e Industrial and Environmental Bioprocesses

ries aims to provide a comprehensive collection of books focusing on bioprocesses in environmental biotechnology. The multidisciplinary content encompasses the chemical cal engineering, industrial microbiology, and energy biosciences, all with a central theme development and circular economy principles. The books align with the Sustainable Goals (SDGs) and offer state-of-the-art information and in-depth knowledge on the r. The book in these series also emphasizes on the application of emerging tools, such arning and artificial intelligence, for the advancement of bioprocesses. While primarily emicians and researchers, the series is valuable for policy planners and industry profes- carefully tailored contents to cater to their specific needs and interests.

Dr Ashok Pandey

shok Pandey is currently Distinguished Scientist at the Centre for Innovation and Research, CSIR-Indian Institute of Toxicology Research, Lucknow, India. His major technological development interests are industrial and environmental biotechnology iosciences, focusing on biomass to biofuels & chemicals, waste to wealth & energy, 5 patents, 120 books, >1000 papers and book chapters, etc. with h index of 138 and ions. Professor Pandey is the recipient of many national and international awards which include Fellow, The World Academy of Sciences (TWAS), India Research itation Award-2023, Clarivate Analytics ; Highly Cited Researcher (Top 1% in the ate Analytics (since 2018–22), Rank #1 in India in Biology in India and Biochemistry he world, Research.Com (2023), Rank #1 in India in Microbiology under Enabling Technologies sector, Elsevier (2021), Rank #1 in India in Biotechnology and #8 in the d University Report (2020-2023), etc.

es: Circular Bioeconomy Perspective
kar, Sunita Varjani, Bhavya B Krishna, Ashok Pandey

ymes in Production of Functional Foods and Nutraceuticals
Rai, Ranjna Sirohi, Luciana Porto de Souza Vandenberghe, Binod Parameswaran

ymes and Metabolites for Health and Well-Being
i, Amit Kumar Rai, Luciana Porto de Souza Vandenberghe, Binod Parameswaran

ird Generation Bioplastics: Production, Application, and Innovation
de Souza Vandenberghe, Ashok Pandey, Ranjana Sirohi, Carlos Ricardo Soccol

als in Aquatic Environments: Toxicity, Monitoring, and Remediation Technologies
Garg, Ashok Pandey, Navish Kataria, Caterina Faggio

ogy and Biotechnological Applications
i, Ashok Pandey, Pau-Loke Show, Binod Parameswaran

chnology
i, Ashok Pandey, Pau Loke Show, Binod Parameswaran

eatment and Management of Sewage Sludge
Awasthi, Zengqiang Zhang, Ashok Pandey

ls in Aquatic Environment: Remediation Technologies and Future Challenges
Garg, Ashok Pandey, Navish Kataria, Caterina Faggio

lysing Enzymes: Basics, Advancements, and Applications
ghania, Anil Kumar Patel, Héctor A. Ruiz and Ashok Pandey

Sanitation and Water Treatment: Concept and Technologies
agi, Ashok Pandey, Patrick Drogui, Bhoomika Yadav, Sridhar Pilli and Jonathan W.C. Wong

tledge.com/Sustainable-Industrial-and-Environmental-Bioprocesses/book-series/SIEB

Decentralized Sanit
Water Treatm

This book provides up-to-date information on the concept
sanitation for wastewater treatment and management. It add
between the understanding of centralized and decentralize
Decentralized Sanitation and Water Treatment: Concept and
principles and technologies involved, and common decentrali
includes topics like septage as a resource, resource recovery,
age, and decentralized sanitation in various countries.

Key Features

- Provides case studies from around the world to explair
  nologies in decentralized sanitation and wastewater tre
- Discusses real examples and cases from rural, urban, a
- Focuses on interdisciplinary approaches of sustainabil
- Covers topics like water recycling, resource recovery, a
- Reviews common treatments and disposal methods of

This book is meant for professionals and researchers work
mental engineering, and ecology.

Sustaina
This book s
industrial an
and biochem
of sustainab
Developmen
subject matt
as machine
targeting aca
sionals, with

Series Edito

Professor A
Translationa
research and
and energy
etc. He has
>81,000 cita
and honors,
Excellence C
world), Clari
and #417 in
and Strategic
world, Stanf

Biotic Resou
Thallada Bha

Microbial En
Amit Kumar

Microbial En
Ranjana Siro

Second and T
Luciana Port

Pharmaceuti
Vinod Kuma

Diatoms: Ec
Archana Tiw

Diatoms Bio
Archana Tiw

Sustainable T
Mukesh Kum

Pharmaceutic
Vinod Kumar

Biomass Hyd
Reeta Rani Si

Decentralized
Rajeshwar D. T

https://www.r

Decentralized Sanitation and Water Treatment

Concept and Technologies

Edited by

Rajeshwar D. Tyagi
BOSK Bioproducts, Québec, Canada
Dongguan University of Technology, Dongguan, China

Ashok Pandey
Centre for Innovation and Translational Research,
CSIR-Indian Institute of Toxicology Research, Lucknow, India

Patrick Drogui
Institut national de la recherche scientifique Centre Eau Terre
Environnement, Université du Québec, Québec, Canada

Bhoomika Yadav
Institut national de la recherche scientifique Centre Eau Terre
Environnement, Université du Québec, Québec, Canada

Sridhar Pilli
National Institute of Technology, Warangal, India

Jonathan W.C. Wong
Research Center for Eco-environmental Engineering,
Dongguan University of Technology, Dongguan, China

CRC Press
Taylor & Francis Group
Boca Raton London New York

CRC Press is an imprint of the
Taylor & Francis Group, an **informa** business

First edition published 2024
by CRC Press
2385 Executive Center Drive, Suite 320, Boca Raton, FL 33431

and by CRC Press
4 Park Square, Milton Park, Abingdon, Oxon, OX14 4RN

ISBN: 9781032443263 (hbk)
ISBN: 9781032463766 (pbk)
ISBN: 9781003381419 (ebk)

DOI: 10.1201/9781003381419

Typeset in Times
by KnowledgeWorks Global Ltd.

Contents

Preface

This book *Decentralized Sanitation and Water Treatment: Concept and Technologies* is dedicated exclusively to the subject of decentralized sanitation and water/wastewater treatment for achieving sustainability. It provides detailed elucidation on the conventional and advanced technologies for decentralized wastewater treatment with the focused themes of water recycling, resource recovery, and the use of sustainable energy sources.

Currently, more than 2 billion people are estimated to lack safe drinking water services, and almost 3.6 billion people do not have safe sanitation systems. Unhygienic practices are widespread thus increasing the ill effects on human health. Thus, there is a strong need to build sustainable mechanisms for water management and wastewater resources. Wastewater systems in developed countries are based on the centralized sewage network and treatment facilities designed to collect and process wastewaters. This concept is challenging due to the reasons of sustainability and difficulties in applying them to under-developed and developing countries with over-population and water scarcity issues. Therefore, a paradigm change is required by adopting the decentralized water systems for wastewater management that would contribute toward providing clean water supply to an increasing population. Such systems have also shown environmental and economic advantages. The goal of sustainability should be applied to reduce the discharges, increase the treatment efficiency, and maximize treated water reuse and by-products recovery. In addition, treatment technologies should be reliable, with low construction costs, easy maintenance, and overall acceptance by the public. The accessibility to water and hygienic sanitation should be increased not by overexploiting existing resources but by improving strategies of reducing, recycling, and reusing as well identifying new water sources like stormwater and reclaimed water.

There is a large knowledge gap existing between the understanding of centralized and decentralized wastewater treatment approaches. Reliable data and comprehensive literature on the decentralized sanitation for wastewater treatment and management are still lacking, which makes this book a useful tool for providing organized and up-to-date information on the various aspects of decentralized wastewater systems in both developing and developed countries. It pays attention to various technologies for reusing wastewater and the potential resource recovery from it.

Based on the above, this book covers as follows: (i) the concept and fundamentals of decentralized treatment toward circular economy and sustainability, as well as (ii) technologies used for decentralized water and wastewater treatment. Each theme covers various chapters that comprehensively illustrate the decentralization model in wastewater treatment and management, thus putting forward all relevant topics in the form of 12 chapters.

We hope that readers will find this book comprehensive and illustrative. All chapters prepared integrate all of the elements related to decentralized sanitation, making it ideal for future research in this field. The book chapters explore the important features for successful decentralized sanitation by discussing the sustainability principles and fundamental concepts, the processes and technologies involved in the concept, and various case studies addressing the decentralized wastewater treatment and management relevant to sustainable development.

We immensely appreciate the excellent work done by the authors in compiling the relevant information on the subject matter. We believe it will be useful to the scientific community, practitioners, and general audience. We also acknowledge the reviewers for their valuable comments, which helped to improve the scientific content and quality of the chapters. We are grateful to the Biotech Research Society, India, for proving us the opportunity to bring out this book under the agreement the Society has with Taylor & Francis. We would like to thank the Taylor & Francis team for their consistent hard work and cooperation in the publication of this book.

Editors
Rajeshwar D. Tyagi
Ashok Pandey
Patrick Drogui
Bhoomika Yadav
Sridhar Pilli
Jonathan W.C. Wong

About the Editors

Professor Rajeshwar D. Tyagi is an internationally recognized professor of biochemical engineering and Distinguished Professor at Dongguan University of Technology, China and Chief Scientific Officer of BOSK Bioproducts. He has served as a professor for 36 years and the Senior Canada Research Chair on Value Added Products from Wastes for 18 years at the University of Québec, Canada. His research on bioconversion and bioprocessing to bioproducts and activity of transfer of technology to the user's sector has received many national and international awards.

Professor Ashok Pandey is currently a Distinguished Scientist at the Centre for Innovation and Translational Research, CSIR-Indian Institute of Toxicology Research, Lucknow, India, and Distinguished Professor, Sustainability Cluster, University of Petroleum and Energy, Dehradun, India. His major research and technological developmental interests are industrial and environmental biotechnology and energy biosciences, focusing on biomass to biofuels and chemicals, waste to wealth and energy, and more.

Professor Patrick Drogui is a professor at the Institut national de la recherche scientifique's Centre Eau Terre Environnement in Canada. He is an internationally recognized expert in electrochemical and membrane technologies and oxidative processes for different applications in water treatment from industrial wastewater to potable water.

Dr. Bhoomika Yadav completed her PhD from Institut national de la recherche scientifique's Centre Eau Terre Environnement. She has strong research skills and works in multidisciplinary areas of biotechnology, microbiology, and biochemical engineering. Her research interests include waste to value transformation, bioflocculants, wastewater treatment, bioplastics, and circular bioeconomy.

Dr. Sridhar Pilli is an assistant professor at the National Institute of Technology Warangal, India. He is an expert in water and wastewater treatment, anaerobic digestion, and waste to energy.

Professor Jonathan W.C. Wong is a Distinguished Professor of the Research Center for Eco-environmental Engineering, Dongguan University of Technology, China. Formerly, he was a professor in the Hong Kong Baptist University, Hong Kong. His major research interests are bioconversion of organic wastes for energy and bioproducts.

Contributors

Anusha Atmakuri
Institut national de la recherche scientifique (INRS)
Centre Eau Terre Environnement (ETE)
Université du Québec
Québec, Canada

Shraddha Chavan
Institut national de la recherche scientifique (INRS)
Centre Eau Terre Environnement (ETE)
Université du Québec
Québec, Canada

Tae Hyun Chung
Department of Civil and Environmental Engineering
University of Alberta
Edmonton, AB, Canada

Bipro Ranjan Dhar
Department of Civil and Environmental Engineering
University of Alberta
Edmonton, AB, Canada

Patrick Drogui
Institut national de la recherche scientifique (INRS)
Centre Eau Terre Environnement (ETE)
Université du Québec,
Québec, Canada

Elsayed Elbeshbishy
Department of Civil Engineering
Toronto Metropolitan University (formerly Ryerson University)
Toronto, Canada

Mohamed El-Qelish
Water Pollution Research Department
National Research Centre
Cairo, Egypt

Hui Guo
Department of Civil Engineering
McMaster University
Hamilton, Ontario, Canada

Rania A. Hamza
Department of Civil Engineering
Toronto Metropolitan University (formerly Ryerson University)
Toronto, Canada

Abir Hamze
Department of Civil Engineering
Toronto Metropolitan University (formerly Ryerson University)
Toronto, Canada

Basem Haroun
Department of Civil Engineering
Toronto Metropolitan University (formerly Ryerson University)
Toronto, Canada
Water Pollution Research Department
National Research Centre
Cairo, Egypt

Amr Ismail
Anaergia Inc.
Burlington, Ontario, Canada

Zexuan Ju
Department of Civil and Environmental Engineering
University of Alberta
Edmonton, AB, Canada

Farokh Laqa Kakar
Brown and Caldwell
Burnaby, British Colombia, Canada

Younggy Kim
Department of Civil Engineering
McMaster University
Hamilton, Ontario, Canada

Sushil Kumar
Institut national de la recherche scientifique (INRS)
Centre Eau Terre Environnement (ETE)
Université du Québec
Québec, Canada

Liwen Luo
Institute of Bioresource and Agriculture
Hong Kong Baptist University
Hong Kong Special Administrative Region, China

Avinash Kumar Reddy Narreddy
Department of Architecture
Dr. YSR Architecture and Fine Arts University
Kadapa, Andhra Pradesh, India

Sri Chandana Panchangam
Department of Civil Engineering
Annamacharya Institute of Technology and Sciences
Kadapa, Andhra Pradesh, India

Sridhar Pilli
Water and Environment Division
Department of Civil Engineering
National Institute of Technology (NIT), Warangal, India

P. Venkatsewara Rao
Water and Environment Division
Department of Civil Engineering
National Institute of Technology Warangal, India

Loka Subramanyam Sarma
Nanoelectrochemistry Laboratory
Department of Chemistry
Yogi Vemana University
Kadapa, Andhra Pradesh, India

Ting Song
School of Environment and Architecture
University of Shanghai for Science and Technology
Shanghai, China

Rajeshwar D. Tyagi
BOSK Bioproducts
Québec, Canada
Research Centre for Eco-Environmental Engineering, Dongguan University of Technology, Dongguan, China

Preethi Vijayarengan
Department of Civil Engineering
Hindustan Institute of Technology and Science
Chennai, India

Jonathan W.C. Wong
Research Centre for Eco-Environmental Engineering, Dongguan University of Technology, Dongguan, China

Suyun Xu
School of Environment and Architecture
University of Shanghai for Science and Technology
Shanghai, China

Manojkumar Y.
Water and Environment Division
Department of Civil Engineering
National Institute of Technology
Warangal, Telangana, India

Bhoomika Yadav
Institut national de la recherche scientifique (INRS)
Centre Eau Terre Environnement (ETE)
Université du Québec
Québec, Canada

Song Yan
Institut national de la recherche scientifique (INRS)
Centre Eau Terre Environnement (ETE)
Université du Québec
Québec, Canada

Mohamed S. Zaghloul
Department of Civil Engineering
Toronto Metropolitan University (formerly Ryerson University)
Toronto, Canada

Basem S. Zakaria
Department of Civil and Environmental Engineering
University of Alberta
Edmonton, AB, Canada

Part I

Concept of Decentralized Treatment toward Sustainability

1 Introduction to Decentralized Treatment

Shraddha Chavan[1], *Anusha Atmakuri*[1], *Sushil Kumar*[1], *Bhoomika Yadav*[1], *Rajeshwar D. Tyagi*[2,3], *Patrick Drogui*[1], *and Jonathan W.C. Wong*[2]
[1] Université du Québec, Québec, Canada
[2] Dongguan University of Technology, Dongguan, China
[3] BOSK Bioproducts, Québec, Canada

1.1 INTRODUCTION

According to Steffen et al. (2015), urban populations have dramatically increased since 1950 (Nations, 2018). As a result, urban areas will shelter more than 80% of the world's gross domestic product. Given the density of people in cities, it is vital to evaluate the techniques used to mitigate the environmental consequences of individuals residing in these places (Shen et al., 2015). In this context, wastewater treatment (WWT) plants (WWTPs) have concentrated on reducing their effectiveness to achieve carbon neutrality. Notably, the process design, infrastructure used, and chemicals or energy demands change significantly between centralized and decentralized WWT (DWWT) systems (DWWTSs) (Samuel et al., 2016). Besides these attributes, an additional distinction between the two approaches is the ability to reuse resources. In centralized systems, effluent is received and processed at a WWTP located outside the covered region; in contrast, the decentralized treatment allows for the reuse of recovered water and biofertilizers in neighboring green and agricultural fields (Ho & Anda, 2006).

In the urban environment of industrialized countries, centralization is frequently the only WWT solution, making it the most widely used strategy for treating wastewater. Decentralization, on the other hand, has become increasingly important in developing economies. In any case, decentralization methods are already globally recognized and acknowledged by both water specialists and legislators. For instance, tiny decentralized WWTPs serve 25% of the US population, primarily in rural areas or where the development of a sewage collecting infrastructure is not financially feasible (Kujawa-Roeleveld & Zeeman, 2006). Compact end-of-pipe decentralized WWTPs are rapidly expanding in the United States due to their relatively reduced expenses and the fact that their installation is perceived in the same way as traditional ones. Similarly, in Colombia, the overhead associated with a single large centralized WWTP has been lowered in favor of two decentralized WWTPs, which might be interpreted as a low level of regional centralization (Bakir, 2001). Numerous factors may influence the advancement of decentralization procedures, including the increasing population in remote regions and developing nations, reduced quality of surface waters, the construction of large building structures in metro regions, the program development of segregated villages, rising shortage of water resources, and the emphasis placed on water recovery and reuse (Kjerstadius et al., 2017).

As demonstrated by Brown et al. (2010), the implementation of decentralized methods is not in opposition to centralized ones. Furthermore, usually, relatively densely inhabited areas have been supplied by a sewage collecting system and one or more centralized WWTPs; therefore,

DOI: 10.1201/9781003381419-2

decentralization would not be a feasible and sustainable economic alternative (Chung et al., 2008). The general approach would be to support a true coexistence between centralized and decentralized systems with various levels of applicability, which appears to be more realistic, especially in the case of great blocks such as commercial centers, hospitals, airports, or new developing areas where treated wastewater reuse might be effectively planned (Lahmouri et al., 2019).

This chapter gives an overview of decentralized WWTSs and examines the studies that explored the amount of decentralization of effluent collection and treatment systems. The purpose of this chapter is to evaluate several technical approaches, highlighting their advantages and drawbacks as well as how they have evolved through time. The primary goal is to assist planners and decision-makers in selecting and implementing the most appropriate methodologies for assessing the decentralization degree of wastewater facilities depending on the objectives established.

1.2 WATER QUALITY AND HEALTH IMPACTS

Low organic strength and high particle organic matter concentration have been the defining characteristics of municipal WWT facilities. Domestic wastewater contamination is a severe issue in developed, developing, and underdeveloped nations and has an impact on the environment, groundwater, and water bodies. Untreated wastewater had contaminated around half of the world's oceans, rivers, and lakes. This has had detrimental consequences on aquaculture and agriculture as well as the destruction of natural resources. The environmental impact of numerous items is currently a topic of great concern among government organizations, municipal workers, residents, and the industrial sector (Estahbanati et al., 2021; Ozgun et al., 2013). These concerns centered on products, production procedures, associated services, transportation, waste management, and environmental impact affect people and the environment. Almost every kind of product that leaves a facility has an impact on the stability of our ecosystem, whether it is home or industrial effluent (Corominas et al., 2020).

The supply of freshwater for agriculture is insufficient around the world, and in most regions, the reuse of either treated or untreated effluent is the primary source of water (Cytryn, 2013; Kumar et al., 2021). Due to resource depletion, global warming, and a lack of freshwater, technologies to produce clean water have drawn attention from all over the world. Nevertheless, the reuse of wastewater is linked to several environmental and health hazards.

1.2.1 Health Issues due to Pathogens

Microorganisms like viruses, bacteria, and protozoa are one of the main sources of waterborne diseases in underdeveloped and developing countries (Akpor & Muchie, 2011). The main source of microbial contamination is human and animal waste (Metcalf & Eddy, 2003). These wastes contain different types of bacteria, viruses, and protozoa, which may later washed off to drinking water supply reservoir (Metcalf & Eddy, 2003). Microbial pathogens are deemed to be important factors causing various waterborne epidemics. World Health Organization (WHO) report indicates that waterborne infection outbreak was top cause of infectious diseases in 132 countries (from 1998 to 2001) (Metcalf & Eddy, 2003). The study suggests that many pathogens in wastewater can cause chronic health diseases with longer term effects such as stomach ulcers and degenerative heart disease. Detection, isolation, and identification of these pathogens in wastewater are challenging and time-consuming process (Metcalf & Eddy, 2003).

Among other microorganisms, viruses are one of the most important and harmful pathogens in the wastewater (Metcalf & Eddy, 2003; Okoh et al., 2007; Paillard et al., 2005). They are hard to detect, more resistant to treatment, highly infectious and need very small dose to cause infections. Hepatitis viruses, adenovirus, and echoviruses are the most common viruses present in wastewater (Metcalf & Eddy, 2003; Paillard et al., 2005). These viruses can cause severe health issues like liver infection (may cause liver failure), eye infection, and meningitis. Another common pathogen in wastewater is bacteria (Akpor & Muchie, 2011; Metcalf & Eddy, 2003). They can cause aide variety of infections and diseases such as diarrhea, skin infection, and dysentery. The most common disease-causing bacteria in the wastewater are *Escherichia coli*, *Salmonella*, *Listeria*, *Campylobacter*, *Vibrio*, etc. (Metcalf & Eddy, 2003). Most of the bacteria present in the wastewater are not harmful. However, pathogenic bacteria may cause health problems like typhoid, pneumonia, intestinal disorders, and dysentery (Akpor & Muchie, 2011). Study indicates that bacteria (*E. coli* and particular strains of *Pseudomonas*) can cause gastroenteritis, which may affect the new-born and can also contribute to gastrointestinal disease epidemics (Akpor & Muchie, 2011).

Protozoa are very common in wastewater and natural water, but most of the protozoa are not harmful (Ingraham & Ingraham, 1998; Metcalf & Eddy, 2003). Pathogenic protozoa can cause milder order gastrointestinal disorders. *Giardia lamblia*, *Cyclospora*, and *Cryptosporidium parvum* are the most common pathogenic protozoa present in wastewater. These protozoa can cause diarrhea and eye infections in immunocompromised person (young children and old people) (Ingraham & Ingraham, 1998; Metcalf & Eddy, 2003).

1.2.2 Health Issues due to Chemicals and Toxic Metals

Wastewater contains chemical pollutants and metals, which are harmful to human health. Some pollutants are considered endocrine disrupting compounds. These compounds if enter the human body can alter the homeostatic and hormonal system. Many studies have indicated that endocrine disrupters can affect metabolism, prostate, breast, liver, lung, thyroid, and reproductive system (Kabir et al., 2015; Kumar et al., 2022; Polyzos et al., 2012). Studies also indicate that some endocrine disrupting compounds can cause diabetes, cancer (testis, prostate, and breast cancer), or may reduce fertility (Kabir et al., 2015). Many authors have reported that endocrine disrupters (dichlorodiphenyltrichloroethane [DDT], dioxins, polychlorinated biphenyls [PCBs], and chlorinated pesticides) can affect the nervous system (Kabir et al., 2015; Polyzos et al., 2012).

Recent studies have identified antibiotic-containing personal care, pharmaceuticals, and human health-related goods as new pollutants and threats to aquatic environments. According to the global reports, antibiotics have been found in our environment in groundwater, sediments, and surface water (Ma et al., 2014; Pruden et al., 2012). Due to improper disposal of antibiotics, water bodies are becoming more and more contaminated, and the concentration of these contaminants is rising daily as a result of population growth, urbanization, industrialization, and poor agricultural practices. A further worry is that prolonged exposure to low levels of antibiotics in the atmosphere, for instance, those found in sewage and surface water, may result in the development of antibiotic-resistant bacteria (Guardabassi et al., 1998). Gene cassettes, transposons, plasmids, and contaminants that are discharged in sewage play crucial roles in the transfer of resistance and assist in maintaining it (Moura et al., 2012).

Wastewater contains huge amount of toxic metals. These toxic metals are normally dissolved in the wastewater. Metals like zinc (Zn), copper (Cu), and manganese (Mn) are essential minerals for human health, but when it exceeds the permissible limit ($Zn < 3000$ µg/L, $Cu < 2000$ µg/L, $Mn < 500$ µg/L), they can adversely affect human health (Azizullah et al., 2011; Crossgrove & Zheng, 2004). Study suggests that high concentration of Mn in drinking water can cause problem in nervous system. In the worst situation it may lead to permanent neurological disorder (Crossgrove & Zheng, 2004). Studies indicate that bone softening disorder has been detected due to high concentration of fluoride in the drinking water system (Ahmad et al., 2003; Azizullah et al., 2011). Another study indicates that people are suffering from skin diseases due to high concentration of arsenic in the drinking water (Azizullah et al., 2011). High concentration of metals like iron (Fe), cadmium (Cd), chromium (Cr), nickel (Ni), lead (Pb), mercury (Hg), and arsenic (As) can lead to severe health complication like cancer, acute gastrointestinal problems, kidney damage, hematopoietic, and digestive problems.

1.3 CENTRALIZED WATER TREATMENT SYSTEM

Rapid urbanization and population expansion have increased the amount of wastewater produced; as a result, sustainable WWT has become more important in recent years. This type of treatment not only deals with recycling and reuse but also with recovery of resources and energy such as nutrients (Verstraete & Vlaeminck, 2011). Conventional WWT must develop into a more resource-oriented strategy due to rising energy costs and the scarcity of nutrient supplies. As a result of this paradigm change, WWTPs are now seen as resource recovery facilities from which a variety of commodities can be generated, including reusable water, energy in various forms (heat, electrical, fuels, etc.), biopolymers, and fertilizer. Novel treatment technologies that demand minimal energy input for carbon and nutrient removal must be designed in order to attain energy neutrality in WWTPs (Scherson & Criddle, 2014).

The centralized (conventional) treatment method, which is depicted in Figure 1.1, involves a combination of the processes of coagulation, flocculation, sedimentation, filtration, and disinfection. It centrally treats water before distributing it through specialized distribution networks (Dhote et al., 2012).

This system comprises a sewer network that carries a variety of wastewater from local houses, small businesses, industrial facilities, institutions, and even storm water runoff to a WWT facility beyond the city or village borders (Arora et al., 2015). The collection and transit subsystems make up the two main components of the sewer network. The WWTPs also include two subsystems, one of which is intended to remove contaminants from the wastewater and the other to transform the substances and their derivatives into a form acceptable for any subsequent uses (e.g., as a soil conditioner in agriculture). The recycled water is released into the nearest aquatic environment (river, estuary, or oceans), where it is anticipated that natural physical, chemical, and biological processes will be efficient and result in the water's final polishing (Kataki et al., 2021; Leigh & Lee, 2019).

A centralized water treatment system can handle all residential, commercial, and industrial needs in urban areas by processing enormous volumes of water quickly (Qu et al., 2019). This method is extensively researched and capable of efficiently eliminating essentially any level of raw water turbidity as well as dangerous organisms like bacterium, viruses, and parasites. Large-scale plants that provide services to extensive municipal or regional service areas are what define it (Hoffmann et al., 2020).

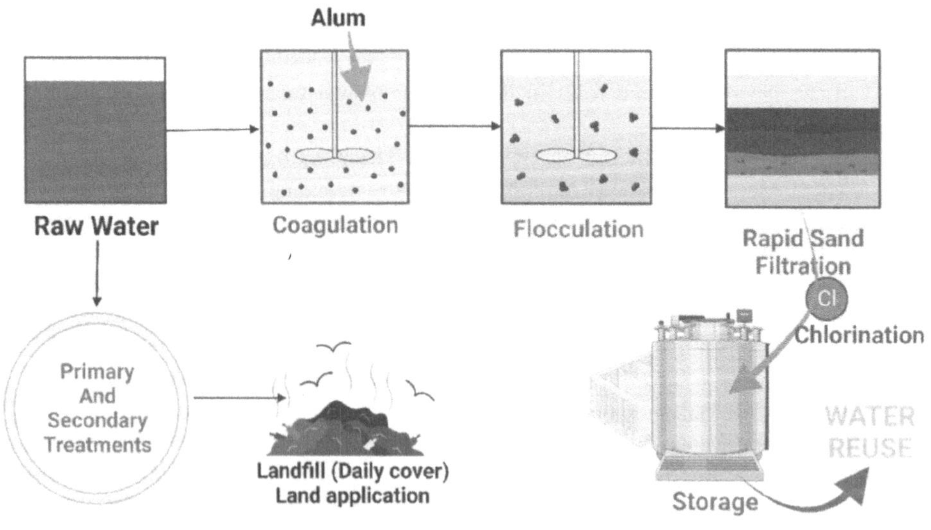

FIGURE 1.1 Typical centralized WWTSs.

Sewage and, in some situations, storm water are collected as part of this treatment, which is the standard method used in many nations. However, a centralized system can have high capital, operational, and maintenance expenditures. It entails the development of water sources, the building of important infrastructure (such as the treatment facility, tank, and water supply main), the use of automated control and surveillance systems, and the presence of on-site operators (Pasciucco et al., 2022).

The primary benefit of centralized WWT method is that one large treatment plant has lower construction and operational expenses than several small-scale plants treating the same urban area (Corominas et al., 2020). Additionally, efficient management over plant operation and effluent control is guaranteed. By adopting a "package plant," wherein treatment units are preassembled at a factory, skid fastened, delivered to the site, and practically equipped to use, smaller cities can cut expenditures. Even yet, some underdeveloped areas would still not be able to afford a centralized treatment system. Therefore, the investment in the accompanying collection sewers, as well as their operation and maintenance (O&M), should be considered while evaluating centralized systems (Nguyen et al., 2019).

Thus, the centralized systems are no longer a viable option due to (Mihai et al., 2022)

- related high collecting system costs;
- soil and groundwater contamination brought on by an undiscovered sewer leak;
- the potential entry of various harmful and resistant materials (pathogens, heavy metals, endocrine disruptors, pharmaceutical residues, etc.) that prevent reuse;
- the variable hydraulic and organic load in the case of combined collecting systems.

Based on integrated, region-specific elements such as availability of land, wastewater qualities, desired produced quality of water, socioeconomic conditions, and municipal and provincial requirements, the best method for WWT should indeed be chosen. The application of the treatment process will be determined in accordance with a number of factors that must be balanced, including the availability of land, capital and operating costs, and sludge production (Gallego-Schmid & Tarpani, 2019). The primary factor to consider when choosing an intense or extensive centralized treatment system is land availability. The following is an example of a practice rule (Molle & Sanchis-Ibor, 2019):

- When there is less than 1 m^2/Phiàng (PE) of available land, intensive systems would be considered.
- Mixed systems (biological secondary treatment accompanied by finishing lagooning, etc.) might be taken into consideration if the attainable area is bigger (up to 5 m^2/PE).
- Extensive systems must be used if area is greater than 6 m^2/PE.

1.4 PROBLEMS WITH CENTRALIZED WATER SYSTEMS

The traditional wastewater management system consisting of wastewater collection and treatment in a central treatment plant has been successfully applied in various areas of industrialized countries. The centralized wastewater collection and treatment has been mostly standardized in mega-cities; however, severe problems relating to the wastewater management in rural areas became apparent. In such areas, the sewer systems do not exist or exist in undeveloped form and the wastewater is not properly treated or not treated at all. The sanitary system in developing countries is mostly inhomogeneous ranging in technological level from pit latrine to biological water treatment plants (Wilderer & Schreff, 2000). Hence, there is a requirement of sanitary infrastructures; however, because the high costs and complexity of construction, O&M, or they require high and reliable water consumption, centralized system may be less suitable for low-income communities, rural areas with low population density, water scarce areas and unreliable water supply areas. DWWTSs are being currently considered by various researchers and authorities as an alternative to centralized systems to be applied in developing countries (Massoud et al., 2009).

Centralized wastewater systems as discussed previously consist of sewer systems that collect wastewater from households, institutions, and small enterprises and transport it to the WWTP located outside the residential areas. In decentralized systems, the wastewater generated is treated close to the point of origin. Also, the length of sewers is quite short as compared to the centralized systems and the collected wastewater flow to the small on-site treatment plants. Decentralization can be considered an economical and ecological alternative to the conventional centralized sanitation systems provided; they produce high-quality effluent at a reasonable price. The DWWTS has many advantages over the traditional ones (Hophmayer-Tokich, 2006). First, the transportation sewers including the lift stations and storage facilities to manage the combined wastewater are not needed resulting in saving associated costs. Second, there are increased opportunities for on-site water reuse and groundwater recharge. The reclaimed water would be more cost-effective as effluent would be available near the point of use, thus reducing the costs for reclaimed water distribution. Third, the failure of one independent unit will not cause the collapse of the whole sanitation system (Wilderer & Schreff, 2000). Fourth, decentralized systems allow for flexibility in management and various processes can be combined to meet treatment goals for small communities. Fifth, they are less water intensive, i.e., the on-site facilities and alternative sewers such as settled sewers reduce the freshwater requirements for transportation. Hence, they are more preferable for communities with improper zoning such as low-density populated rural or peri-urban areas (Massoud et al., 2009).

1.5 DEVELOPMENT SOLUTIONS: NEED FOR DECENTRALIZED WATER TREATMENT

A decentralized system – point-of-entry (POE) or point-of-use (POU) – installed at the individual home or business can be used to achieve potable water in the circumstances of centralized community treatment system being unavailable. POEs treat the raw water prior to its entrance into the property or home, while POUs are installed to treat water where needed, for example, at kitchen and bathroom taps (Massoud et al., 2009).

Usually, in underdeveloped locations facing substantial deficiencies in financial and technical support, POEs/POUs might be considered the lone option. At present, sustainability has become a prime issue of wastewater management. However, the systems offered for sustainable management are expensive enough that a developing country cannot adopt (Capodaglio, 2017). The use of conventional WWT and sewer system for rural communities is not only expensive in terms of provision of services but O&M as well. Finally, in the absence of the needed technical and funding assistance, the implementation of centralized systems is not plausible, thereby pushing them toward decentralized systems (Massoud et al., 2009).

POEs/POUs might be inexpensive compared to centralized systems, as they waive large initial capital investments and minimize operating costs along with maintenance costs. However, their treatment capacity and ability to deal with multiple contaminants might be limiting factors. In addition, they need full community buy-in and monitoring.

Decentralized water treatment offers the flexibility needed to ensure a steady supply of potable water during unpredictable water shortages and weather. In addition to these, its potential for water reuse and possibility to reduce long-term treatment costs attract more attention. Decentralized water treatment, including desalination, might be ideal solution for variety of residential, commercial, and industrial settings, comprising

- small communities;
- school campuses;
- commercial developments;
- hotels and resorts;

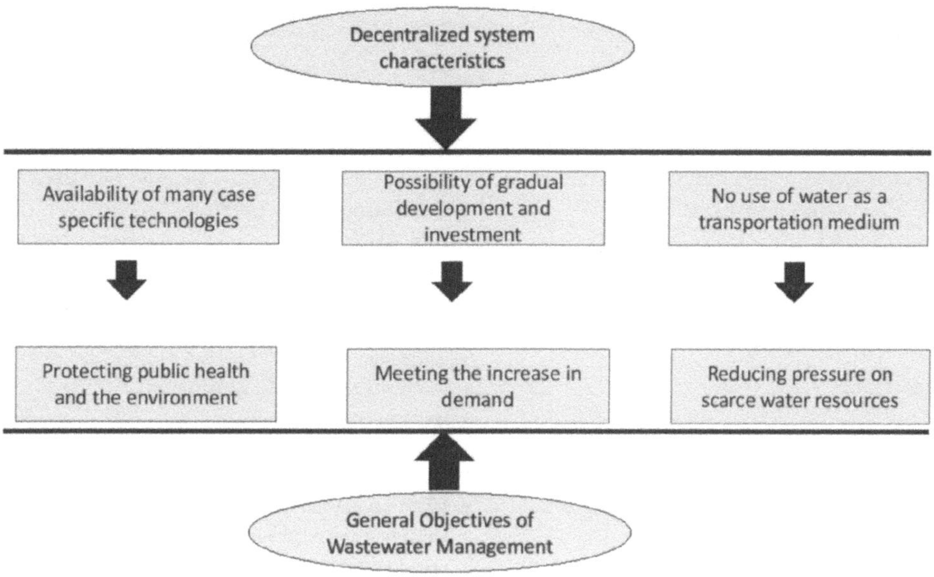

FIGURE 1.2 Decentralized system characteristics making it an attractive option for WWT.

- golf courses;
- industrial plants;
- mining, or oil and gas camps;
- construction sites.

Figure 1.2 illustrates the characteristics of decentralized water systems, which can be compared against the general objectives of wastewater management being fulfilled.

Decentralized systems pose an advantage of easy adaptation to conditions at local level of urban areas, along with expanding their capacity in agreement with population growth. This approach hence facilitates reuse of water and treatment by-products, such as nutrients, sludge, and energy (Capodaglio, 2017).

The implementation of decentralized systems comprises varied planning methods, wherein design, feasibility, and implementation need to be performed by independent service areas, with contexts and whose solutions must respond to their individual needs, considering the heterogeneity that occurs within a same urban center, where social, environmental, geographic, economic, and technological conditions can vary widely (Bernal & Restrepo, 2012).

Along with the abovementioned advantages, a notable addition is the lessening of pollutant amount received by the water bodies post-exiting the plant, keeping in mind that the flows treated by decentralized systems are lower, hence, facilitating pollutant dilution and plummeting its environmental impact on parameters such as diluted oxygen, chemical oxygen demand, and biochemical oxygen demand (Bernal et al., 2021; Massoud et al., 2009).

1.6 REQUIREMENTS/STANDARDS FOR DECENTRALIZED SOLUTIONS

The various key indicators of decentralization in wastewater management are related to technological, economic, social, environmental, and institutional aspects, the importance of which ranges from centralized to decentralized approaches.

Table 1.1 lists the various aspects and variables of the key indicators of decentralized wastewater management system. The success of a decentralized management program for urban areas will depend largely on institutional support as, even though the system's installation, operation, and

TABLE 1.1
Combined Indicators of DWWTSs

Aspect	Variable	Indicator
Technological	Technological features	• Reuse potential • Sewer coverage • Centralization level
	Area features	• Area availability • Quality goals • Area topography • Distance to the treatment point
	Reuse type	• Agriculture • Urban development, street washing • Aquaculture • Energy recovery
Economic and institutional	Costs	• Investment in treatment systems • Investment in sewage systems • Operation and maintenance (O&M) of treatment system • O&M in sewage system
	Planning	• Institutional support for decentralized systems • Interest of service companies and O&M • Peri-urban and expansion areas without wastewater treatment coverage
Environmental	Environmental impact	• Water availability • Nutrient recycling • Sludge production • Noise/smell/insects/landscaping
Social	Community	• Community participation in system management and O&M • Reuse acceptation
	Demographics	• Population size • Population distribution • Population density

maintenance costs might be diminished, a risk of failing to continue operations in the long term due to financial crises stays, especially since micro local governments usually lack the resources required for investing in infrastructure improvements to provide better services.

The coverage and the existence of sewer systems are a key indicator, especially as, in the case of incomplete sewer coverage, the planning of decentralized wastewater management systems is aided by step-by-step planning aimed at escalating sewer and treatment coverage (Leigh & Lee, 2019; Roefs et al., 2017).

Based on technological aspects, three prime indicators were identified: reuse potential, sewage system coverage, and the potential for combining centralized and decentralized systems. A thorough analysis of these factors is crucial for project planning using a decentralized approach (Capodaglio, 2017).

For social aspects, the most pertinent indicators are environmental awareness, reuse acceptance, community participation in system management, and O&M. The systems' success will be impacted by acceptance from the population associated with access to information, environmental education, and water culture (van Afferden et al., 2015).

Among the pertinent environmental indicators are the removal of contaminants and pathogens, conservation of natural resources, and water availability. The chief goal of water treatment systems is to mitigate or minimize impacts on natural resources and on the ecosystem services they provide (Bernal et al., 2021).

System resilience is basically the degree to which the system minimizes the magnitude of the failures caused by exceptional circumstances when rendering services during its lifetime. Currently, system resilience has become a key aspect – pertinent in implementation and planning of the treatment system based on decentralization or centralization levels. System resilience will also help in determining the success rate of the system under harsh weather conditions and external stressors (Bernal et al., 2021).

1.7 THE CURRENT SITUATION IN DEVELOPING REGIONS – OVERVIEW

Based on a study by Muzioreva et al. (2022) offering critical evaluation of use of decentralized technologies in developing countries, 53 papers about decentralized wastewater systems over 20 years in developing countries were identified. The focus on the topic is still developing, with momentum rising slowly. The plethora of technologies considered decentralized wastewater systems and the variety of elements that comprise the systems illustrate the flexibility of decentralized technologies, especially in developing nations. However, the presence of constructed wetlands and septic tanks as the common types of decentralized wastewater technologies in developing countries meant that there were not many high-rate technologies studied in the technology aspect (Brixa et al., 2010).

Despite the growing work and interest, gaps are still pertinent in the developing nations and their policies to adopt DWWTSs. One of the key factors to be corrected in these nations is their policy, as usually government policies have a relevant impact and can change the course of any WWT method.

Some examples of issues faced by developing countries can be summarized below. Caribbean countries despite their proximity to the United States face problems in the case of WWTSs. Even now, there are only two WWTPs, Sistema Central and María del Carmen, within the City of Havana, as compared to none in the year 2005 (de Freitas Alves & Westbrook12, 2016).

Another such instance can be the case of Zimbabwe. Until recently there has been little, if any, concern over revamping let alone improving wastewater management system in Zimbabwe's urban areas given the dominance and institutionalized waterborne system (Chirisa et al., 2017).

In Southern Africa, countries illustrate varying outlines with respect to centralized and decentralized wastewater systems. For instance, Kenya uses conventional WWTSs, inadequate and non-functional due to higher costs of O&M (Chirisa et al., 2017; Water, 2006).

Windhoek is one of the cities in Southern Africa that has presented meaningful progress in relation to the implementation and management of decentralized systems (Sönmez et al., 2011). In Zimbabwe, cases of decentralized wastewater management system are primarily in the rural areas where pit latrines are commonly used. According to Nhapi, examples of such a system are in places such as Gweru, Redcliff, Nemanwa, and Mupandawana (Nhapi, 2004a, 2004b, 2009). In Gweru, the decentralized wastewater system started in 1994 (Nhapi, 2009).

Based on the various studies done over the years on the use of DWWTSs in developing nations, some key issues and risks involved can be zeroed in Table 1.2.

As far as the implementation level of decentralized WWTSs in developing countries is concerned, efforts are being made to implement decentralized WWTSs based on the community's needs. For the successful implementation or adoption of decentralized wastewater systems, developing countries need to have proper regulations and institutional frameworks that comprise sanitation policies commensurate with decentralized technologies (Muzioreva et al., 2022).

1.7.1 Rural Areas

Despite the continuing global urbanization phenomena, half of the global population still lives in rural settings. As a result, many investigations are done for decentralized wastewater management as solutions for communities located in rural or peri-urban settings (Capodaglio, 2017). Decentralized management is primarily centered on the improved sanitation idea that emerged as a viable and cost-effective alternative for rural WWT, with the possibility for recovering energy, water reuse,

TABLE 1.2
Issues, Potentialities, Opportunities, Risks, and Strategies of DWWTSs

Issues	Potentialities	Risks	Opportunities	Strategies
Design	Unsophisticated design	Poorly designed system	Availability of engineers	Training of engineers on decentralized wastewater systems, capacity building for local authority
Financing	Reduction in financial cost	Financial risk in investments	Ratable property in the city	Tapping of low-cost methods, negotiating with government on grants for infrastructure development, levying taxes on households
Public health	Reduction in public health threats	Public health risks in case of poor designs	Availability of city health departments	Setting standards to ensure public health, review of public health legislation and policies

and fuel savings. Vacuum sewage systems (VSS) are a type of decentralization which is a long time, technical solution for wastewater collection. The core principle of VSS is that wastewater is transported by air pressure rather than gravity. VSS have two key benefits over traditional gravity systems: they use substantially lesser water than gravity systems and wastewater transit is not limited to following hydraulic grade lines (Averill & Heinke, 1974). Vacuum sewers are capable of fulfilling the fundamental objective of limiting human contact with fecal contaminated water while providing comprehensive ecological and health safety, decreasing seepage and odors, and achieving power and savings in operating costs (Panfil et al., 2013). Moreover, for rural sewage treatment, an integrated decentralization comprising source segregation, a traditional septic tank, a built wetland a subsurface drip irrigation system can be a potential solution based on considerable studies (Terryn et al., 2014). However, additional investigation into the economic and ecological viability of water reuse, resource recoveries, and electricity generation is required for areas where conventional centralization plants are currently in existence (Elawwad et al., 2014).

1.7.2 Urban and Peri-Urban Areas

In peri-urban areas (settlements near large urban areas but without proper services and facilities) and some urban areas in low-income countries, conventional centralized sanitation systems for WWT have generally failed to address the needs of communities for wastewater management (Bernal et al., 2021). For example, in the peri-urban areas of Xochimilco, Mexico City, a large number of residents have centralized waste collection service; however, the separation and recycling of solid waste is uncommon. There are various households that do not make use of centralized facilities or are not connected to such services, thus, leading the solid wastes to be burnt or discarded on streets and in environment, causing pollution (Nanninga et al., 2012). Similarly, the sanitation situation in the peri-urban areas of this region differs by location. For example, houses located near urban areas are connected to the water supply and use flush toilets that discharge the wastewater into canals or septic tanks that are poorly constructed and regulated. Whereas, households in remote areas use poor flush toilets or pit latrines that are poorly constructed or defecate and urinate openly in the environment.

A comparative study performed in ten peri-urban areas in the peri-urban interface of five metropolitan areas, i.e., Chennai, Mexico, Caracas, Dar es Salaam, and Cairo, revealed the lack of access to adequate water and sanitation system to not only the low-income households in peri-urban areas but also other income households. However, the peri-urban residents are more vulnerable than higher income households as they lack financial and political means to improve their situation with respect to sanitation services (Allen et al., 2006).

Arguments in favor of decentralized sanitation systems for communities in rural, peri-urban, and some urban areas have been advocated by many researchers (Capodaglio, 2017). A decade ago, around 25% of the population in the United States was using DWWTSs (Capodaglio, 2017). In remote settings, the centralized WWT would cause unsustainable long-term economical burdens; therefore, decentralized systems are alternative solutions as they can be built exactly to fulfill the requirements of the community and can be expanded later depending on their needs.

1.8 CHALLENGES IN THE SYSTEM

The primary problems to the adoption of decentralization in urban settings were recognized and categorized based on an examination of the state of the art and reported occurrences (Angelakis et al., 2003). Figure 1.3 depicts these concerns.

The ultimate purpose of a treatment process is environmental sustainability. There is a special implication for decentralization in the setting of substantial pollution issues caused by wastewater without processing or inadequate treatment connected with conventional approaches. On the contrary, in a decentralized arrangement, the fluxes at any site would be limited, meaning that any error would cause minimal ecological harm (Nhapi, 2004a). The smaller collection pipes would be put at the subsurface and could be directed more freely, resulting in low ecological damage. Typically, large communities support centralized infrastructure, whereas the viability of a decentralized system is dependent on the variables such as the acceptability of the group and the neighboring population where the system is situated. This acceptability is supported by awareness of climate change, which is connected to information accessibility and ecological studies (Libralato et al., 2012).

Most of the economic implications in traditional centralized structures are well-recognized to be connected to the construction and operation of the sewage collection infrastructure. In contrast, most decentralization expenses are associated with the treatment facility (Howard et al., 2010). Each enterprise must undergo an economical study to determine the expenses related to the environmental advantages and associated with the technical system (Mankad & Tapsuwan, 2011). The integration of decentralized systems may reduce the amount of capital required for wastewater management, as well as save water supplies and financial assets. However, they have the possibility

FIGURE 1.3 The key challenges in DWWTSs.

of failing to operate in long term and relying on funds from higher governmental levels to support infrastructure advancements (Parkinson & Tayler, 2003).

The "zero-discharge" and/or "closing-the-loop" approaches directed at reprocessing the processed effluent for agro-based or municipal reuse are common characteristics of all the decentralization choices. In this context, the selected decentralized technology can provide water quality based on the successive use of the raw sewage, and the facilities implemented in a decentralized scheme should enable reusability (Bernal & Restrepo, 2012). Finally, an additional major technical concern is the ability to merge centralized and decentralized strategies, which would permit the deployment of decentralized treatment facilities. In the case of remote treatment facilities, even if merely for solid sludge processing, decentralized systems might be incorporated into the present centralized model (Libralato et al., 2012).

In the planning category of DWWT, six main challenges were identified: urban area plan, geographic location and land use, enhancement of urban development projects, institutional improvement, organizational changes, and political changes. All these fundamental difficulties are linked to the type of water management in a metropolis, in which a centralized strategy dominates with traditional "end-of-the-pipe" alternatives that have yet to demonstrate the predicted effectiveness in reducing pollution. In this regard, it is essential to combine urban planning with water resource development using geographic and temporal distribution and land use to establish the threshold of decentralization and assess the opportunities for reutilization within every field. In regard to the problem of decentralization, the significance of restructuring the governmental and organizational structure, as well as reinforcing the potential of organizations in the water and hygiene areas, could be a vital tool to guarantee the long-term viability of these choices and their articulation with already functioning centralized strategies (Burkhard et al., 2000).

1.9 CONCLUSIONS AND PERSPECTIVES

The socioeconomic and environmental benefits of decentralization are becoming vital aspects, particularly in peri-urban regions in which effluent accumulation and/or treatment is unavailable.

Considering advancements in both decentralized and centralized technology, it is now feasible to generate effluent water of the required quality. The task is to use treated wastewater, which in a way will lead to a sustainable future. Considering the studies in the chapter, it can be concluded that (I) DWWT will eventually become an essential component of forthcoming techniques to maintain and handle the country's water supplies; (II) the advancement of new techniques will revolutionize the on-site wastewater management; and (III) DWWT will attain similar rights with centralized treatment with impactful quality standards, application requirements, and management methods.

Modern decentralized infrastructure includes a range of approaches, technologies, and systems, such as (I) ultra-high water-use efficiency fixtures and appliances, along with in-building source separation; (II) small-diameter wastewater collection and conveyance networks; (III) reactor-based and landscape-based treatment unit operations; (IV) methods for monitoring system performance and ensuring its effectiveness; and (V) management systems designed to ensure long-term sustainability.

Decentralized graywater treatment has emerged as a frequent approach for WWT and water reclamation, effectively reducing the high demand for tap water in tourist destinations. Nevertheless, achieving a fully closed urban water cycle is just one facet of the circular economy concept. To realize its full potential, decentralized WWT using membrane technologies must also prioritize resource recovery as a crucial objective. Resource recovery scenarios with decentralized WWT have shown positive environmental outcomes, even with high energy consumption in the operation of the anaerobic membrane bioreactor system, resulting in an approximately 9% reduction in global warming potential.

Addressing future challenges in the DWWT field involves creating databases to measure gaseous emissions, developing correlations for energy consumption at different scales, and conducting life cycle assessment (LCA) studies in situations of environmental and political concern. For instance, comparing regular operation with a WWT scenario where the centralized facility has exceeded its capacity due to heavy rainfall or population growth could be an example of such studies.

REFERENCES

Ahmad, M., Ahmad, B., Nawazish-i-Hussain, S., Mahmood, S. 2003. Clinical investigations of skeletal fluorosis in children of Manga Mandi in Pakistan. *Pakistan Journal of Pharmaceutical Sciences*, **16**(2), 9–11.

Akpor, O., Muchie, B. 2011. Environmental and public health implications of wastewater quality. *African Journal of Biotechnology*, **10**(13), 2379–2387.

Allen, A., Dávila, J.D., Hofmann, P. 2006. *Governance of Water and Sanitation Services for the Peri-Urban Poor: A Framework for Understanding and Action in Metropolitan Regions*, The Development Planning Unit, University College London.

Angelakis, A., Bontoux, L., Lazarova, V. 2003. Challenges and prospectives for water recycling and reuse in EU countries. *Water Science and Technology: Water Supply*, **3**(4), 59–68.

Arora, M., Malano, H., Davidson, B., Nelson, R., George, B. 2015. Interactions between centralized and decentralized water systems in urban context: A review. *Wiley Interdisciplinary Reviews: Water*, **2**(6), 623–634.

Averill, D., Heinke, G. 1974. Vacuum sewer systems and their possible Canadian applications. *Canadian Journal of Civil Engineering*, **1**(1), 50–61.

Azizullah, A., Khattak, M.N.K., Richter, P., Häder, D.-P. 2011. Water pollution in Pakistan and its impact on public health—A review. *Environment International*, **37**(2), 479–497.

Bakir, H.A. 2001. Sustainable wastewater management for small communities in the Middle East and North Africa. *Journal of Environmental Management*, **61**(4), 319–328.

Bernal, D.P., Restrepo, I. 2012. Key issues for decentralization in municipal wastewater treatment. *12th Edition of the World Wide Workshop for Young Environmental Scientists (WWW-YES-2012)-Urban Waters: Resource or Risks?* HAL-ENPC.

Bernal, D., Restrepo, I., Grueso-Casquete, S. 2021. Key criteria for considering decentralization in municipal wastewater management. *Heliyon*, **7**(3), e06375.

Brixa, H., Koottatepb, T., Frydc, O., Laugesend, C.H. 2010. Appropriate and Sustainable Wastewater Management in Developing Countries by the use of Constructed Wetlands. *Proceedings of the 3rd IASTED African Conference*. p. 175.

Brown, V., Jackson, D., Khalifé, M. 2010. 2009 Melbourne metropolitan sewerage strategy: A portfolio of decentralised and on-site concept designs. *Water Science and Technology*, **62**(3), 510–517.

Burkhard, R., Deletic, A., Craig, A. 2000. Techniques for water and wastewater management: A review of techniques and their integration in planning. *Urban Water*, **2**(3), 197–221.

Capodaglio, A.G. 2017. Integrated, decentralized wastewater management for resource recovery in rural and peri-urban areas. *Resources*, **6**(2), 22.

Chirisa, I., Bandauko, E., Matamanda, A., Mandisvika, G. 2017. Decentralized domestic wastewater systems in developing countries: The case study of Harare (Zimbabwe). *Applied Water Science*, **7**, 1069–1078.

Chung, G., Lansey, K., Blowers, P., Brooks, P., Ela, W., Stewart, S., Wilson, P. 2008. A general water supply planning model: Evaluation of decentralized treatment. *Environmental Modelling & Software*, **23**(7), 893–905.

Corcoran, E., Nellemann, C., Baker, E., Sos, R., Osborn, D., & Savelli, H. (2010). Sick water? The central role of wastewater management. *Sustainable development: A rapid response assessment, UNEP, UN-HABITAT, GRID-Arendal, 88.*

Corominas, L., Byrne, D.M., Guest, J.S., Hospido, A., Roux, P., Shaw, A., Short, M.D. 2020. The application of life cycle assessment (LCA) to wastewater treatment: A best practice guide and critical review. *Water Research*, **184**, 116058.

Crossgrove, J., Zheng, W. 2004. Manganese toxicity upon overexposure. *NMR in Biomedicine: An International Journal Devoted to the Development and Application of Magnetic Resonance In Vivo*, **17**(8), 544–553.

Cytryn, E. 2013. The soil resistome: The anthropogenic, the native, and the unknown. *Soil Biology and Biochemistry*, **63**, 18–23.

de Freitas Alves, N.S., Westbrook12, A. 2016. Havana's Wastewater Treatment Plants: Changes Over Time and Estimate of Replacement Cost. *Annual Proceedings*. The Association for the Study of the Cuban Economy.

Dhote, J., Ingole, S., Chavhan, A. 2012. Review on wastewater treatment technologies. *International Journal of Engineering Research & Technology*, **1**(5), 1–10.

Elawwad, A., Ragab, M., Abdel-Halim, H. 2014. Vacuum sewerage system in developing regions and the impact on environmental management. *4th International Conference on Environmental Pollution and Remediation, August 11.*

Estahbanati, M.K., Kumar, S., Khajvand, M., Drogui, P., Tyagi, R.D. 2021. Environmental Impacts of Recovery of Resources from Industrial Wastewater. in: *Biomass, Biofuels, Biochemicals*, Elsevier, pp. 121–162.

Gallego-Schmid, A., Tarpani, R.R.Z. 2019. Life cycle assessment of wastewater treatment in developing countries: A review. *Water Research*, **153**, 63–79.

Guardabassi, L., Petersen, A., Olsen, J.E., Dalsgaard, A. 1998. Antibiotic resistance in *Acinetobacter* spp. isolated from sewers receiving waste effluent from a hospital and a pharmaceutical plant. *Applied and Environmental Microbiology*, **64**(9), 3499–3502.

Ho, G., Anda, M. 2006. Centralised versus decentralised wastewater systems in an urban context: the sustainability dimension. *2nd IWA Leading Edge on Sustainability in Water-Limited Environments. IWA Publishing*, 81–89.

Hoffmann, S., Feldmann, U., Bach, P.M., Binz, C., Farrelly, M., Frantzeskaki, N., Hiessl, H., Inauen, J., Larsen, T.A., Lienert, J. 2020. A research agenda for the future of urban water management: Exploring the potential of nongrid, small-grid, and hybrid solutions. *Environmental Science & Technology*, **54**(9), 5312–5322.

Hophmayer-Tokich, S. 2006. *Wastewater Management Strategy: Centralized v. Decentralized Technologies for Small Communities*, Center for Clean Technology and Environmental Policy, p. 27.

Howard, G., Bartram, J., World Health Organization. 2010. *Vision 2030: The Resilience of Water Supply and Sanitation in the Face of Climate Change: Technical Report.* World Health Organization.

Ingraham, J.L., Ingraham, C.A. 1998. *Introducción a la microbiología. II*, Reverté.

Kabir, E.R., Rahman, M.S., Rahman, I. 2015. A review on endocrine disruptors and their possible impacts on human health. *Environmental Toxicology and Pharmacology*, **40**(1), 241–258.

Kataki, S., Chatterjee, S., Vairale, M.G., Sharma, S., Dwivedi, S.K. 2021. Concerns and strategies for wastewater treatment during COVID-19 pandemic to stop plausible transmission. *Resources, Conservation and Recycling*, **164**, 105156.

Kjerstadius, H., Saraiva, A.B., Spångberg, J., Davidsson, Å. 2017. Carbon footprint of urban source separation for nutrient recovery. *Journal of Environmental Management*, **197**, 250–257.

Kujawa-Roeleveld, K., Zeeman, G. 2006. Anaerobic treatment in decentralised and source-separation-based sanitation concepts. *Reviews in Environmental Science and Bio/Technology*, **5**(1), 115–139.

Kumar, S., Mostafazadeh, A.K., Kumar, L.R., Tyagi, R., Drogui, P., Brien, E. 2022. Advancements in laundry wastewater treatment for reuse: A review. *Journal of Environmental Science and Health, Part A*, **57**(11), 927–946.

Kumar, S., Talan, A., Boyle, K., Ormeci, B., Drogui, P., Tyagi, R.D. 2021. Water Recycling: Economic and Environmental Benefits. in: *Biomass, Biofuels, Biochemicals*, Elsevier, pp. 91–120.

Lahmouri, M., Drewes, J.E., Gondhalekar, D. 2019. Analysis of greenhouse gas emissions in centralized and decentralized water reclamation with resource recovery strategies in Leh Town, Ladakh, India, and potential for their reduction in context of the water–energy–food nexus. *Water*, **11**(5), 906.

Leigh, N.G., Lee, H. 2019. Sustainable and resilient urban water systems: The role of decentralization and planning. *Sustainability*, **11**(3), 918.

Libralato, G., Ghirardini, A.V., Avezzù, F. 2012. To centralise or to decentralise: An overview of the most recent trends in wastewater treatment management. *Journal of Environmental Management*, **94**(1), 61–68.

Ma, L., Li, B., Zhang, T. 2014. Abundant rifampin resistance genes and significant correlations of antibiotic resistance genes and plasmids in various environments revealed by metagenomic analysis. *Applied Microbiology and Biotechnology*, **98**(11), 5195–5204.

Mankad, A., Tapsuwan, S. 2011. Review of socio-economic drivers of community acceptance and adoption of decentralised water systems. *Journal of Environmental Management*, **92**(3), 380–391.

Massoud, M.A., Tarhini, A., Nasr, J.A. 2009. Decentralized approaches to wastewater treatment and management: Applicability in developing countries. *Journal of Environmental Management*, **90**(1), 652–659.

Metcalf, E.E., Eddy, H. 2003. *Wastewater Engineer: Treatment Disposal, Reuse*, McGraw, p. 191.

Mihai, F.-C., Gündoğdu, S., Markley, L.A., Olivelli, A., Khan, F.R., Gwinnett, C., Gutberlet, J., Reyna-Bensusan, N., Llanquileo-Melgarejo, P., Meidiana, C. 2022. Plastic pollution, waste management issues, and circular economy opportunities in rural communities. *Sustainability*, **14**(1), 20.

Molle, F., Sanchis-Ibor, C. 2019. Irrigation policies in the Mediterranean: Trends and challenges. *Irrigation in the Mediterranean*, **22**, 279–313.

Moura, A., Pereira, C., Henriques, I., Correia, A. 2012. Novel gene cassettes and integrons in antibiotic-resistant bacteria isolated from urban wastewaters. *Research in Microbiology*, **163**(2), 92–100.

Muzioreva, H., Gumbo, T., Kavishe, N., Moyo, T., Musonda, I. 2022. Decentralized wastewater system practices in developing countries: A systematic review. *Utilities Policy*, **79**, 101442.

Nanninga, T.A., Bisschops, I., López, E., Martínez-Ruiz, J.L., Murillo, D., Essl, L., Starkl, M. 2012. Discussion on sustainable water technologies for peri-urban areas of Mexico City: Balancing urbanization and environmental conservation. *Water*, **4**(3), 739–758.

Nations, U. 2018. *The World's Cities in 2018—Data Booklet*. New York: Department of Economic and Social Affairs, Population Division.

Nguyen, T.T., Ngo, H.H., Guo, W., Wang, X.C., Ren, N., Li, G., Ding, J., Liang, H. 2019. Implementation of a specific urban water management – Sponge city. *Science of the Total Environment*, **652**, 147–162.

Nhapi, I. 2004a. A framework for the decentralised management of wastewater in Zimbabwe. *Physics and Chemistry of the Earth, Parts A/B/C*, **29**(15–18), 1265–1273.

Nhapi, I. 2004b. Potential for the use of duckweed-based pond systems in Zimbabwe. *Water SA*, **30**(1), 115–120.

Nhapi, I. 2009. The water situation in Harare, Zimbabwe: A policy and management problem. *Water Policy*, **11**(2), 221–235.

Okoh, A.I., Odjadjare, E.E., Igbinosa, E.O., Osode, A.N. 2007. Wastewater treatment plants as a source of microbial pathogens in receiving watersheds. *African Journal of Biotechnology*, **6**(25), 2932–2944.

Ozgun, H., Dereli, R.K., Ersahin, M.E., Kinaci, C., Spanjers, H., van Lier, J.B. 2013. A review of anaerobic membrane bioreactors for municipal wastewater treatment: Integration options, limitations and expectations. *Separation and Purification Technology*, **118**, 89–104.

Paillard, D., Dubois, V., Thiebaut, R., Nathier, F., Hoogland, E., Caumette, P., Quentin, C. 2005. Occurrence of *Listeria* spp. in effluents of French urban wastewater treatment plants. *Applied and Environmental Microbiology*, **71**(11), 7562–7566.

Panfil, C., Mirel, I., Szigyarto, I., Isacu, M. 2013. Technical, economical, social and ecological characteristics of vacuum sewage system. *Environmental Engineering & Management Journal (EEMJ)*, **12**(5), 1017–1022.

Parkinson, J., Tayler, K. 2003. Decentralized wastewater management in peri-urban areas in low-income countries. *Environment and Urbanization*, **15**(1), 75–90.

Pasciucco, F., Pecorini, I., Iannelli, R. 2022. Planning the centralization level in wastewater collection and treatment: A review of assessment methods. *Journal of Cleaner Production*, **375**, 134092.

Polyzos, S.A., Kountouras, J., Deretzi, G., Zavos, C., Mantzoros, C.S. 2012. The emerging role of endocrine disruptors in the pathogenesis of insulin resistance: A concept implicating nonalcoholic fatty liver disease. *Current Molecular Medicine*, **12**(1), 68–82.

Pruden, A., Arabi, M., Storteboom, H.N. 2012. Correlation between upstream human activities and riverine antibiotic resistance genes. *Environmental Science & Technology*, **46**(21), 11541–11549.

Qu, J., Wang, H., Wang, K., Yu, G., Ke, B., Yu, H.-Q., Ren, H., Zheng, X., Li, J., Li, W.-W. 2019. Municipal wastewater treatment in China: Development history and future perspectives. *Frontiers of Environmental Science & Engineering*, **13**(6), 1–7.

Roefs, I., Meulman, B., Vreeburg, J.H., Spiller, M. 2017. Centralised, decentralised or hybrid sanitation systems? Economic evaluation under urban development uncertainty and phased expansion. *Water Research*, **109**, 274–286.

Samuel, S., Lim, M., Hsien, C., Ho, A., Schraudolph, U., White, S. 2016. Blue buildings: Decentralized and integrated management of water from 'source-to-source, at source'. *Water Practice and Technology*, **11**(3), 601–609.

Scherson, Y.D., Criddle, C.S. 2014. Recovery of freshwater from wastewater: Upgrading process configurations to maximize energy recovery and minimize residuals. *Environmental Science & Technology*, **48**(15), 8420–8432.

Shen, Y., Linville, J.L., Urgun-Demirtas, M., Mintz, M.M., Snyder, S.W. 2015. An overview of biogas production and utilization at full-scale wastewater treatment plants (WWTPs) in the United States: Challenges and opportunities towards energy-neutral WWTPs. *Renewable and Sustainable Energy Reviews*, **50**, 346–362.

Sönmez, N., Tengnäs, A., McGrath, V. 2011. *Design of Sustainable Sanitation Systems for IDPA Corail Camp in Haiti*. The École des Mines de Nantes.

Steffen, W., Richardson, K., Rockström, J., Cornell, S.E., Fetzer, I., Bennett, E.M., Biggs, R., Carpenter, S.R., De Vries, W., De Wit, C.A. 2015. Planetary boundaries: Guiding human development on a changing planet. *Science*, **347**(6223), 1259855.

Terryn, I.C.C., Lazar, I., Nedeff, V., Lazar, G. 2014. Conventional vs. vacuum sewerage system in rural areas—An economic and environmental approach. *Environmental Engineering and Management Journal*, **13**(8), 1847–1859.

van Afferden, M., Cardona, J.A., Lee, M.-Y., Subah, A., Müller, R.A. 2015. A new approach to implementing decentralized wastewater treatment concepts. *Water Science and Technology*, **72**(11), 1923–1930.

Verstraete, W., Vlaeminck, S.E. 2011. ZeroWasteWater: Short-cycling of wastewater resources for sustainable cities of the future. *International Journal of Sustainable Development & World Ecology*, **18**(3), 253–264.

Water, U. 2006. Kenya National Water Development Report: A WWAP Case Study Prepared for the 2nd UN World Water Development Report: Water, a Shared Responsibility (2006). Un-Water/Wwap/2006/12. Nairobi.

Wilderer, P.A., Schreff, D. 2000. Decentralized and centralized wastewater management: A challenge for technology developers. *Water Science and Technology*, **41**(1), 1–8.

2 Circular Economy and Sustainable Development

Manojkumar Y.[1], *Sridhar Pilli*[1], *P. Venkatsewara Rao*[1], *and Rajeshwar D. Tyagi*[2]
[1] National Institute of Technology, Warangal, India
[2] Dongguan University of Technology, Dongguan, China

2.1 INTRODUCTION

The problems associated with waste and wastewater management are complicated and dynamic due to urbanization, climate change, and population expansion. The concentration of urban trash necessitates the use of the proper infrastructure, transportation, and disposal facilities. Financing, inadequate household waste separation, a lack of secure landfills, illegal dumping, burning, and street litter are additional issues that prevent nations from achieving sustainability. Other issues include low public awareness, insufficient instruction in proper waste management techniques, and weak formal institutions (Kumar et al. 2017; Raab et al. 2021).

Water that has had an adverse anthropogenic (human) effect is referred to as wastewater. It can come from a variety of sources, including household, commercial, industrial, agricultural, surface runoff or stormwater, and sewage input or infiltration (Cosgrove & Loucks 2015). Ordinarily, untreated sewage deposited into a body of water may naturally clean itself through stream cleaning and self-purification. However, sewage discharge has increased at a pace that is significantly greater than the rate of natural purification as a result of population growth and widespread urbanization. The water body becomes eutrophicated as a result of the extra nutrients produced, gradually lowering the quality of the water.

Operational effectiveness is always of the highest significance in treatment facilities, and this has always been a driving force behind innovation. Although there have been significant improvements recently in the creation of efficient technology, difficulties still exist. One of the biggest costs of running a wastewater treatment is energy usage. The biological treatment process typically accounts for between 50 and 60% of plant energy use in the treatment of municipal wastewater. Operators of wastewater treatment plants must be professionals with the necessary training and certification. The waste produced during physical, chemical, and biological treatment is known as sludge. The disposal of extra sludge generated during the treatment process is a significant environmental concern for wastewater treatment. Wastewater treatment demands a footprint and is costly to construct and occupy substantial land areas.

The following are the harms brought about by improper wastewater management and the cost of inactivity.

- Increasing direct healthcare costs and indirect costs, due to an increase in disease or fatality from ingesting polluted water.
- Farmers lose money owing to lost production days.
- Increase in treatment costs of drinking water and industrial wastewater from productivity sectors.
- The tourism business depends on ecologically appealing locations; poor water quality turns away visitors and reduces income.

DOI: 10.1201/9781003381419-3

- Nutrients leak into the subsoil and contaminate groundwater.
- The usage of wastewater containing hazardous wastes can lead to an increase in waterborne illnesses and environmental deterioration.

In order to safeguard public health and avoid polluting water sources, wastewater management must be carried out effectively. Water utilization, weather, and sewage system condition are the key determinants of the produced wastewater quantity in the region.

2.2 RESOURCE AND WASTE MANAGEMENT CHALLENGES

Urbanization and population expansion are the major causes of high resource utilization and waste generation. Growing resource exploitation entails significant environmental and health hazards, including resource depletion, air, water, and soil pollution, climate change, and biodiversity loss. Managing urban garbage is expensive. When it comes to financing, expensive and sophisticated garbage operations must compete with other priorities, including access to clean water and other utilities, education, and healthcare. Ineffective garbage collection, ununiformed consumer behaviour, market-related factors, technological hurdles, complicated design issues, and hazardous nature cause challenges in waste management systems. Recycling is a process that involves both quantity and quality, but the quality of downcycling generally won't meet the criteria.

Wastewater from major cities is frequently discharged without treatment directly into rivers or the ocean, causing pollution and endangering the well-being of both people and ecosystems. When resources are lost, crucial raw materials that are essential to the operation of important industrial applications and industries are also lost.

Waste management statistics are crucial for developing policy and planning for the local context but are inaccurate. Future waste generation forecasts are generally imprecise. Weaker regulations and illegal exports are additional issues. The imprecision of current frameworks for evaluating material circularity makes it challenging to track the development of the circular economy.

2.3 WHAT IS CIRCULAR ECONOMY (CE)?

The "take-make-use-dispose" linear economic system is unsustainable since it relies heavily on the usage of non-renewable resources and harmful energy sources. Natural and circular economy systems were adopted throughout the pre-industrial era, whereas the linear economy was first implemented during the industrial era. It is our responsibility to move towards the circular economy. The circular economy has developed as a workable alternative to the linear economy, removing waste from the system, optimizing resource utilization, using commodities at their best value, using renewable energy sources, and replenishing natural resources continuously (Mallory et al. 2020).

The circular economy is defined as the actions that replace the end-of-life concept by reducing, recycling, and recovering materials in production, distribution, and consumption processes (Ellen Macarthur Foundation). A circular economy attempts to retain goods, parts, and materials at their greatest usability and worth at all times. It is restorative and regenerative by design. The more garbage is recycled and reused in an industrial plant, the more it conforms to the circular economy idea while being less damaging to the environment and more profitable. The circular economy is a regenerative system in which resource input, waste, pollution, and energy leakage are minimized by long-lasting design, maintenance, repair, reuse, re-manufacturing, refurbishing, recycling, and upcycling (Morseletto et al. 2022).

In a circular economy, resources are used less often; materials, goods, and services are redesigned to use less resources; and "waste" is recovered and used as a resource to create new goods and materials. Within the context of sustainable waste management, circularity is valued. The circular economy has the ability to advance social justice, enhance economics, and save the environment when it is thoughtfully and broadly constructed. Social fairness is fundamental to sustainability. Communities that are already vulnerable may be disproportionately impacted by how to acquire, use, and dispose of the resources (United States Environmental Protection Agency [USEPA] 2022).

The circular economy equips us with the means to simultaneously address pressing socioeconomic needs and the threat of biodiversity loss posed by climate change. It empowers us to increase resilience, prosperity, and employment while reducing greenhouse gas emissions, waste, and pollution. In a circular economy, goods are made with the intention of recycling the resources from which they were manufactured after they have reached the end of their useful lives. In the circular economy, it's important to reduce waste, material, and energy usage.

The principles of circular economy are as follows:

- Eliminate waste and pollution.
- Circulate products and materials.
- Regenerate nature.

The circular economy places a lot of focus on waste management. The waste hierarchy provides a hierarchy of waste management practices, starting with prevention and moving via energy recovery, preparation for reuse, recycling, and landfilling to disposal. This rule attempts to promote choices that have the best effects on the environment as a whole. The way the waste is collected and managed can result in high recycling rates and the economic recovery of valuable materials, or it can result in an ineffective system where the majority of recyclable waste is disposed of in landfills or burned, potentially harming the environment and causing significant economic losses. All trash should be taken into account, whether it is produced by homes, enterprises, industries, mines, or the building industry (European Commission 2015).

2.4 SUSTAINABLE DEVELOPMENT IN WASTE AND WATER TREATMENT TOWARDS THE CIRCULAR ECONOMY

Development that "meets the demands of the present without compromising the potential of future generations to satisfy their own needs" is referred to as sustainable development (United Nations [UN] 2015). Sustainability refers to the potential for the ecosystem and human civilization to coexist and is the ability to continue existing continuously. Only when the demand for productive land (ecological footprint) is equal to or lower than the supply of productive land available can environmental sustainability be achieved (biocapacity). Economic, environmental, and social sustainability are the three pillars of sustainable development.

The "5R" idea of a closed-cycle economy may be used to summarize the essence of sustainable development, which has direct economic and technical aid (Morseletto et al. 2022).

- Restriction of energy and material consumption
- Replacement of non-renewable resources with renewables
- Recovery of the necessary components from recycled waste
- Recycling
- Reuse of products

Water is essential to human existence and is closely connected to both the environment and the economy (Chen 2016). Regardless of its significance, over 80% of the wastewater produced by civilization is disposed of in the environment without being recycled or treated. Due to the rising population and a more prosperous economy, there is an ongoing increase in the amount of wastewater generated and its overall pollutant load. Untreated wastewater pollution is a threat to human health. The role of sustainable development in waste and water treatment is portrayed in Figure 2.1.

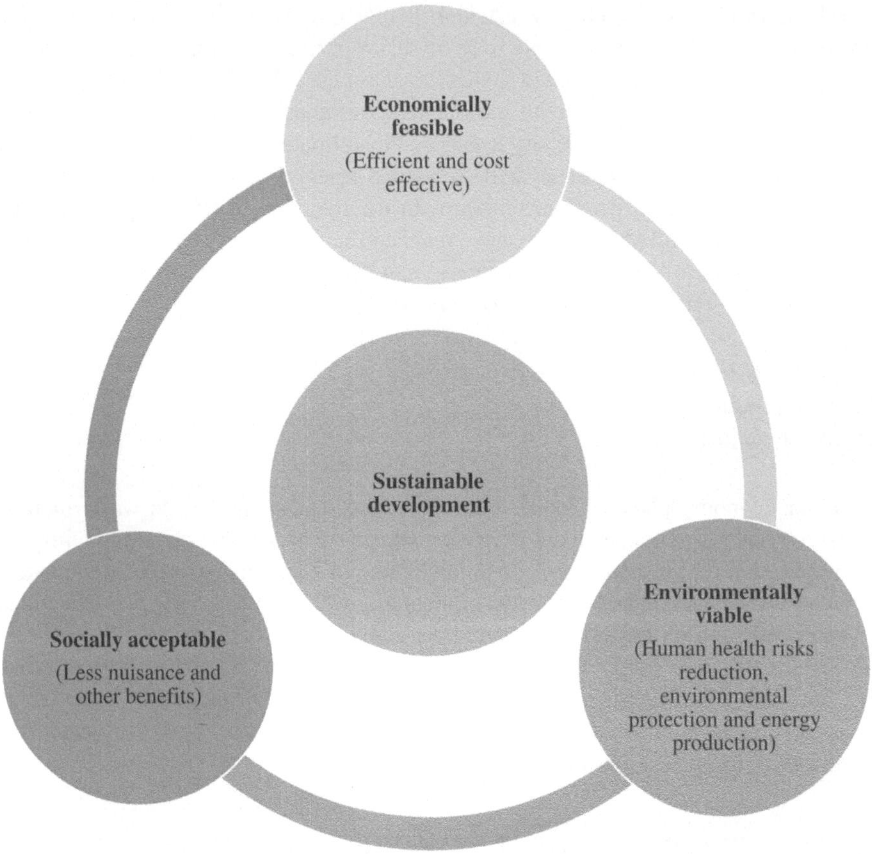

FIGURE 2.1 Sustainable development in waste and water treatment.

Wastewater may be a resource if it is managed appropriately (Morseletto et al. 2022).

- Its nutrients may be used to produce crops, which is very advantageous for farming communities.
- It is a resource that is renewable and might be recovered after usage.
- The usage of wastewater will significantly reduce the overstretching of communal potable water.
- The reclaimed water can be used for toilets and urinals in public buildings, fire protection, etc.
- The reclaimed water can be used for irrigation of public parks, athletic fields, recreation centres, school yards, and playing fields.
- Additionally, treated wastewater can be recycled in the industry (for instance, in cooling towers) and for artificial aquifer replenishment and restoration of native ecologies.
- Rarely, it is also used to increase the supply of drinking water.
- Not all water-intensive activities need high-quality water. So, minimal treatment is required to reuse wastewater.

2.5 ROLE OF DECENTRALIZED SANITATION IN THE CIRCULAR ECONOMY

Decentralized sanitation and reuse, which takes a multidisciplinary perspective, discuss the most advanced sanitation technology now available and positions public sanitation in a global context. It demonstrates how these technologies may be used to combine the treatment of residential wastewater and waste in order to optimize resource recycling in daily life. Decentralized sanitation and reuse focus on the trade-off between dependability and minimum water wastage and offer technical solutions for on-site collection and delivery of concentrated waste streams. A wide range of sustainable decentralized approaches, both low-tech and high-tech, are available to treat concentrated (black water) and diluted (grey water) streams. Decentralized sanitation and reuse will be invaluable resources for treating wastewater (Zeemann & Lettinga 2001).

Decentralization seems like a reasonable way to address the sustainability issues with wastewater management systems since it concentrates on local resource recycling and reuse as well as on-site wastewater treatment. Decentralized solutions often tend to be consistent with regional water usage and reuse needs, where locally treated water might improve agricultural production or be utilized as a substitute for supplying water of a potable quality for appropriate purposes. Social, economic, and environmental factors should be considered while examining the sustainability of technology (Capodaglio 2017).

Most developing nations rely on on-site wastewater systems, including septic tanks, drain fields, and on-site treatment units, to handle wastewater generated in locations without access to centralized sewage networks. To achieve universal access to sanitation, decentralized alternatives must be created that, by cutting off the most deprived areas from sewage networks and centralized treatment, provide them access to the bare necessities of health and dignity. In isolated rural regions and slums where the construction of centralized treatment facilities or sewage networks is impractical, they offer an excellent alternative. They provide communities with a way to benefit from self-management and the circular economy.

The decentralized approach becomes a viable option to take into account in small communities or growing residential areas where bioenergy production can be improved by recovering organic matter in separate streams, or where building a sewer network to connect to a centralized facility may be technologically or financially impractical. The major goal is to compare the hybrid-decentralized configuration to the centralized treatment scenario in order to assess the environmental and economic profile for effective wastewater management and resource recovery. The decentralized treatment gives the opportunity to value digestate streams as fertilizer sources for horticultural or decorative crops nearby in addition to water reclamation. The environmental profile results demonstrate that, although it demands energy in the toilets, the new scenarios in which energy and nutrients are recovered from black water result in a benefit for the environment. Given that the suggested topologies are adaptable systems that enable resource recovery close to the site of generation, they are suitable for incorporating wastewater treatment programmes into the circular economy (Estévez et al. 2022).

In Europe advancements like recycling energy and resources, new holistic communities will be built; instead of treating wastewater, wastewater treatment plants will operate as energy and resource factories. The holistic perspective encompasses source separation, decentralized systems, new energy solutions, integrated or novel ways to combat climate change, water reuse, food and resource utilization from wastewater, and new treatment objectives (China Europe Water Platform 2020).

Less than 80% of people were linked to public urban wastewater treatment systems, and there were 3,278 municipal wastewater treatment facilities operating in Poland. Different technologies are used in wastewater treatment facilities as a result of decentralization, and environmental, economic, and social factors influence the pace of change and the adoption of a circular economy paradigm in the Polish wastewater industry (Kacprzak & Kupich 2021).

Jamestown in the United States is a tiny island settlement that relies on individual sewage systems and private wells for drinking water. Groundwater quality was being impacted by poorly maintained on-site wastewater systems on small lots with high seasonal water tables. According to studies, 32% of the local wastewater treatment facilities are to blame for the microbial and nutrient issues in private water wells. A law was passed in Jamestown requiring regular inspections of each wastewater system. A High Groundwater Table District also directs future construction to safeguard the quality of drinking water. Therefore, 50 systems have been the target of town-initiated repair and replacement efforts since 2003 (USEPA 2012).

In New York, the lake and groundwater supply the drinking water to 20,000 inhabitants of the Keuka Lake watershed. Nearly all of the inhabitants of the watershed likewise rely on decentralized, closely spaced wastewater systems that drain to the soil for treatment. Testing, however, showed that poorly maintained individual on-site systems were polluting drinking water wells and adding excessive amounts of germs to the lake. A regional watershed cooperative comprising eight towns put in place consistent permits and inspection mechanisms to find and fix or replace broken treatment facilities. As a result, Keuka Lake has some of the best water quality in the Finger Lakes area (USEPA 2012).

More than 1,000 relatively tiny lots were created in the California subdivision of Auburn Lake Trails throughout the 1970s and 1980s as a recreational location near Auburn Lake in a region with shallow, low-permeability soils and steep hills. When planners realized that the area's soils could not properly handle the waste to safeguard water supplies after construction was complete, they suggested constructing a centralized sewage collection and treatment system. Residents, however, objected to it as being too expensive. The Georgetown Divide Public Utility District (PUD) has been permitted by the community to develop and operate individual and clustered wastewater systems with conventional and advanced treatment. The PUD created a method where maintenance and monitoring schedules are determined by the kind of system and where the required performance levels for treatment systems are linked to health and environmental risk (USEPA 2012).

For the 800 residents of Peña Blanca, New Mexico, outdated, abandoned, or nonexistent wastewater infrastructure constituted a threat to public health. Open cesspools and seepage pits discharged waste onto irrigation canals and yards. According to surveys, 86% of individual wastewater systems need maintenance or replacement. A planned consolidated sewer system that would have cost $3.1 million was rejected by the locals. The water and sanitation district is now in charge of operating and managing both the renovated and new facilities after the community decided to repair or replace 133 of the 185 treatment systems that were already in existence. For the 133 households served by repaired or new systems, the decentralized wastewater alternative was less expensive than anticipated compared to central sewage treatment. There are no longer any cesspool discharges or sewage surfacing in the neighbourhood (USEPA 2012).

For individual systems on lots with a minimum size of three acres, Missouri enacted more rigorous public health rules in 1995. Property owners in Phelps County were required to update each of their systems in order to adhere to those laws. The fact that local lenders would not offer loans on homes that were not in line with state regulations served as more evidence that processes needed to be upgraded. The water and sanitation district is in charge of operating and managing the renovated and new facilities after the community decided to replace or repair 133 of the 185 treatment systems that were already in existence. Nitrate levels in the project region were 1 mg/L or less after construction, which is far lower than the 10 mg/L guideline for groundwater used for drinking (USEPA 2012).

In Agra's Kachpura village, a decentralized wastewater treatment system has been in operation since 2010. The three-chamber septic tank receives the wastewater. It moves to nine chambered baffled anaerobic reactors after primary treatment. The programme's objective was to enhance the slum neighbourhoods' sanitary conditions. The system handles around 50 kilolitres per day (KLD) of the total wastewater it collects from five slum clusters via a shared drain (NIUA 2016).

Since 2005, Bangalore has had a decentralized wastewater treatment system operating at the neighbourhood level for the Beedi Workers colony. Wastewater moves through an anaerobic baffled reactor and planted gravel bed before treated water is finally collected in a collection tank. The colony's occupants use the generated biogas for cooking (NIUA 2016).

Decentralized wastewater management is essential to the circular economy since it is economically feasible, environmentally sound, and widely accepted. Focus has to be shifted from regulatory governance to infrastructure, taking into account the elements essential to the long-term viability of decentralized wastewater management. Decentralized sanitation functions as a closed loop and

TABLE 2.1
Characteristics and Impact of Decentralized Systems

S. No.	Decentralized Systems Characteristics	Impact on Parameters
1	Capital is high, and operation and maintenance costs are low	Economical (cost-effective)
2	Biogas production, emissions reduction, reuse and recycling of wastewater	Environmental (sustainable and less polluting)
3	Though compliance with regulations, people's acceptance is comparatively less	Socio-cultural (less acceptable)
4	Effluents free from pollutants, pathogens, and hazardous substances	Hygiene and human health (risk and disease reduction)
5	Government and stakeholders support individuals through knowledge and equipment transfer	Technology (easy and functional)
6	Discomfort due to odour, noise, land use, etc.	Aesthetics (may affect)

offers an appealing substitute for traditional sanitation (Lourenço & Nunes 2020). Table 2.1 shows the circular economy parameters fulfilled by the decentralized water treatment systems.

The advantages of the circular economy may be developed and put into practice with the help of decentralized systems. Waste is given value, and the system as a whole encourages engagement in other local sectors like agriculture with fertilizers or biogas for electricity and transportation. They contribute to community cohesiveness and empowerment by promoting participation, promoting self-management, and encouraging the sharing of information. They are an essential connection in providing everyone with sustainable sanitation. Figure 2.2 illustrates the various available decentralized treatment technologies.

2.6 FACTORS FOR SHIFTING TOWARDS THE CIRCULAR ECONOMY

Due to resource depletion, the rising world population, and an increase in waste, newer and more sustainable forms of consumption are practised. The parameters which influence the circular economy are organization, finance, technology, knowledge, supply chain, market, government, society, and the environment. The circular economy's initial objective is to positively affect natural systems

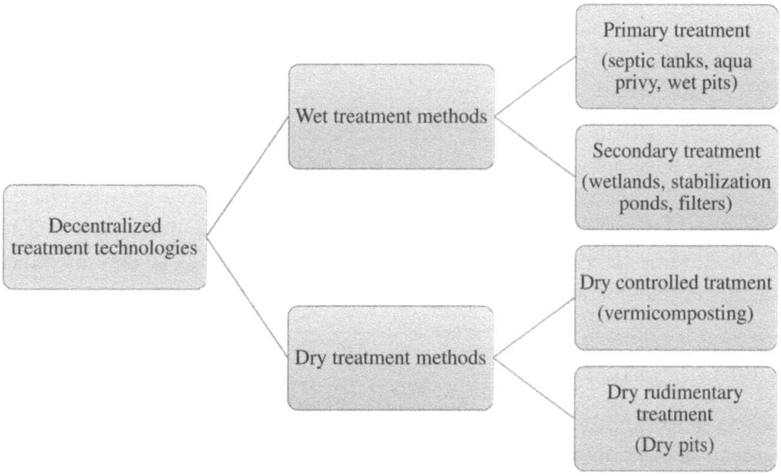

FIGURE 2.2 Types of decentralized treatment technologies.

without overloading or depleting them. A circular economy produces less greenhouse gas emissions, ensures the sustainability of the land, air, and water, and protects natural areas.

2.6.1 Less Greenhouse Gases

Circular economy calculations show that only 38% of greenhouse gas emissions are produced during the supply and use of goods and services, while 62% of global greenhouse gas emissions (excluding those from land use and forestry) are caused by the extraction, processing, and production of goods to meet societal needs. The circular economy, a regenerative system that reduces resource input and waste, emissions, and energy leakage by limiting, shutting, and narrowing energy and material loops, is therefore a key component of the answer to climate change. Therefore, the circular economy will mitigate global greenhouse gas emissions (Lehtoranta et al. 2022; United Nations Climate Change [UNCC] 2019).

2.6.2 Vital Soil, Air, and Water Bodies

Circularity in the economy produces essential ecosystems, including soil, air, and water bodies. These ecosystems offer products like clean drinking water, pollination, and productive farmland in addition to services like cleaning. In a linear economy, these services eventually become overloaded by the continual dumping of toxins or product withdrawal, respectively. The soil, air, and water bodies continue to be robust and productive if these goods are employed in a cycle and the services are not burdened by hazardous compounds.

2.6.3 Conservation of Nature Reserves

Nature reserves suffer from the exploitation of raw resources and the disposal of garbage. These natural regions are crucial for maintaining ecosystem functions, as well as our natural and cultural legacy. At the time, the major focus of many governments and organizations is safeguarding nature from resource exploitation and garbage disposal (Lehtoranta et al. 2022).

2.6.4 Substantial Resource Savings

In principle, the circular economy requires no new virgin raw materials because 100% of all raw materials are totally recycled. Because all existing used items must be recycled, achieving this situation will take a very long period.

2.6.5 Economic Growth

The circular economy would undoubtedly achieve economic development. A skilled workforce is necessary for the development, manufacture, and maintenance of circular goods, which will result in an increase in employment. There are several benefits of circularity for the economy. The global economy would gain $2 trillion by 2050 from better resource management (United Nations Environment Programme [UNEP] 2017). The price of raw resources will significantly drop, fostering innovation and employment.

2.6.6 Growth of Employment

Labour is valued more highly in a circular economy than in raw commodities. These occupations will increase as a result of labour-intensive recycling and high-quality repairs, local product take-back jobs, new businesses created via innovation and the service economy, and new business models.

2.6.7 Innovation Stimulus

Innovative ideas built on a fresh way of thinking are challenged by circular economics. This leads to fresh perspectives, multidisciplinary collaboration between manufacturers, designers, and recyclers, and, ultimately, sustainable inventions.

2.6.8 Changing Demand

Circularity changes consumer demand, which reduces raw material use, reduces waste output, and modifies manufacturing.

2.7 RESOURCE-ORIENTED SANITIZATION TOWARDS THE CIRCULAR ECONOMY

A normative objective that necessitates cost information is a circular economy approach to sanitation. Understanding the costs to build, run, and maintain resource-oriented systems is necessary to implement a circular economy approach to sanitation. However, the lack of information on the costs of resource-oriented systems in particular and urban sanitation in general limits the potential to advance sustainable sanitation in low- and middle-income nations. A deliberate and fair shift towards circular economy methods to sanitation is supported by policy and planning that is informed by the examination of overall costs, financial perspectives, and reuse specifics. Reuse-focused sanitation, regenerative sanitation, sustainable sanitation, resource recovery and reuse, and ecological sanitation are some of the labels used to describe sanitation strategies that include reuse. Reuse connected to sanitation has several cross-sectoral advantages, such as partially offsetting treatment costs, offering substitutes for expensive or imported inputs, and facilitating resource access for low-income communities (Carrard et al. 2021).

A resource orientation can enhance a sanitation service's overall financial profile and partially offset the needed public investment. The benefits of resource-focused sanitation are widely known and extensive. Resource-oriented systems need to be studied further and put into practice at the municipal or community level. As evidenced by literature linking circular economy concepts with sanitation to date, implementing such systems at scale necessitates a shift in perspective away from "whether" resource-oriented systems can achieve cost recovery or encourage investment and towards "how" can effectively finance and manage systems where reuse is the necessary end-point of the sanitation chain (Carrard et al. 2021).

The transition to a circular economy necessitates strategic infrastructure investments as well as policy coherence, coordination, and stakeholder participation across sectors and governance levels. Resources like water, energy, and nutrients are recovered from organic waste streams like faeces, wastewater, and food waste through the use of products like compost, biogas, and recycled water in resource-oriented sanitation and waste management systems (Ddiba et al. 2020).

2.8 CIRCULAR ECONOMY: PRACTICAL CHALLENGES WITH RESPECT TO DECENTRALIZATION

The most significant challenges of decentralized systems are the lack of financial resources and legislation to support implementing the circular economy in the water sector. Other challenges are limited awareness of the benefits, public perception, health concerns, lack of technology readiness, high technology, energy costs for wastewater treatment, lack of experienced manpower, lack of promoting or supporting policies, lack of awareness about water scarcity, low public acceptance of water reuse, the unclear economic benefits, lack of process design, the lack of government support and encouragements, the lack of policies and protocols, lack of infrastructure, lack of involvement of stakeholders and citizens, data uncertainty, lack of expertise, lack of coordination and knowledge, etc. (Liu et al. 2021).

Public involvement in waste management raises serious challenges because residents often don't take responsibility for their waste. To create effective and sustainable waste management systems, it is essential to raise community knowledge and alter people's attitudes about waste (Kumar et al. 2017).

Decentralized systems typically demand local users to be more informed and involved than centralized ones. Local stakeholders are typically more proactive when contemplating these systems, and the choice to install a decentralized solution to wastewater treatment demands is typically made or addressed at the local level. When their goals and benefits are made apparent to consumers, as is typically the case with other environmental projects, they may be extremely favourably received (Capodaglio 2017).

To help with the transition to a fully circular economy of water and wastewater treatment, policymakers should focus on three primary issues: normative, governance, and implementation obstacles.

Normative challenges are situations when a new law or changes to current legislation are required. The use of enabling mechanisms, such as a system of rewards and incentives for adopting circular economy efforts, such as nutrient recovery, or the use of disincentives, like taxation, to discourage wasteful linear processes, should be taken into account when developing normative features (Morseletto et al. 2022).

Roles and duties for the development and implementation of circular economy initiatives are among the governance problems. In light of regional specifics and contextual variables, governments and water authorities are responsible for establishing the most efficient water governance processes and arrangements (such as centralized/decentralized; public, private, or mixed solutions) for the circular economy. All necessary parties must be involved for a smooth transition to the circular economy, especially when rethinking a system (Morseletto et al. 2022). Barriers and chances for systematic use of the circular economy are examples of implementation issues. Further pilot phase studies have to be conducted to achieve many circular economy solutions which need sufficient funding to be adopted widely (Morseletto et al. 2022).

2.8.1 Other Major Challenges of Decentralized Systems

- The toughness of systemic change promotion
- Economic difficulties (a circular economy may be unprofitable in the short term)
- Inadequate markets (lack of required goods and facilities, competition, market knowledge, and/or incentives)
- Poor legislation and/or execution (imperfect regulation)
- Social aspects (insufficient knowledge and skills related to the circular economy)
- The inability to get sufficient money
- The absence of uniform practices

2.9 CONCLUSIONS AND PERSPECTIVES

The potential benefits of using wastewater as a resource are thus tremendous. Wastewater that is safely handled offers accessible, long-lasting sources of water, energy, nutrients, and other recoverable components. Current methods of wastewater management result in ineffective nutrient recovery and reuse, which can have negative effects on the ecosystem. To develop sustainable sanitation services towards a circular economy where wastewater is realized as a useful resource rather than a burden, decentralized wastewater treatment systems could be helpful. Through water conservation and effective usage, the circular economy may support sustainable water management, which helps achieve many other sustainable development objectives. In this situation, circular economy tactics are essential to the development and implementation of sustainable water treatment systems. Water supply and sanitation systems, as well as the water utilities that run them, can be sustained by recovering resources from wastewater facilities in the form of electricity, reusable water, biosolids, and other resources, such as nutrients. Adopting circular economy concepts in wastewater processing has a number of benefits, one of which is resource recovery and reuse, which may change sanitation from being an expensive service to one that is self-sustaining and contributes value to the economy. In order for circular economy plans and solutions to be successfully implemented, favourable circumstances are necessary. Water usage and energy usage have already been significantly decreased because of decentralized water and wastewater infrastructure. The world community has acknowledged the promise of circular economy methods to sanitation to solve both human and environmental issues and achieve sustainable development. In the circular economy, there is a chance to make sure that investments are done in the most effective and sustainable manner possible as cities continue to expand. Future urban development calls for strategies that reduce resource use and concentrate on resource recovery, in line with the circular economy's guiding principles. Increasing ambition and taking real action in the direction of more widespread adoption and optimization of circular economy methods for sanitation can be aided by accumulating data regarding the costs of resource-oriented sanitation systems in actual practice.

REFERENCES

Capodaglio, Andrea G., "Integrated, decentralized wastewater management for resource recovery in rural and peri-urban areas." *Resources* 6, no. 2 (2017): 22.

Carrard, Naomi, Nilanthi Jayathilake, and Juliet Willetts, "Life-cycle costs of a resource-oriented sanitation system and implications for advancing a circular economy approach to sanitation." *Journal of Cleaner Production* 307 (2021): 127135.

Chen, G., "Challenges in water resources and wastewater treatment." *The International Journal of Water and Wastewater Treatment* 2, no. 2 (2016). http://dx.doi.org/10.16966/2381-5299.e102.

China Europe Water Platform, Circular Economy of Municipal Wastewater Treatment Plants in Europe. 26.06.2020. https://www.cewp.eu/sites/default/files/2021-05/CEWP_Whitepaper_Circular%20Economy%20of%20Water%20in%20Waste%20Water%20Treatment_final.pdf.

Cosgrove, William J., and Daniel P. Loucks, "Water management: Current and future challenges and research directions." *Water Resources Research* 51, no. 6 (2015): 4823–4839.

Ddiba, Daniel, Kim Andersson, Steven H.A. Koop, Elisabeth Ekener, Göran Finnveden, and Sarah Dickin, "Governing the circular economy: Assessing the capacity to implement resource-oriented sanitation and waste management systems in low-and middle-income countries." *Earth System Governance* 4 (2020): 100063.

Ellen Macarthur Foundation. https://ellenmacarthurfoundation.org/topics/circular-economy-introduction/overview (referred as on date 10.11.2022).

Estévez, Sofía, Sara González-García, Gumersindo Feijoo, and María Teresa Moreira, "How decentralized treatment can contribute to the symbiosis between environmental protection and resource recovery." *Science of the Total Environment* 812 (2022): 151485.

European Commission, 2015. Communication from the Commission to the European Parliament, the Council, the European Economic and Social Committee and the Committee of the Regions. Closing the loop – An EU Action Plan for the Circular Economy. https://eur-lex.europa.eu/resource.html?uri=cellar:8a8ef5e8-99a0-11e5-b3b7-01aa75ed71a1.0012.02/DOC_1&format=PDF (referred as on date 08.11.2022).

Kacprzak, Malgorzata J., and Iwona Kupich, "The specificities of the circular economy (CE) in the municipal wastewater and sewage sludge sector—Local circumstances in Poland." *Clean Technologies and Environmental Policy* (2023): 25, 519–535.

Kumar, Sunil, Stephen R. Smith, Geoff Fowler, Costas Velis, S. Jyoti Kumar, Shashi Arya, Rakesh Kumar, and Christopher Cheeseman, "Challenges and opportunities associated with waste management in India." *Royal Society Open Science* 4, no. 3 (2017): 160764.

Lehtoranta, Suvi, Vuokko Laukka, Brenda Vidal, Elisangela Heiderscheidt, Heini Postila, Ritva Nilivaara, and Inga Herrmann, "Circular economy in wastewater management—The potential of source-separating sanitation in rural and peri-urban areas of Northern Finland and Sweden." *Frontiers in Environmental Science* (2022): 97.

Liu, Qinglan, Longjian Yang, and Miying Yang, "Digitalisation for water sustainability: Barriers to implementing circular economy in smart water management." *Sustainability* 13, no. 21 (2021): 11868.

Lourenço, N., and L. M. Nunes, "Review of dry and wet decentralized sanitation technologies for rural areas: Applicability, challenges and opportunities." *Environmental Management* 65, no. 5 (2020): 642–664.

Mallory, Adrian, Daniel Akrofi, Jenica Dizon, Sourav Mohanty, Alison Parker, Dolores Rey Vicario, and Sharada Prasad et al., "Evaluating the circular economy for sanitation: Findings from a multi-case approach." *Science of the Total Environment* 744 (2020): 140871.

Morseletto, Piero, Caro Eline Mooren, and Stefania Munaretto, "Circular economy of water: Definition, strategies and challenges." *Circular Economy and Sustainability* (2022): 2(4), 1463–1477.

NIUA, 2016. Handbook on Decentralised Waste Water Treatment Module, 2016. Delhi, India. https://niua.org/khub/wp-content/uploads/2017/05/DEWATS7.pdf.

Raab, Katharina, Giulia Tolotti, and Ralf Wagner, 2021. Challenges in Solid Waste Management: Insights into the Disposal Behavior of Suburban Consumers in Guatemala City. 3, 683576. https://www.frontiersin.org/articles/10.3389/frsc.2021.683576/full#:~:text=In%20suburban%20areas%20of%20Guatemala,suburban%20areas%20lag%20far%20behind.

United Nations (UN), 2015. Sustainable Development Goals. The Sustainable Development Agenda – United Nations Sustainable Development (referred as on date 08.11.2022).

United Nations Climate Change (UNCC), 2019. Circular Economy Is Crucial for Climate Protection – Patricia Espinosa. https://unfccc.int/news/circular-economy-is-crucial-for-climate-protection-patricia-espinosa (referred as on date 08.11.2022).

United Nations Environment Programme (UNEP), 2017. Smarter Use of Resources Can add $2 Trillion Annually to Global economy (unep.org) (referred as on date 07.11.2022).

United States Environmental Protection Agency (USEPA), 2012. Case Studies of Individual and Clustered (Decentralized) Wastewater Management Programs. https://www.epa.gov/sites/default/files/2015-06/documents/decentralized-case-studies-2012.pdf.

United States Environmental Protection Agency (USEPA), 2022. https://www.epa.gov/recyclingstrategy/what-circular-economy#EPA (referred as on date 10.11.2022).

Zeemann, Grietje, and Gatze Lettinga, eds. *Decentralised Sanitation and Reuse: Concepts, Systems and Implementation.* IWA, 2001.

3 Sanitation and Sustainable Water Management Concepts in Rural, Urban, and Peri-Urban Areas

Song Yan[1], *Patrick Drogui*[1], *Rajeshwar D. Tyagi*[2,3], *and Jonathan W.C. Wong*[3]
[1] Université du Québec, Québec, Canada
[2] BOSK Bioproducts, Québec, Canada
[3] Dongguan University of Technology, Dongguan, China

3.1 INTRODUCTION

Sanitation and sustainable water management are essential components for the provision of clean and safe water in rural, urban, and peri-urban areas. Access to safe water and proper sanitation is a fundamental human right, as recognized by the United Nations General Assembly in 2010 (UN 2010). However, despite the significant progress that has been made in improving water and sanitation infrastructure in many parts of the world, there are still millions of people who lack access to these basic necessities (Messerli et al. 2019; WHO 2019).

Despite the global trend of increasing urbanization, about half of the world population still lives in rural areas. In the European Union (EU), about 30% of the population of former CEE (Central and Easter European) countries (42 million people) live in settlements of less than 2000, while this percentage is lower than 20% in the western part. Many other areas of the world show a still preponderantly rural or peri-urban (settlements in proximity of a large urban areas, but without the proper services and facilities of such areas) character. A large part of this population is still waiting for proper sanitation systems or is aiming to improve the efficiency of existing ones and scale-up environmental protection and resources recovery. Goal 6 of the United Nations Sustainable Development Goals ("Clean Water and Sanitation") foresees achieving access to adequate and equitable sanitation and hygiene for all by 2030 and, within the same time frame, halving the proportion of untreated wastewater and substantially increasing recycling and safe reuse globally. Still, at the moment, at least 1.8 billion people globally use sources of drinking water that are fecally contaminated, over 1.7 billion people currently live in river basins where water use exceeds recharge, and more than 80% of wastewater from human activities is discharged without any pollution removal. Collection and treatment of wastewater have a huge impact on the environment and economy, both at the local and global levels (UN 2010; WHO 2017, 2019).

Sustainable water management, on the other hand, involves the management of water resources in a way that is environmentally sound, socially equitable, and economically viable, and that meets the needs of present and future generations (WHO 2017).

In rural areas, where access to safe water and sanitation services is often limited, sustainable water management and sanitation initiatives are essential for improving health outcomes and enhancing economic development. In many rural communities, access to clean water and proper sanitation is a key factor in reducing the incidence of water-borne diseases such as cholera, typhoid,

DOI: 10.1201/9781003381419-4

and diarrhea. These diseases are major causes of morbidity and mortality in many developing countries, particularly among children under the age of five (WHO 2017).

In urban areas, where the population is concentrated and the demand for water and sanitation services is high, sustainable water management and sanitation initiatives are critical for reducing the burden on existing infrastructure and ensuring that these services are accessible to all residents. In many urban areas, there is a significant gap between the demand for water and sanitation services and the capacity of existing infrastructure to meet these needs. This can lead to overcrowding, poor hygiene, and the spread of water-borne diseases (WHO 2017).

Peri-urban areas, which lie on the outskirts of urban areas and are characterized by rapid population growth, often face unique challenges in providing safe and sustainable water and sanitation services. These areas are often underserved by existing infrastructure and may lack access to clean water and proper sanitation facilities. In addition, peri-urban areas are often subject to environmental degradation, which can exacerbate existing water and sanitation challenges (Capodaglio 2017).

The provision of safe and sustainable water and sanitation services in rural, urban, and peri-urban areas is a complex and multifaceted challenge. There are many factors that contribute to the difficulties faced by these areas in providing these essential services. In many rural, urban, and peri-urban areas, there is a lack of adequate infrastructures for the provision of water and sanitation services, which can include a lack of piped water supply, inadequate wastewater treatment facilities, and poor sanitation infrastructure. Financing for water and sanitation infrastructure is often limited in these areas, making it difficult to invest in the necessary infrastructure and services. Rapid population growth in urban and peri-urban areas can put a strain on existing water and sanitation infrastructure, leading to overcrowding, poor hygiene, and the spread of water-borne diseases. Climate change can exacerbate existing water and sanitation challenges, particularly in rural areas where access to water is often dependent on rainfall patterns (Capodaglio 2017; WHO 2019).

This chapter provides a comprehensive overview of the concepts of sanitation and sustainable water management in rural, urban, and peri-urban areas. Table 3.1 gives the example summary of various technologies for water/wastewater treatment in rural, urban, and peri-urban areas. It covers a range of topics, including water supply and decentralized sanitation, wastewater treatment, and the challenges and strategies associated with sanitation for domestic, industrial, hospital, and agricultural wastewater treatment. Additionally, the chapter explores the reuse of treated wastewater as a means of promoting sustainability. The chapter concludes by summarizing the key challenges faced by these areas in providing safe and sustainable water and sanitation services and discusses various strategies and initiatives being implemented to address these challenges.

3.2 WATER SUPPLY AND DECENTRALIZED SANITATION IN RURAL, URBAN, AND PERI-URBAN AREAS

3.2.1 Drinking Water Supply

Access to safe and reliable drinking water is essential for human health and well-being, yet over 2 billion people worldwide lack access to safe drinking water. In many cases, those affected are living in rural, urban, and peri-urban areas where centralized water supply systems may not be practical or cost-effective. In recent years, decentralized drinking water supply systems have emerged as a promising alternative to traditional centralized systems, particularly in areas where access to safe drinking water is a challenge (UN 2010; WHO 2017, 2019).

Decentralized drinking water supply systems are typically small-scale, community-based systems that rely on a variety of water sources, treatment technologies, and distribution methods. These systems can be implemented in a range of settings, from rural communities with no existing water supply infrastructure to peri-urban areas with unreliable or unsafe water sources (UN 2010; WHO 2017, 2019).

One common decentralized drinking water supply system is the use of household-level water treatment and storage systems, such as point-of-use filters and household-level rainwater

TABLE 3.1
Examples of Various Technologies for Water/Wastewater Treatment in Rural, Urban, and Peri-Urban Areas

SN	Water/WW	Technology	After Treatment or Treatment Efficiency	If Reuse	References
1	Graywater	Constructed wetlands	To remove 97% of the TSS, 98% of the BOD_5, 98% of ammonia nitrogen, and 99.9% of the total coliform. Study points out that the system required an investment of only about Rs. 35,000 (US$500) and the family is able to save about 400 L of water per day.	The system is good enough to treat all the graywater generated in the house with seven members and the treated water is used for non-consumptive uses such as flushing, gardening, and washing.	Tuladhar, Shrestha and Shrestha (2008)
2	Domestic wastewater	Reed bed systems	The reduction in the concentrations of chemical oxygen demand (COD), total N, total P, and fecal coliform (FC) bacteria ranged from 80 to 99%		Kern and Idler (1999)
3	Mixture of domestic and agricultural wastewater	Reed bed systems	Efficiency ranged between 13 and 99%.		Kern and Idler (1999)
4	Industrial wastewater (combined wastewater and dye wash water)	Sequential batch operation under aerobic and aerobic-anoxic microenvironments	76.2% of organic fraction, 73.3% of phosphate, and 81.6% of nitrate removal. Based on these studies, a closed-loop decentralized effluent treatment system was designed to ensure zero-liquid discharge (ZLD).		Kopperi et al. (2023)
5	Mixed animal and human wastewaters	Biological processes	Removal of an important amount of noxious pollutants with an average efficiency of 80–98% (COD basis).	The production of 1200 m^3/day of biogas as an energy supplier and a source of CO_2 for microalgae culture in a high-rate algal pond. It can be possible to recover the digested sludge as biofertilizer and low-grade water useful for crop irrigation. The proposed integrated system also provides the recovery of 8.3 $g/m^2/d$ of algal biomass suitable for animal consumption.	Rovirosa et al. (1995)

(Continued)

TABLE 3.1 (*Continued*)
Examples of Various Technologies for Water/Wastewater Treatment in Rural, Urban, and Peri-Urban Areas

SN	Water/WW	Technology	After Treatment or Treatment Efficiency	If Reuse	References
6	Agricultural wastewater	Constructed wetlands (CW)	Average concentration reductions were: BOD_5 81%, TSS 83%, TKN 75%, NH_4^+-N 76%, NO_3^--N 42%, and TP 64%. Average log reductions for *Escherichia coli* and fecal coliforms were 1.63 and 1.93, respectively.		Rozema et al. (2016)
7	Hospital wastewater	Constructed wetland at Dhulikhel Hospital	Initial tests in 1997: to remove 98% of total suspended solids (TSS), 98% of BOD5, 96% of COD, and 99.9% of total coliforms. It also removed 80% of the ammonia nitrogen and 54% of phosphate. Follow-up monitoring in 2003 showed that the plant was still removing 96% of BOD5 and 93% of TSS and COD.		Tuladhar, Shrestha, and Shrestha (2008)

harvesting systems. These systems can be highly effective in reducing microbial and chemical contaminants in drinking water and are often preferred by households due to their low cost and ease of use (Clasen, Brown and Collin 2006).

In rural areas, decentralized drinking water supply systems can also include community-scale systems, such as hand pumps or gravity-fed systems that draw water from nearby sources such as wells or springs. These systems can be highly effective in remote areas with limited water supply infrastructure but may require ongoing maintenance and management to ensure long-term sustainability (Parkinson and Tayler 2003).

In peri-urban areas, decentralized drinking water supply systems can include community-scale systems that draw water from nearby sources. They can also include small-scale systems that use rainwater harvesting or treated wastewater for non-potable uses such as irrigation or toilet flushing. These systems can be effective in reducing demand on centralized water supply systems, while also providing a reliable source of water for households (Parkinson and Tayler 2003).

In urban areas, decentralized drinking water supply systems are often used in informal settlements or slums where existing water supply infrastructure may be inadequate or inaccessible (Van der Hoek et al. 2001).

While decentralized drinking water supply systems offer many benefits, their implementation and operation do come with certain complexities. Ensuring the long-term sustainability of the system requires ongoing maintenance and management, which can be particularly difficult in areas with limited resources or institutional capacity. In such cases, the involvement of community-based organizations or other stakeholders can be necessary to ensure effective management (Van der Hoek et al. 2001). In addition to maintenance and management, the safety and quality of the water supply must also be a priority, particularly in areas where water sources may be contaminated or unreliable. Careful selection of water sources, appropriate treatment, and distribution methods are crucial to ensure that the water is safe for consumption (Van der Hoek et al. 2001).

It is also important to ensure that decentralized drinking water supply systems are socially and economically sustainable and do not perpetuate existing inequalities or create new ones. Community engagement and participation are critical in this regard, as is an understanding of the broader social and economic context in which the system is being implemented (Goverdhanam 2014; Van der Hoek et al. 2001).

Despite these complexities, there is increasing interest in decentralized drinking water supply systems as a promising alternative to traditional centralized systems. These systems can be more cost-effective and sustainable in many cases, particularly in areas where access to safe drinking water is a challenge. By carefully planning, implementing, and managing decentralized drinking water supply systems, communities can benefit from improved access to safe drinking water (Goverdhanam 2014; Mankad and Tapsuwan 2011; Peter-Varbanets et al. 2009).

3.2.2 Sustainable Wastewater Management Concepts

The environmental impact of sewer infrastructures was found to be significant, surpassing both the construction and operation of wastewater treatment plants (WWTPs) in several categories. Eco-innovation is critical in reducing the environmental impact of urban water systems (UWSs) and enhancing their sustainability in economic, environmental, and social terms. Decentralized wastewater management systems have been proposed as a solution for rural or peri-urban communities by various advocates. In addition to the unaffordability of centralized treatment for remote and low-income regions, decentralized wastewater management is an excellent solution to address sustainability challenges. These systems can be built to meet current needs and can be expanded as needs arise. Even in developed countries, small decentralized WWTPs served 25% of the population over a decade ago. Since most of the urban water infrastructure in developed countries is near or beyond its useful design lifespan, the switch to smaller, local cluster-based systems is a sensible and financially sound solution. The gradual loss of the densely concentrated settlement character of cities, which is expanding into the countryside, is also a significant driving factor behind decentralization (UN 2010; WHO 2017, 2019).

Overall, decentralization offers a practical, cost-effective, and sustainable solution for managing wastewater in rural, urban, and peri-urban areas. With Goal 6 of the United Nations Sustainable Development Goals focusing on achieving access to adequate and equitable sanitation and hygiene for all by 2030, it is essential to promote and implement eco-innovation strategies and decentralized wastewater management systems to achieve this global target (UN 2010; WHO 2017, 2019).

In rural areas, sanitation is often inadequate, with a lack of access to toilets and proper waste management systems. To address this issue, decentralized sanitation systems such as composting toilets and decentralized WWTPs can be used. These systems are low-cost, easy to maintain, and do not require a connection to a centralized sewer system (UN 2010; WHO 2017, 2019).

In urban areas, sanitation is usually better, with access to toilets and centralized sewer systems. However, these systems can be overwhelmed in areas with high population density, leading to wastewater pollution and public health concerns. Decentralized WWTPs can be used to supplement the centralized system and reduce the load on the existing infrastructure (UN 2010; WHO 2017, 2019).

In peri-urban areas, sanitation is a combination of rural and urban systems. They may have access to centralized sewer systems but also rely on decentralized systems, such as septic tanks and composting toilets, to manage their waste (UN 2010; WHO 2017, 2019).

Therefore, both water supply and sanitation in rural, urban, and peri-urban areas can be improved through the use of decentralized systems. These systems can provide communities with access to clean water and proper waste management, improving public health and quality of life (UN 2010; WHO 2017, 2019).

3.3 WASTEWATER TREATMENT INRURAL, URBAN, AND PERI-URBAN AREAS

3.3.1 Sanitation for Domestic Wastewater Treatment

3.3.1.1 Black Water

Black water, which is the wastewater from toilets, contains significant amounts of organic matter and pathogens that can pose a health risk to humans and the environment if not properly treated. Conventional centralized wastewater treatment systems are not always feasible or affordable in rural, urban, and peri-urban areas, hence the need for decentralized treatment systems that can be tailored to local conditions and can be managed by local communities (Capodaglio 2017).

Firstly, they can be designed and implemented to suit local conditions, including the availability of land, water, and other resources, the level of technology required, and the local culture and social norms (Capodaglio 2017). This flexibility allows for a more customized and appropriate solution to black water management. Secondly, decentralized systems can be less expensive than centralized systems, as they require less infrastructure and can often use natural processes for treatment, such as soil infiltration or wetlands. Thirdly, decentralized systems can enhance the capacity of local communities to manage their own wastewater, thereby contributing to their empowerment and the development of local institutions. Black water treatment can also provide a solution for areas with limited access to water resources, such as arid regions or areas with low rainfall. In these areas, the treated black water can be used for irrigation or other non-potable uses, reducing the strain on freshwater resources. In addition, black water treatment can also contribute to reducing the pollution of rivers and lakes, which often serve as sources of drinking water in rural areas (Capodaglio 2017; Fendel et al. 2022; Lakho et al. 2021).

Although black water decentralized treatment systems offer potential benefits, their implementation also poses several obstacles. One significant obstacle is the absence of technical expertise and knowledge among local communities and institutions. This can lead to incorrect installation, operation, and maintenance of the systems, resulting in system failure and posing environmental and health risks. Inadequate financial resources and the absence of sustainable funding mechanisms can also impede the implementation and operation of decentralized systems. Additionally, social and cultural obstacles, such as communities' reluctance to change their sanitation practices or the stigmatization of handling human waste, can hinder the adoption and success of decentralized systems (Parkinson and Tayler 2003; Tilley et al. 2014; Tooke 2012).

To overcome the obstacles of black water decentralized treatment, several potential solutions have been proposed (Otoo and Drechsel 2018). Firstly, capacity-building and knowledge-sharing programs can be implemented to train and educate local communities and institutions on the installation, operation, and maintenance of the systems. This can be achieved through community-led initiatives, partnerships with universities or research institutes, and the involvement of local non-governmental organizations (NGOs) and government agencies. Secondly, innovative financing mechanisms, such as microfinance schemes, public-private partnerships, and community-based organizations, can be used to provide the necessary financial resources and sustain the operation and maintenance of the systems. Thirdly, awareness-raising campaigns can be launched to overcome social and cultural barriers and to promote the benefits of decentralized systems for public health and the environment. This can involve the participation of local leaders, opinion-makers, and media outlets (Otoo and Drechsel 2018).

Several case studies have been reported on the successful implementation of black water decentralized treatment systems in rural, urban, and peri-urban areas (Fendel et al. 2022). For instance, in India, biogas received partial support from stakeholders' perspectives, with co-composting of black water and wet waste being the most popular, underlined by numerous positive arguments packaged as indicators. The current situation in South India shows a focus on these mature technologies and the majority sees great potential in co-compost in particular. The most difficult perceived scenario in comparison was that of "Rural Urban Nutrient Partnership" (RUN)'s high-tech solution, with co-recycling generally raising more concerns than mono-recycling. Nevertheless, combinations with alternative and even more environmentally friendly concepts are seen, especially for the future once the pressing problems have been solved. Thus, RUN can be a good, decentralized option for prosperous new developments if the focus is on high compatible sustainability and recycling rates of valuable materials, closed loops, and reduced wastewater. It is agreed that there is no ultimate technology or concept. The choice depends on the circumstances and location but currently shows a clear preference (Fendel et al. 2022; Lakho et al. 2021).

3.3.1.2 Graywater

Graywater is wastewater generated from domestic activities such as dishwashing, laundry, and bathing, while blackwater consists of toilet water. Graywater can be classified as low or high load based on its organic strength, with low-load graywater excluding more polluted kitchen and laundry wastewater (Friedler 2004; Winward et al. 2008a, 2008b).

Water used for activities like laundry, dishwashing, and bathing, which is not flushed down the toilet, can be recycled and used for purposes like irrigation or other non-drinking purposes. However, its reuse requires proper treatment to remove contaminants (suspended and dissolved solids, oils grease, surfactants, etc.) and ensure safety for human health and the environment. Decentralized graywater treatment systems have emerged as an alternative to traditional centralized treatment systems, especially in rural, urban, and peri-urban areas where centralized systems are often not feasible or affordable (Filali et al. 2022; Lakho et al. 2021).

Graywater is an important source of water for sustainable consumption in buildings, with significant development potential (Santos et al. 2012). Compared to combined municipal wastewater, graywater has lower organic pollutants and pathogen content and is therefore particularly suitable for on-site treatment (Eriksson et al. 2002; Revitt et al., 2011).

Graywater accounts for 50–80% of the total domestic water use, making it the largest potential source of water savings in domestic residences (Al-Hamaiedeh and Bino 2010; Christova-Boal et al. 1996; Eriksson et al. 2002). Graywater can contribute to this cause as it is still water, makes up the largest volume of waste flow from households, and has low pathogen content, and low nutrient content that can be beneficially used for crop irrigation (Mandal et al. 2011; WHO 2006). In dwellings and residential buildings, treated graywater reuse can represent a 40–47% reduction in total water consumption (Santasmasas et al. 2013; Santos et al. 2012).

Various technologies have been developed to treat graywater, including membrane bioreactor (MBR) (Jefferson et al. 2001; Lesjean and Gnirss 2006), rotating biological contactor (Friedler et al. 2005; Nolde 1999), and constructed wetland (CW) (Dallas and Ho 2005). Graywater treatment system designs depend on the site conditions and graywater characteristics, including the quantity to be treated and reuse applications (Mandal et al. 2011). However, the major difficulty in treating graywater is the large variation in its composition, and the reused graywater must meet four criteria: hygienic safety, aesthetics, environmental tolerance, and technical and economic feasibility (Al-Hamaiedeh and Bino 2010; DHWA 2002). All these biological systems have been capable of meeting a 10 mg/L biological oxygen demand (BOD) standard, although the level of suspended solids and microorganism removal varies between them (Winward et al. 2008a). Consistently detected indicator bacteria in graywater (coliforms, *Escherichia coli*, and enterococci) indicate the potential for enteric pathogenic bacteria (e.g., salmonella, campylobacter), protozoa (e.g., cryptosporidium, giardia), and viruses (e.g., rotavirus, norovirus) to persist in graywater (Ottoson and Stenström 2003; Winward et al. 2008a).

Rural areas in developing countries often face challenges of water scarcity and contaminated groundwater, due to limited access to water treatment options. A decentralized graywater treatment system can be an effective solution to address these challenges. A study (Subramanian et al. 2020) evaluated the performance of a decentralized graywater treatment and reuse system implemented in a government-managed school in rural India for over 12 months. The system treated handwash and kitchen wash wastewater separately, utilizing pre-treatment screens and grease traps, slow sand biofiltration combined with anaerobic sludge bioreactor, aeration, and a final ozone-based disinfection stage. The treated water was used for toilet-flushing in the school. The system achieved high removal efficiencies, including 99% for turbidity, 98% for total suspended solids (TSSs), 66% for nitrate, 73% for total phosphorus (TP), 98% for BOD (five days), 96% for chemical oxygen demand, and >99.99% for fecal coliform (FC), demonstrating the potential for effective and economical decentralized graywater treatment in rural India (Subramanian et al. 2020).

One of the commonly used technologies for decentralized graywater treatment is CWs. CWs are designed to mimic natural wetland systems, and they use a combination of physical, chemical, and biological processes to remove contaminants from graywater. Studies have shown that CWs can effectively remove pollutants from graywater, such as organic matter, nutrients, and pathogens (Sijimol and Joseph 2021). However, the effectiveness of the treatment process depends on various factors, such as the design and operation of the wetland, the composition of the graywater, and the climatic conditions of the region.

Another technology that has shown promising results for graywater treatment is MBRs. MBRs use a combination of biological processes and membrane filtration to remove contaminants from graywater. Studies have demonstrated that MBRs can achieve high removal rates for pollutants such as organic matter and nutrients (Sijimol and Joseph 2021). However, MBRs are often associated with high energy consumption and operating costs, which can be a barrier to their widespread adoption in decentralized systems (Filali et al. 2022; Lakho et al. 2021; Sijimol and Joseph 2021).

Electrochemical treatment techniques like electrocoagulation (EC), electrooxidation (EO), and electroflotation are useful for treating graywater (Barışçı and Turkay 2016; Karichappan, Venkatachalam and Jeganathan 2014; Nasr, Ateia and Hassan 2016). By EC using aluminum electrodes, under the optimal conditions (J = 12.5 mA/cm^2, t = 30 min, and l = 0.5 cm), pollutants removal was: COD_{total} = 52.8%, $COD_{soluble}$ = 31.4%, BOD_{total} = 32.8%, $BOD_{soluble}$ = 27.6%, SS = 64.6%, TN = 30.1%, and TP = 13.6%. The consumed electrical-energy recorded 4.1 kWh/m^3 with an operating cost 0.25 US \$/m^3 (Nasr et al. 2016). When graywater was treated by the EC process, the highest COD removal was obtained with the Al–Fe–Fe–Al hybrid combination. Current density of 1 mA/cm^2 provided the highest efficiency (Barışçı and Turkay 2016).

Other technologies that have been used for decentralized graywater treatment include sand filters, activated sludge systems, and trickling filters. Sand filters use sand beds to remove contaminants from graywater, while activated sludge systems use aeration and microbial processes to break down pollutants. Trickling filters use a combination of physical, chemical, and biological processes to remove contaminants from graywater. Studies have shown that these technologies can effectively remove pollutants from graywater, but their effectiveness depends on various factors such as the design and operation of the system, the composition of the graywater, and the climatic conditions of the region (Filali et al. 2022; Sijimol and Joseph 2021).

While decentralized graywater treatment systems offer several benefits, such as reduced water consumption and lower energy requirements, there are also several challenges associated with their implementation. One of the main challenges is the lack of awareness and technical expertise among local communities and institutions. This can result in improper installation, operation, and maintenance of the systems, leading to system failure and environmental and health hazards. Additionally, the lack of financial resources and sustainable funding mechanisms can hinder the implementation and operation of decentralized systems. Furthermore, social and cultural barriers, such as the reluctance of communities to change their water use habits, can impede the adoption and success of decentralized systems (Lakho et al. 2021; Sijimol and Joseph 2021).

To overcome these challenges, it is important to develop appropriate policies and regulations to support the implementation and operation of decentralized graywater treatment systems. This includes providing technical assistance and training to local communities and institutions, developing sustainable financing mechanisms, and promoting awareness and education about the benefits of decentralized systems. Moreover, it is essential to establish monitoring and evaluation mechanisms to ensure the proper functioning and performance of the systems (Sijimol and Joseph 2021).

Therefore, decentralized graywater treatment systems offer a promising alternative to traditional centralized systems for the treatment and reuse of graywater in rural, urban, and peri-urban areas (Filali et al. 2022).

Another challenge associated with graywater decentralized treatment is the lack of proper regulations and guidelines. Without clear guidelines, there is a risk of unregulated and potentially harmful practices. Therefore, it is important to develop regulations and guidelines that promote the safe and effective use of graywater treatment systems.

Despite these challenges, there have been successful examples of decentralized graywater treatment systems in various contexts. In rural areas, simple and low-cost systems such as gravel filters, CWs, and soil infiltration systems have been implemented (Capodaglio et al. 2017). In urban areas, systems such as graywater recycling and reuse have been implemented in individual households and larger buildings, such as apartment complexes and hotels (Capodaglio et al. 2017).

Graywater exhibits highly variable characteristics, with amounts ranging from 50 to 80% of the wastewater volume produced by households (Albalawneh and Chang 2015). All types of graywater display good biodegradability, as indicated by COD:BOD_5 ratios that range from 0.31 to 0.71. While most countries apply the same standards for reclaimed municipal wastewater as they do for graywater, some have established specialized standards for graywater reuse. Graywater treatment technologies are classified into physical, chemical, biological, and natural systems, or a combination of these. Physical treatment processes alone are insufficient for graywater treatment, while chemical treatment processes are suitable for low-strength graywater treatment systems in single households, as their treatment performance is not affected by the variability in the strength and flow of the graywater. CWs are considered the most environmentally friendly and cost-effective technology for graywater treatment and reuse. Finally, the study proposes a possible scheme for graywater recycling for agricultural irrigation reuse purposes (Albalawneh and Chang 2015).

Decentralized graywater treatment systems have the potential to provide various benefits in addition to improving sanitation and reducing water usage. For example, the reuse of graywater can reduce the demand for freshwater resources and improve water security, particularly in areas where water scarcity is an issue (Albalawneh and Chang 2015). Furthermore, the reuse of graywater for irrigation can improve soil quality and crop yields (Albalawneh and Chang 2015).

The implementation of decentralized graywater treatment systems offers a promising alternative to traditional centralized wastewater treatment systems in rural, urban, and peri-urban areas. These systems have the potential to improve sanitation, reduce water usage, and provide various additional benefits. However, there are challenges associated with the implementation and operation of these systems, including the lack of technical expertise and knowledge, financial resources, sustainable funding mechanisms, social and cultural barriers, and the lack of proper regulations and guidelines. Therefore, it is important to address these challenges and develop strategies to promote the safe and effective use of decentralized graywater treatment systems (Albalawneh and Chang 2015; Lakho et al. 2021).

3.3.2 Sanitation for Industrial Wastewater Treatment

Sanitation for industrial wastewater treatment is a critical aspect of environmental protection and public health. Industrial wastewater is characterized by high levels of pollutants, which can cause significant environmental damage and pose a risk to human health if not treated properly. In many rural, urban, and peri-urban areas, centralized wastewater treatment systems are either non-existent or inadequate to handle the volume and complexity of industrial wastewater (Frontiers 2016). In such situations, decentralized wastewater treatment systems offer a promising alternative for managing industrial wastewater in a safe and sustainable manner (Victor et al. 2008).

Industrial (for example: mining industries, forestry industries, agriculture and farming, tourism industry, fishing industries, and manufacturing industries) wastewater is one of the most polluting types of wastewater due to its high concentration of organic and inorganic pollutants, heavy metals, and toxic substances. To prevent the contamination of natural water resources and protect public health, it is essential to properly manage and treat industrial wastewater before discharge into the environment. Sanitation technologies have been widely applied in industrial wastewater treatment to remove pollutants and reduce the risk of environmental pollution.

One of the most common sanitation technologies used in industrial wastewater treatment is the use of aerobic and anaerobic biological treatment systems. These systems rely on the natural ability of microorganisms to break down organic matter in the wastewater, transforming it into less harmful substances such as carbon dioxide, water, and biomass. The performance of biological treatment systems depends on several factors such as the type and concentration of pollutants, temperature, pH, and oxygen levels, and the hydraulic retention time. Some of the most commonly used biological treatment systems for industrial wastewater include activated sludge systems, sequencing batch reactors, and anaerobic digesters (Kopperi et al. 2023).

In addition to biological treatment, physical and chemical treatment technologies are also commonly used in industrial wastewater treatment. These technologies involve the use of physical and chemical processes such as sedimentation, filtration, adsorption, coagulation, electro-technology, and oxidation to remove pollutants from the wastewater. Physical and chemical treatment systems can be used alone or in combination with biological treatment systems to achieve the desired level of wastewater treatment (Kopperi et al. 2023).

Decentralized wastewater treatment systems for industrial wastewater treatment can be implemented at various scales, including household, community, and industrial levels. At the household level, on-site wastewater treatment systems, such as septic tanks, can effectively treat wastewater from small-scale industrial activities. Community-level decentralized treatment systems, such as CWs or decentralized package treatment plants, can be used to treat wastewater from multiple industrial (for example: mining industries, forestry industries, agriculture and farming, tourism industry, fishing industries, and manufacturing industries) sources. At the industrial level, on-site treatment systems, such as activated sludge systems or MBRs, can be used to treat wastewater from specific industrial processes (Frontiers 2016; Victor et al. 2008).

Decentralized treatment systems offer several advantages over centralized systems for industrial wastewater treatment. Firstly, they can be implemented quickly and cost-effectively, particularly in areas where centralized infrastructure is not available or is prohibitively expensive to build. Decentralized systems can also be designed to meet specific treatment requirements, such as the removal of heavy metals or organic pollutants, which may not be possible with traditional centralized systems. Additionally, decentralized systems can be tailored to the specific needs of individual industries, reducing the risk of cross-contamination and ensuring optimal treatment performance (Frontiers 2016; Victor et al. 2008).

However, despite the potential benefits, there are also several challenges associated with the implementation of decentralized treatment systems for industrial wastewater. One of the major challenges is the lack of technical expertise and knowledge among local communities and institutions. This can result in improper installation, operation, and maintenance of the systems, leading to system failure and environmental and health hazards. Additionally, the lack of financial resources and sustainable funding mechanisms can hinder the implementation and operation of decentralized systems. Furthermore, social and cultural barriers, such as the reluctance of industries to adopt new sanitation practices or the stigmatization of handling industrial waste, can impede the adoption and success of decentralized systems (Frontiers 2016).

Several studies (Frontiers 2016; Victor et al. 2008) have demonstrated the effectiveness of decentralized treatment systems for industrial wastewater treatment in various settings. For example, in a study conducted in India, decentralized package treatment plants were installed in several small- and medium-scale industries to treat their wastewater. The treatment plants were designed to remove organic pollutants and heavy metals, and the treated wastewater was used for irrigation purposes. The study found that the treatment plants were effective in removing pollutants from the wastewater, and the treated wastewater was safe for use in agriculture (Frontiers 2016; Victor et al. 2008).

Similarly, in a study conducted in Brazil, a decentralized treatment system was implemented to treat wastewater from a pig farming facility. The system consisted of an anaerobic digester and a CW and was designed to remove organic matter and nitrogen from the wastewater. The study found that the treatment system was effective in reducing the concentration of pollutants in the wastewater, and the treated wastewater met the discharge standards for agriculture and environmental purposes (Victor et al. 2008).

In India, decentralized handlooms have been a traditional practice, and the Siripuram village, known for Pochampally/Ikat work, was investigated as a case study for a sustainable solution (Kopperi et al. 2023). The village's primary occupation is weaving and dyeing, with 65% of the population solely dependent on it, generating an average wash water of 127 KLD from dyeing operations (Kopperi et al. 2023). A topographical survey using an aerial drone was conducted to understand the drainage pattern, elevations, and contours. Lab scale studies were also done on the combined wastewater and dye wash water using sequential batch operation under aerobic and aerobic-anoxic microenvironments, with the latter proving effective in organic and nutrient removal. The treatment studies showed 76.2% of organic fraction, 73.3% of phosphate, and 81.6% of nitrate removal. Based on these studies, a closed-loop decentralized effluent treatment system was designed to ensure zero-liquid discharge (ZLD) (Kopperi et al. 2023).

The use of decentralized systems for industrial wastewater treatment offers a promising alternative to traditional centralized treatment methods, particularly in rural, urban, and peri-urban areas. These systems can effectively treat and reuse wastewater, reducing water demand and decreasing the pollution load on receiving water bodies. Additionally, decentralized systems offer economic and social benefits, such as creating local job opportunities, increasing community involvement in wastewater management, and reducing health risks associated with traditional sanitation practices (Kopperi et al. 2023).

Despite the potential benefits, there are also several challenges associated with the implementation of decentralized industrial wastewater treatment systems. These include technical and knowledge gaps among local communities and institutions, lack of financial resources, and social and cultural barriers. Therefore, the success of decentralized systems requires the involvement of all stakeholders, including local communities, industries, and government agencies. Capacity-building and awareness-raising programs are necessary to overcome these challenges and ensure the sustainable operation of decentralized systems.

Future research should focus on improving the efficiency and reliability of decentralized systems, particularly in developing countries where the need for industrial wastewater treatment is high. Additionally, there is a need for the development of appropriate regulations and policies to support the implementation and operation of decentralized industrial wastewater treatment systems. Such policies should address issues related to institutional arrangements, financial mechanisms, and operation and maintenance requirements (Kopperi et al. 2023).

3.3.3 Sanitation for Hospital Wastewater Treatment

Sanitation is a critical aspect of healthcare facilities as it ensures a hygienic environment for patients, healthcare workers, and the general public. In hospitals, wastewater is generated from various activities, including patient care, cleaning, and medical procedures, which can contain hazardous substances such as pathogens, pharmaceuticals, and chemical disinfectants. If not treated properly, hospital wastewater can have adverse effects on human health and the environment. Therefore, proper sanitation for hospital wastewater decentralized treatment is crucial in rural, urban, and peri-urban areas to ensure that the wastewater is treated before it is discharged into the environment.

CWs are commonly utilized for decentralized wastewater treatment and in small communities. However, the discharge of effluent into small watercourses can result in pharmaceutical residues that pose a threat to the aquatic ecosystem. A study (Auvinen et al. 2017) described the performance of a pilot-scale aerated subsurface-flow CW treating municipal and hospital wastewater. Especially the effect of active aeration on the removal of selected pharmaceuticals was assessed. The removal of metformin and valsartan was significantly increased when continuous aeration was applied (99 ± 1% vs. 68 ± 32% for metformin and 99 ± 1% vs. 17 ± 19% for valsartan), although the microorganisms can adapt to degrading metformin also in anoxic conditions. At the hospital site, high concentrations of pharmaceuticals were measured in the influent. Atenolol and bisoprolol were efficiently removed (>75 and >50%, respectively), although the effluent concentrations of these compounds remained high (up to 0.5 μg/L). Intermittent aeration (50%) was found to be as effective as continuous aeration in removing the selected pharmaceuticals (Auvinen et al. 2017).

In the context of hospital wastewater treatment, the treatment process typically involves several stages, including primary treatment, secondary treatment, and tertiary treatment. Primary treatment involves the removal of solids and grease from the wastewater, while secondary treatment involves the removal of organic matter and nutrients. Tertiary treatment involves the removal of pathogens and other harmful substances to meet discharge standards. The treated water can then be reused for non-potable purposes such as irrigation, toilet flushing, or industrial processes (Auvinen et al. 2017).

Several types of decentralized wastewater treatment systems (DEWATS) have been developed for hospital wastewater treatment, including anaerobic digestion (Azizan, Yuzir and Abdullah 2021), aerobic treatment (Alsubih et al. 2021), CWs (Shrestha et al. 2001), and MBRs (Kovalova et al. 2012). Each system has its advantages and disadvantages, depending on the local conditions and treatment requirements. For example, anaerobic digestion is effective for high-strength wastewater and produces biogas, which can be used for energy generation. Aerobic treatment is suitable for low-strength wastewater and produces less sludge than anaerobic digestion. CWs are effective for removing nutrients and provide additional benefits such as biodiversity and aesthetic value. MBRs are efficient for producing high-quality treated water but require a significant amount of energy for operation (Auvinen et al. 2017; Kovalova et al. 2012; Shrestha et al. 2001).

In rural areas, DEWATS can be implemented at the household or community level to treat hospital wastewater. For example, in a study conducted in a rural hospital in Ghana, a DEWATS system was developed to treat the hospital wastewater using anaerobic digestion followed by CWs (Auvinen et al. 2017; Kovalova et al. 2012; Shrestha et al. 2001).

The Swiss pilot study provided valuable insight into the most appropriate technologies for decentralized wastewater treatment in hospitals (McArdell et al. 2011). The results showed that biological treatment of hospital wastewater was feasible, as demonstrated by the successful one-year operation of an MBR without any issues related to nitrification inhibition. While biological treatment was a crucial first step, it alone was insufficient for adequately eliminating pharmaceuticals. Additional advanced treatment steps were necessary for effective removal of pharmaceuticals. Ozonation and adsorption to powdered activated carbon (PAC) were two promising technologies for achieving effective elimination of pharmaceuticals.

As hospital wastewater contains high levels of pathogens, the treatment process should aim to eliminate them. While MBR is a good barrier, an additional disinfection step is recommended. Ozone treatment is advantageous because it kills pathogens at the same time. PAC treatment needs to be followed by disinfection, such as UV. Hospital wastewater is a major contributor to the release of antibiotic-resistant and multi-resistant bacteria, which need to be eliminated (McArdell et al. 2011).

Hospital wastewater treatment can meet the standard parameters for wastewater effluents that are required to be directly discharged into receiving water bodies. However, the remaining high concentrations of X-ray contrast media containing halogen iodide make adsorbable organic halogen compounds (AOX) problematic. X-ray contrast media is a pharmaceutical class that accounts for the highest load in hospital wastewater, and its overall load remains high even after post-treatment. Measurements from the site showed that AOX was still higher than requirements (effluent concentration requirement: CH: 0.08 mg/L, D: 0.05–0.1 mg/L), even after PAC treatment (McArdell et al. 2011).

For iodinated X-ray contrast media (ICM), particularly anionic ICM, direct phototransformation through UV was found to be more efficient than treatment with PAC or ozone. Advanced oxidation processes, such as ozone/UV or UV/H_2O_2, can be used to eliminate this class of pharmaceuticals, despite their high energy consumption. Since contrast media are excreted entirely in urine within 24 hours, separate urine collection (e.g., through roadbags) or treatment can also be considered a strategy to reduce pharmaceutical loads in the aquatic environment (McArdell et al. 2011).

3.3.4 Sanitation for Agricultural Wastewater Treatment

Agricultural wastewater is a significant source of pollution, particularly in rural and peri-urban areas, where centralized wastewater treatment systems are often not available or feasible. DEWATS can provide an effective solution to treat agricultural wastewater, allowing for the reuse of treated wastewater for irrigation and other purposes (Dordio and Carvalho 2013; Kern and Idler 1999; Lydecker and Drechsel 2010; Nguyen et al. 2007).

CWs are attracting renewed attention as a viable phytotechnology for treating agricultural wastewaters and removing specific pollutants, particularly recalcitrant ones (Dordio and Carvalho 2013). The performance of CW mesocosms was investigated (Dordio and Carvalho 2013) using light expanded clay aggregates (LECA) as the bed substrate and planted with *Phragmites australis* for the treatment of olive mill wastewater (OMW), swine wastewater (SW) contaminated with oxytetracycline, and water contaminated with the herbicide MCPA (2-methyl-4-chlorophenoxyacetic acid). Both wastewaters (OMW and SW) had high initial organic matter content and TSSs, which were efficiently removed by the system with efficiencies exceeding 80%. The removal of polyphenols in OMW and nitrogen compounds in SW also demonstrated similar or higher efficiencies in comparison with other treatment systems in the literature. While the antibiotic oxytetracycline was completely removed from SW in unplanted LECA beds, planted beds allowed for significantly faster removal. With regard to water contaminated with MCPA, the results demonstrated that LECA has a high sorption capacity for this herbicide, with removal efficiencies ranging from 56 to 97%. Overall, substantially higher pollutant removal efficiencies were achieved when plants were used (up to 28% higher). These findings indicate that CWs with LECA as a substrate may represent a suitable option for agricultural wastewater treatment (Dordio and Carvalho 2013).

In recent times, the escalating agro-industrial activities and the inadequacy of traditional WWTPs to eradicate persistent organic contaminants are posing a potential global risk to the environment (Theodorakopoulos et al. 2023). Among the advanced water treatment technologies proposed for mitigating this issue, the development and optimization of a ground-breaking hybrid photocatalytic nanofiltration reactor (PNFR) prototype has emerged as a notable solution that achieves synergistic effects between photocatalytic degradation activity and size exclusion capacity for micropollutant molecules (Theodorakopoulos et al. 2023). These features can be simultaneously endowed to numerous membrane monoliths. The physicochemical and photoinduced decontamination properties of titania materials were first assessed in the powder form, followed by the structural and morphological characterization of the titania-modified membrane monoliths obtained. The PNFR unit can function at high water recovery and low pressures, displaying promising removal efficiencies against Acetamiprid (ACT) and Thiabendazole (TBZ) pesticides and enabling the recycling of 15 m^3/day of actual agro-wastewater. The results obtained are very promising, showcasing the integration of titania photocatalysts in a photocatalytic membrane reactor as a viable technological solution for the purification of agricultural wastewater (Theodorakopoulos et al. 2023).

Treating agro-industrial wastewaters can prevent noxious pollutants from further deteriorating the ecosystem (Rovirosa et al. 1995). By constructing and implementing integrated systems for treating such waste, effluents can be reused, producing useful by-products (Rovirosa et al. 1995). Biological processes used for treating 300 m^3/day of mixed animal and human wastewaters can also generate 1200 m^3/day of biogas, serving as an energy source and a CO_2 supplier for microalgae culture in a high-rate algal pond. Additionally, the digested sludge can be recovered as biofertilizer and low-grade water suitable for crop irrigation. The proposed integrated system can also recover 8.3 g/m^2/d of algal biomass suitable for animal consumption and remove a significant amount of noxious pollutants with an average efficiency of 80–98% (COD basis) (Rovirosa et al. 1995).

Over the course of one year, the effectiveness and seasonal variations of two newly established reed bed systems were observed (Kern and Idler 1999). In one of the systems, which was designed to treat domestic wastewater, the reduction in concentrations of chemical oxygen demand (COD), total nitrogen (N), total phosphorus (P), and FC bacteria ranged from 80 to 99%. In the second system, which treated a mixture of domestic and agricultural wastewater from a cheese dairy, the treatment efficiency ranged between 13 and 99%. The removal rates for domestic wastewater did not exhibit any seasonal patterns, whereas the mixture of domestic and agricultural wastewater displayed increased removal rates during the summer. The abundance of bacteria involved in nitrogen metabolism was estimated using the most probable number (MPN) test. For both reed bed systems, the MPN values of proteolytic bacteria, ammonifiers, nitrifiers, and denitrifiers per gram of dry weight of the gravel substrate were in the range of 10^4–10^6, 10^5–10^7, 10^1–10^5, and 0–10^3, respectively. Based on the composition of the microbial community, it is necessary to improve the conditions for both nitrification and denitrification to increase the efficiency of nitrogen removal (Kern and Idler 1999).

CWs are a promising treatment option for agricultural wastewater, but their ability to function effectively in cold climates is still being evaluated as they rely on microbial and plant activity (Rozema et al. 2016). To assess their performance and identify design considerations specific to the region, a review (Rozema et al. 2016) of CWs in Eastern Canada and the Northeastern USA was conducted. The review synthesized data from 21 studies evaluating 25 full-scale wetlands. Data was also separated seasonally to assess the effects of climate on treatment performance (Rozema et al. 2016). The wastewater parameters considered were five-day biochemical oxygen demand (BOD_5), TSSs, *E. coli*, FCs, total Kjeldahl nitrogen (TKN), ammonia/ammonium (NH_3/NH_4^+-N), nitrate-nitrogen (NO_3^--N), and TP. Average concentration reductions were as follows: BOD_5 81%, TSS 83%, TKN 75%, NH_4^+-N 76%, NO_3^--N 42%, and TP 64%. Average log reductions for *E. coli* and FCs were 1.63 and 1.93, respectively. Average first order areal rate constants (k_a, m/y) were as follows: BOD_5 6.0 m/y, TSS 7.7 m/y, *E. coli* 7.0 m/y, FCs 9.7 m/y, TKN 3.1 m/y, NH_4^+-N 3.3 m/y, NO_3^--N 2.5 m/y, and TP 2.9 m/y. In general, CWs effectively treated a variety of agricultural wastewaters, regardless of season (Rozema et al. 2016).

3.4 REUSE OF TREATED WASTEWATERS TOWARD SUSTAINABILITY

Water scarcity is a growing global concern, especially in rural, urban, and peri-urban areas where access to clean water is limited. The use of decentralized wastewater treatment systems (DEWATS) has been increasingly promoted as a solution to improve sanitation, protect public health, and reduce water pollution. In addition, the reuse of decentralized treated wastewater can provide a sustainable source of water for various purposes such as irrigation, aquaculture, and groundwater recharge (Lopez et al. 2006).

As part of a strategic R&D project that began in April 2002, membrane filtration, simplified treatments, storage reservoirs, and CWs technologies were being investigated at the field scale to evaluate their effectiveness for treating municipal effluents for agricultural reuse (Lopez et al. 2006). The following main results have been recorded so far: microbial quality of treated effluents using membrane filtration was higher than that of local well-water used for irrigation; simplified treatment by skipping biological processes and irrigating olive trees with effluents resulted in a 50% increase in yield by saving the agronomic potential of organic matter and nutrients present in urban wastewater; storage reservoirs achieved the Italian limits for TSS, BOD_5, COD, and nutrients concentrations for WW agricultural reuse; CWs had an average efficiency of 85, 65, 75, 42, and 32% for TSS, BOD_5, COD, TN, and TP removal, respectively (Lopez et al. 2006).

The development of regulations and technology needed to efficiently and safely reuse municipal wastewater in irrigated agriculture in Mexico was discussed (Jiménez 2005). The regulations allow for the presence of nutrients and organic matter in wastewater at appropriate concentrations, while requiring the removal of pathogens, particularly helminth ova, before reuse. Custom research was conducted to develop technology that could treat wastewater to meet the new definition of effluent quality. They examined the Tula Valley case, which had the largest compact irrigation area using wastewater in the world, to illustrate the process. The more than 100 years of experience in reusing wastewater in the Tula Valley have allowed for the study of the short- and long-term effects. The technology developed is called APT (advanced primary treatment) and comprises a coagulation-flocculation process combined with high-rate sedimentation. The paper concludes that with appropriate regulations and wastewater treatment technology, it may be possible to safely reuse wastewater in agriculture at an affordable cost, resulting in a win-win situation for both farmers and public health (Jiménez 2005).

Decentralized wastewater treatment is crucial for sustainable water management because it facilitates the reuse of effluent (and nutrients) for irrigation or as service water in households. MBRs are capable of producing effluents of bathing water quality. Septic tanks can be retrofitted to MBR units and package MBR plants for wastewater or graywater treatment are available (Meuler, Paris and Hackner 2008). Decentralized treatment and reuse systems for domestic wastewater or graywater are also feasible for hotels, condominiums, apartment, or office complexes. This paper presents the effluent qualities of various decentralized MBR applications. The high effluent quality enables infiltration even in sensitive areas or reuse for irrigation, toilet flushing, and cleaning purposes in households. As these systems allow the reuse of treated water and the possibility to design them for carbon reduction only, they can ideally and easily help to close water and nutrient loops (Meuler et al. 2008).

Graywater has lower organic pollutants and pathogen content compared to combined municipal wastewater and is therefore particularly suitable for on-site treatment and reuse (Eriksson et al. 2002; Revitt et al. 2011). Graywater production and reuse potential depend on several factors, including building activity. In dwellings and residential buildings, treated graywater reuse can represent a 40–47% reduction in total water consumption (Santos et al. 2012).

The reuse of treated wastewater has many environmental, economic, and social benefits (Albalawneh and Chang 2015; Jiménez 2005; Lakho et al. 2021; Lopez et al. 2006; López-Morales and Rodríguez-Tapia 2019; Meuler et al. 2008; Santos et al. 2012; Sijimol and Joseph 2021). First, it reduces the demand for freshwater resources, which are becoming increasingly scarce in many parts of the world. Second, it helps to reduce water pollution by treating wastewater before it is discharged into water bodies. Third, it promotes sustainable agriculture, aquaculture, and other economic activities by providing a sustainable source of water. Fourth, it helps to improve public health by reducing the risk of waterborne diseases associated with untreated wastewater (Albalawneh and Chang 2015; Jiménez 2005; Lakho et al. 2021; Lopez et al. 2006; López-Morales and Rodríguez-Tapia 2019; Meuler et al. 2008; Santos et al. 2012; Sijimol and Joseph 2021).

However, the reuse of treated wastewater also poses some challenges and risks. The most significant risk is related to the presence of pathogens and contaminants in the treated wastewater, which can pose health risks to humans and animals if not properly treated and monitored. The risk can be mitigated by implementing appropriate treatment processes and monitoring programs to ensure that the treated wastewater meets the required quality standards for its intended reuse (Albalawneh and Chang 2015; Jiménez 2005; Lakho et al. 2021; Lopez et al. 2006; López-Morales and Rodríguez-Tapia 2019; Meuler et al. 2008; Santos et al. 2012; Sijimol and Joseph 2021).

Another challenge is the lack of awareness and acceptance of the reuse of treated wastewater among the public, which can lead to social and cultural barriers. This can be addressed by promoting awareness and education campaigns to increase public understanding of the benefits and risks associated with the reuse of treated wastewater.

Overall, the reuse of decentralized treated wastewaters can provide a sustainable source of water for various purposes such as agriculture, aquaculture, and groundwater recharge in case of water scarcity in rural, urban, and peri-urban areas. The implementation of DEWATS systems can effectively treat wastewater and provide a reliable source of water for non-potable purposes such as irrigation, livestock watering, and toilet flushing. However, it is important to ensure that appropriate treatment methods are used to minimize health risks and that the reuse of treated wastewater complies with relevant regulations and guidelines (Albalawneh and Chang 2015; Jiménez 2005; Lakho et al. 2021; Lopez et al. 2006; López-Morales and Rodríguez-Tapia 2019; Meuler et al. 2008; Santos et al. 2012; Sijimol and Joseph 2021).

3.5 CONCLUSIONS AND PERSPECTIVES

Decentralized sanitation and sustainable water management are crucial for achieving sustainable development goals and ensuring access to clean water and sanitation for all. These concepts offer unique advantages, including improved water quality, reduced demand for freshwater resources, and reduced environmental impacts in rural, urban, and peri-urban areas.

In rural areas, decentralized sanitation systems provide a cost-effective and sustainable solution to improve access to sanitation facilities that can be tailored to meet the specific needs of a community. In urban areas, decentralized systems can reduce the burden on central WWTPs, improve water quality, and provide a more sustainable approach to water management. In peri-urban areas, decentralized systems can manage wastewater from agriculture and livestock production, reducing the environmental impacts of these activities.

Despite the many benefits, challenges exist, including the lack of access to finance and technical expertise, and the need to increase awareness and education about decentralized systems and sustainable water management practices.

To fully realize the benefits of decentralized systems, increased investment, awareness, education, and continued research and development are necessary. Remote monitoring and control systems will make it easier to manage decentralized systems and ensure their long-term sustainability.

The integration of decentralized systems with circular economy approaches and renewable energy systems provides additional opportunities for achieving sustainability goals. Reusing treated wastewater for irrigation and other non-potable uses has the potential to significantly reduce freshwater demand and increase the resilience of communities to climate change.

Addressing the challenges requires a multifaceted approach, involving increased investment in infrastructure, community participation, public-private partnerships, and the use of appropriate technologies. By implementing these strategies and initiatives, it is possible to improve access to safe and sustainable water and sanitation services and promote health and economic development in these areas.

The future of decentralized sanitation and sustainable water management looks promising, with significant potential for improving access to clean water and sanitation, increasing agricultural productivity, and reducing environmental impacts. Continued investment, research, and education will be a key to realizing these benefits and ensuring a sustainable future for all.

REFERENCES

Albalawneh, Abeer, and Tsun-Kuo Chang. 2015. "Review of the greywater and proposed greywater recycling scheme for agricultural irrigation reuses." *International Journal of Research – GRANTHAALAYAH* 3:16–35. doi: 10.29121/granthaalayah.v3.i12.2015.2882.

Al-Hamaiedeh H., and M. Bino. 2010. "Effect of grey water reuse in irrigation on soil and plants." *Desalination* 256 (1):115–19.

Alsubih, Majed, Roohul Abad Khan, Nadeem A. Khan, Amadur Rahman Khan, and Gaurav Sharma. 2021. "Performance evaluation of aerobic fluidized bed bioreactor coupled with tube-settler for hospital wastewater treatment." *Journal of Environmental Chemical Engineering* 9 (5):105896.

Auvinen, Hannele, Iva Havran, Laurens Hubau, Lize Vanseveren, Wilhelm Gebhardt, Volker Linnemann, Dion Van Oirschot, Gijs Du Laing, and Diederik P. L. Rousseau. 2017. "Removal of pharmaceuticals by a pilot aerated sub-surface flow constructed wetland treating municipal and hospital wastewater." *Ecological Engineering* 100:157–164. doi: 10.1016/j.ecoleng.2016.12.031.

Azizan, Nur Alyaa Zahida, Ali Yuzir, and Norhayati Abdullah. 2021. "Pharmaceutical compounds in anaerobic digestion: A review on the removals and effect to the process performance." *Journal of Environmental Chemical Engineering* 9 (5):105926. doi: 10.1016/j.jece.2021.105926.

Barışçı, Sibel, and Ozge Turkay. 2016. "Domestic greywater treatment by electrocoagulation using hybrid electrode combinations." *Journal of Water Process Engineering* 10:56–66. doi: 10.1016/j.jwpe.2016.01.015.

Capodaglio, Andrea G. 2017. "Integrated, decentralized wastewater management for resource recovery in rural and peri-urban areas." *Resources* 6 (2):22.

Capodaglio, A. G., A. Callegari, D. Cecconet, and D. Molognoni. 2017. "Sustainability of decentralized wastewater treatment technologies." *Water Practice and Technology* 12 (2):463–477.

Clasen, Thomas F., Joseph Brown, and Simon M. Collin. 2006. "Preventing diarrhoea with household ceramic water filters: Assessment of a pilot project in Bolivia." *International Journal of Environmental Health Research* 16 (03):231–239.

Dallas, S., and G. Ho. 2005. "Subsurface flow reedbeds using alternative media for the treatment of domestic greywater in Monteverde Costa Rica, Central America." *Water Science and Technology* 52:119–28. doi: 10.2166/wst.2005.0358.

DHWA (Department of Health Western Australia). 2002. Draft guidelines for the reuse of greywater in Western Australia. Perth, Australia.

Dordio, A., and A. J. P. Carvalho. 2013. "Constructed wetlands with light expanded clay aggregates for agricultural wastewater treatment." *Science of The Total Environment* 463–464:454–461. doi: 10.1016/j.scitotenv.2013.06.052

Eriksson, E., K. Auffarth, M. Henze, and A. Ledin, A. 2002. "Characteristics of grey wastewater." *Urban Water* 4:85–104.

Fendel, Veronika, Martin Kranert, Claudia Maurer, Gabriela Garcés-Sánchez, Jingjing Huang, and Girija Ramakrishna. 2022. "Stakeholder assessment on closing nutrient cycles through co-recycling of biodegradable household kitchen waste and black water between rural and urban areas in South India." *Recycling* 7 (4):49.

Filali, Hanen, Narcis Barsan, Dalila Souguir, Valentin Nedeff, Claudia Tomozei, and Mohamed Hachicha. 2022. "Greywater as an alternative solution for a sustainable management of water resources—A review." *Sustainability* 14 (2):665.

Friedler, E. 2004. "Quality of individual domestic greywater streams and its implication on on-site treatment and reuse possibilities." *Environmental Technology*, 25 (9): 997–1008.

Friedler, Eran, R. Kovalio, and Noah I. Galil. 2005. "On-site greywater treatment and reuse in multistorey buildings." *Water Science and Technology : A Journal of the International Association on Water Pollution Research* 51 (10):187–94.

Frontiers, UNEP. 2016. Report: Emerging Issues of Environmental Concern. United Nations Environment Programme, Nairobi.

Goverdhanam, Suraj. 2014. "Decentralized water systems for sustainable and reliable supply."

Jefferson, B., J. E. Burgess, A. Pichon, J. Harkness, and S. J. Judd. 2001. "Nutrient addition to enhance biological treatment of greywater." *Water Research* 35 (11):2702–10. doi: 10.1016/s0043-1354(00)00553-4.

Jiménez, Blanca. 2005. "Treatment technology and standards for agricultural wastewater reuse: A case study in Mexico." *Irrigation and Drainage* 54 (S1):S23–S33. doi: 10.1002/ird.183.

Karichappan, Thirugnanasambandham, Sivakumar Venkatachalam, and Prakash Maran Jeganathan. 2014. "Optimization of electrocoagulation process to treat grey wastewater in batch mode using response surface methodology." *Journal of Environmental Health Science and Engineering* 12:1–8.

Kern, Jürgen, and Christine Idler. 1999. "Treatment of domestic and agricultural wastewater by reed bed systems." *Ecological Engineering* 12 (1):13–25. doi: 10.1016/S0925-8574(98)00051-2.

Kopperi, Harishankar, Manupati Hemalatha, Boda Ravi Kiran, J. Santhosh, and S. Venkata Mohan. 2023. "Sustainable consideration for traditional textile handloom cluster/village in pollution abatement – A case study." *Environmental Pollution* 324:121320. doi: 10.1016/j.envpol.2023.121320.

Kovalova, Lubomira, Hansruedi Siegrist, Heinz Singer, Anita Wittmer, and Christa S. McArdell. 2012. "Hospital wastewater treatment by membrane bioreactor: Performance and efficiency for organic micropollutant elimination." *Environmental Science & Technology* 46 (3):1536–1545.

Lakho, Fida Hussain, Hong Quan Le, Flor Mattheeuws, Wouter Igodt, Veerle Depuydt, Joachim Desloover, Diederik P. L. Rousseau, and Stijn WH Van Hulle. 2021. "Decentralized grey and black water reuse by combining a vertical flow constructed wetland and membrane based potable water system: Full scale demonstration." *Journal of Environmental Chemical Engineering* 9 (1):104688.

López-Morales, Carlos A., and Lilia Rodríguez-Tapia. 2019. "On the economic analysis of wastewater treatment and reuse for designing strategies for water sustainability: Lessons from the Mexico Valley Basin." *Resources, Conservation and Recycling* 140:1–12.

Lopez, A., A. Pollice, A. Lonigro, S. Masi, A. M. Palese, G. L. Cirelli, A. Toscano, and R. Passino. 2006. "Agricultural wastewater reuse in southern Italy." *Desalination* 187 (1):323–334. doi: 10.1016/j.desal.2005.04.091.

Lydecker, Mary, and Pay Drechsel. 2010. "Urban agriculture and sanitation services in Accra, Ghana: The overlooked contribution." *International Journal of Agricultural Sustainability* 8 (1–2):94–103.

Mandal, D., P. Labhasetwar, S. Dhone, A. S. Dubey, G. Shinde, S. Wate. 2011. "Water conservation due to greywater treatment and reuse in urban setting with specific context to developing countries." *Resources Conservation and Recycling* 55:356–61. doi: 10.1016/j.resconrec.2010.11.001.

Mankad, Aditi, and Sorada Tapsuwan. 2011. "Review of socio-economic drivers of community acceptance and adoption of decentralised water systems." *Journal of Environmental Management* 92 (3):380–391.

McArdell, Christa S., Lubomira Kovalova, Hansruedi Siegrist, C. Kienle, R. Moser, and T. Schwartz. 2011. "Input and elimination of pharmaceuticals and disinfectants from hospital wastewater." *Final project report. Eawag: Das Wasserforschungs-Institut des ETH-Bereichs, Dubendorf, Suiza.*

Messerli, Peter, Endah Murniningtyas, Parfait Eloundou-Enyegue, Ernest G. Foli, Eeva Furman, Amanda Glassman, Gonzalo Hernández Licona, Eun Mee Kim, Wolfgang Lutz, and J.-P. Moatti. 2019. "Global sustainable development report 2019: the future is now – Science for achieving sustainable development." United Nations; New York. September 11, 2019. https://sustainabledevelopment.un.org/gsdr2019.

Meuler, S., Stefania Paris, and T. Hackner. 2008. "Membrane bio-reactors for decentralized wastewater treatment and reuse." *Water Science and Technology: A Journal of the International Association on Water Pollution Research* 58:285–94. doi: 10.2166/wst.2008.356.

Nasr, Mahmoud, Mohamed Ateia, and Kareem Hassan. 2016. "Artificial intelligence for greywater treatment using electrocoagulation process." *Separation Science and Technology* 51 (1):96–105.

Nguyen, Anh Viet, Nga Thuy Pham, Thang Huu Nguyen, Antoine Morel, and Karin Tonderski. 2007. "Improved septic tank with constructed wetland, a promising decentralized wastewater treatment alternative in Vietnam." International Conference of the National Onsite Wastewater Recycling Association (NOWRA) and International Water Association.

Nolde, E. 1999. "Greywater reuse systems for toilet flushing in multi-storey buildings — over ten years experience in Berlin." *Urban Water* 1:274–84.

Otoo, Miriam, and Pay Drechsel. 2018. *Resource recovery from waste: Business models for energy, nutrient and water reuse in low-and middle-income countries*, Routledge.

Ottoson, J., and T. A. Stenström. 2003. "Faecal contamination of greywater and associated microbial risks." *Water Research* 37 (3):645–55. doi: 10.1016/s0043-1354(02)00352-4.

Parkinson, Jonathan, and Kevin Tayler. 2003. "Decentralized wastewater management in peri-urban areas in low-income countries." *Environment and Urbanization* 15 (1):75–90.

Peter-Varbanets, Maryna, Chris Zurbrügg, Chris Swartz, and Wouter Pronk. 2009. "Decentralized systems for potable water and the potential of membrane technology." *Water Research* 43 (2):245–265.

Revitt, D. M., E. Eriksson, E. Donner. 2011. "The implications of household greywater treatmentand reuse for municipal wastewater flows and micropollutant loads." *Water Research* 45:1549–60.

Rovirosa, N., E. Sánchez, F. Benítez, L. Travieso, and A. Pellón. 1995. "An integrated system for agricultural wastewater treatment." *Water Science and Technology* 32 (12):165–171. doi: 10.1016/0273-1223(96)00151-5.

Rozema, Eric R., Andrew C. VanderZaag, Jeff D. Wood, Aleksandra Drizo, Youbin Zheng, Ali Madani, and Robert J. Gordon. 2016. "Constructed wetlands for agricultural wastewater treatment in northeastern North America: A review." *Water* 8 (5):173.

Santasmasas, Carme, Miquel Rovira, Frederic Clarens, and César Valderrama. 2013. "Grey water reclamation by decentralized MBR prototype." *Resources, Conservation and Recycling* 72:102–107. doi: 10.1016/j.resconrec.2013.01.004.

Santos, C., F. Taveira-Pinto, C. Y. Cheng, and D. Leite. 2012. "Development of an experimental system for greywater reuse." *Desalination* 285:301–305. doi: 10.1016/j.desal.2011.10.017.

Shrestha, R. R., R. Haberl, J. Laber, R. Manandhar, and J. Mader. 2001. "Application of constructed wetlands for wastewater treatment in Nepal." *Water Science and Technology* 44 (11–12):381–386. doi: 10.2166/wst.2001.0855.

Sijimol, M. R., and S. Joseph. 2021. "Constructed wetland systems for greywater treatment and reuse: A review." *International Journal of Energy and Water Resources* 5 (3):357–369.

Subramanian, P. S. Ganesh, Anjali V. Raj, Priyanka Jamwal, Stephanie Connelly, Jagadeesh Yeluripati, Samia Richards, Rowan Ellis, and Lakshminarayana Rao. 2020. "Decentralized treatment and recycling of greywater from a school in rural India." *Journal of Water Process Engineering* 38:101695.

Theodorakopoulos, George V., Michalis K. Arfanis, José Antonio Sánchez Pérez, Ana Agüera, Flor Ximena Cadena Aponte, Emilia Markellou, George Em. Romanos, and Polycarpos Falaras. 2023. "Novel pilot-scale photocatalytic nanofiltration reactor for agricultural wastewater treatment." *Membranes* 13 (2):202.

Tilley, Elizabeth, Lukas Ulrich, Christoph Lüthi, Philippe Reymond, and Christian Zurbrügg. 2014. "Compendium of sanitation systems and technologies."

Tooke, M. A. 2012. "Decentralized wastewater treatment: A sensible solution." *US EPA* 8.

Tuladhar, Bhushan, Prajwal Shrestha, and Rajendra Shrestha. 2008. "Decentralised wastewater management using constructed wetlands." Beyond *construction*: Use *by all*: A collection of case studies from sanitation and hygiene promotion practitioners in South Asia, 86–94. https://www.ircwash.org/sites/default/files/Tuladhar-2008-Decentralised.pdf

UN. 2010. *General assembly. Resolution adopted by the general assembly.* United Nations. https://documents-dds-ny.un.org/doc/UNDOC/GEN/N09/479/35/PDF/N0947935.pdf?OpenElement.

Van der Hoek, Wim, Flemmming Konradsen, Jeroen H. J. Ensink, Muhammad Mudasser, and Peter K. Jensen. 2001. "Irrigation water as a source of drinking water: Is safe use possible?" *Tropical Medicine & International Health* 6 (1):46–54.

Victor, Reginald, Richard Kotter, Geoff O'Brien, Mit Mitropoulos, and Gabriel Panayi. 2008. *WHO guidelines for the safe use of wastewater, excreta and greywater*, volumes 1–4. Taylor & Francis.

WHO. 2006. Guidelines for the Safe Use of wastewater, excreta and Greywater. Volume I: Policy and Regulatory Aspects. WHO, UNEP, 2006.

WHO. 2017. *Progress on drinking water, sanitation and hygiene: 2017 update and SDG baselines.*

WHO. 2019. *Progress on household drinking water, sanitation and hygiene 2000-2017: special focus on inequalities*: World Health Organization.

Winward, G. P., L. M. Avery, R. Frazer-Williams, M. Pidou, P. Jeffrey, T. Stephenson, and B. Jefferson. 2008b. "A study of the microbial quality of greywater and an evaluation of treatment technologies for reuse." *Ecological Engineering* 32:187–97.

Winward, G. P., L. M. Avery, T. Stephenson, B. Jefferson. 2008a. Chlorine disinfection of greywater for reuse: effect of organics and particles. *Water Research* 42:483–491.

Part II

Technologies for Decentralized Treatment

4 Common Decentralized Treatment and Disposal Methods

Sri Chandana Panchangam[1], Loka Subramanyam Sarma[2], Preethi Vijayarengan[3], and Avinash Kumar Reddy Narreddy[4]

[1] Annamacharya Institute of Technology and Sciences, Kadapa, India

[2] Yogi Vemana University, Kadapa, India

[3] Hindustan Institute of Technology and Science, Chennai, India

[4] Dr. YSR Architecture and Fine Arts University, Kadapa, India

4.1 INTRODUCTION

Decentralized wastewater treatment systems (DWWTSs) are a feasible alternative for tackling challenges such as connectivity of waste generation points to the central treatment systems, increasing loads in accordance with the population, addressing the issue of improper sanitation and generating reclaimable or reusable water for non-potable purposes, sustainable and less spatial footprint (Bajpai et al, 2019; Libralato et al, 2012; Singh et al, 2019; Varma et al, 2022). Moreso, pliability of DWWTSs toward unforeseen external factors is welcoming feature for their applicability toward stresses caused by epidemics or disasters (Juan-Garcia et al, 2017; Kohler et al, 2020). Technologies developed so far are targeted for the treatment of either wastewater or sludge (NIUA, 2016) or even for organic waste (Biswas et al, 2021). Treatment of wastewater in decentralized system is achieved by primary, secondary and tertiary treatments. The objective of primary treatment is to remove solids by settling along with the removal of floating material (Sharma et al, 2022). The goal of secondary treatment is to meet the effluent standards for non-potable purposes (Arora et al, 2016). When the secondary treatment cannot meet the standards which are not possible in all the cases, tertiary treatment options are chosen to polish the secondary effluents (Philip et al, 2019). The technological options in DWWTSs are segregated as natural treatment systems, aerobic and anaerobic systems, and a combination of them (Bernal et al, 2021; Singh et al, 2015). Natural treatment involves treatment by a combination of earthen material, plants and microorganisms; thus, these systems are economical, highly functional in tropical regions and sustain load fluctuations (Singh et al, 2015). The disadvantages include long retention times, greater land requirement, generation of bad odors due to the hotter climate. Nevertheless, they are highly suitable in developing countries (Starkl et al, 2013) and that natural treatment systems are more suitable as decentralized treatment options (Bernal et al, 2021).

Decentralized treatment systems are usually more demanding and required in urban settlements with large scope in fluctuation in population density. Domestic wastewater is generated from activities such as cooking, washing, gardening, toilet flushing and shower waters. Domestic wastewater consists of 70% organic content, while the remaining is inorganic in nature (Huang et al, 2010). The wastewater generated from dwellings can be classified as gray water and black water (Estevez et al, 2022). Another potential application of DWWTSs or techniques is for the treatment of hospital effluents, which consists of pathogens, infectious microorganisms and other contaminants that impact human health (Kamliya et al, 2023; Sharma et al, 2022). DWWTSs can treat wastewaters

DOI: 10.1201/9781003381419-6

originated from both domestic and industrial origins up to a capacity ranging from 1 to 1000 m^3/day (Singh et al, 2019).

4.2 PRIMARY TREATMENT METHODS

The primary treatment methods that are available include sedimentation tanks, septic tanks, or simple bio-digesters (Varma et al, 2022). Septic tank is simplest and oldest available technique, also called as cesspool. They are installed as standalone treatment units for households in developing countries for treating the wastewater through settlement of solids and also anaerobic digestion to certain extent (Capodaglio et al, 2017; Massoud et al, 2009; Moussavi et al, 2010). The functioning of septic tank is effective in hotter climate and the organic removal efficacy is about 50%. A septic tank is not efficient in removing pathogens, and thus, additional or augmented treatment is required for achieving good quality effluents (Singh et al, 2015). Septic tanks are a common choice in the developing countries due to their suitability to the local conditions. A septic tank may consist of a maximum of three sections (Varma et al, 2022), usually being two (Mahon et al, 2022). One chamber usually separates organic solids, solids and scum through the process of sedimentation and floatation (Nasr et al, 2013). Because of the presence of solids and mechanisms occurring in the tank such as accumulation and storage of solids (Siegrist, 2017), followed by their digestion, three layers are seen in the septic tank. The three layers include sludge, liquid and scum layers from bottom to top (Al Momani, 2015). However, the efficiency of a septic tank is challenging because of the escape of unsettleable solids rich in microbial population for the digestion of sludge from the tank (Saeed et al, 2021). The cost burden in maintenance and failure of the septic tank is mostly associated with the improper desludging operation (Mahon et al, 2022). The desludging of septic tanks is usually recommended once in two to six years for typical sized tanks. The depth of sludge deposition in the septic tank shall affect the effluent solids concentration and the dimensions of the tank (Gray, 1995; Lossing et al, 2010). The material used for construction of septic tanks is usually concrete, though other materials such as tires were used, concrete remains standard choice based on evaluation in terms of technical and cost (Martins et al, 2021).

Waste stabilization ponds are another old form of primary treatment technique predominantly used as a decentralized treatment, and these ponds can be of aerobic, facultative or anaerobic type. Waste stabilization ponds are simple in design, have long retention times, economical, easy to construct and operate, low maintenance cost and skill, not energy intensive and can handle wide hydraulic loading rates (Bajpai et al, 2019; del Castillo et al, 2022). However, waste stabilization ponds require larger areas which make them less preferable to septic tanks. Other primary treatments include settlers or sedimentation tanks for the removal of solids due to retention by settling mechanism. It can be seen from Table 4.1 that the performance of primary treatment is not sufficient to meet the effluent standards or reuse standards and hence require further treatment in the form of secondary phase. Therefore, primary treatment methods are usually used in conjunction with other methods as elaborated in the on-site treatment methods section.

4.3 SECONDARY TREATMENT METHODS

A better effluent quality can be obtained through secondary treatment by the removal of total solids and achieve water standards useful for non-potable purposes (Arora et al, 2016) basically by enhancing contact between activated biomass and fresh wastewater (Saeed et al, 2021; Singh et al, 2019). A large variety of decentralized secondary treatment units include constructed wetlands (CWs), anaerobic baffled reactors (ABRs), horizontal gravity filters, anaerobic filters, anaerobic and facultative pond systems, anaerobic reactors (AR), upflow anaerobic sludge blanket (UASB) reactors, membrane bioreactors (MBRs), biomass concentrator reactor (BCR), moving bed biofilm reactor (MBBR), lagoons, suspended growth systems, attached growth systems and sequencing batch reactor (SBR). Choice of a treatment unit depends on site-specific conditions.

TABLE 4.1
Performance of Various Primary Decentralized Treatment Techniques

	Removal (%)						
Treatment Technique	**BOD**[a]	**COD**	**TSS**	**TKN**	**TN**	**TP**	**Reference**
Waste stabilization pond	75	70	60	–	51	51	Rodrigues (2009)
	50.6	48.9	44.3	–	–	–	Ghazy and El-Senousy (2008)
	6.7	16.3	–	–	–	18.1	Avsar et al (2008)
	94	–	63	–	–	–	Papadopoulos and Tsihrintzis (2011)
	38	42	–	77.4	–	76	Del Nery et al (2013)
	11.7	17.7	–	–	–	52	
	20	21.4	10.64	–	–	–	Hayati et al (2013)
	33	21.3	15.14	–	–	–	
	56.5	49.3	56	–	–	–	
Septic tank	68.4	65.3	65.3	–	26.8	29.3	Nasr & Mikhaeil (2013)
	58	61	32	20	–	–	Koottatep et al (2020)
Septic tank with solar-heated water system	83	88	71	35	–	–	Koottatep et al (2020)

[a] BOD: Biological oxygen demand, COD: chemical oxygen demand, TSS: total suspended solids, TKN: total kjeldahl nitrogen, TN: total nitrogen, TP: total phosphorous.

4.3.1 Construction Wetlands (CWs)

CWs are prominent engineered secondary treatment units for the treatment of wastewater originated from various sources such as domestic, industrial and agricultural origins. CWs are popular in recent years because of robustness, low capital cost, simplicity in design and operation, capacity to withstand shock loads and load fluctuation (Capodaglio et al, 2017; Zhang et al, 2015). Operational efficacy and stability of the CWs are limited due to the clogging of substrate particularly for the high organic loaded influents eventually become harmful to the macrophytes in the CW unit (Langergraber et al, 2003).

CWs can be classified (Figure 4.1) into basically two types depending on the type of flow. They are free-water surface flow (SFCW) and subsurface flow (SSFCW) (Capodaglio et al, 2017). SFCWs have open water surfaces and can be further divided in to three types of wetlands owing to the

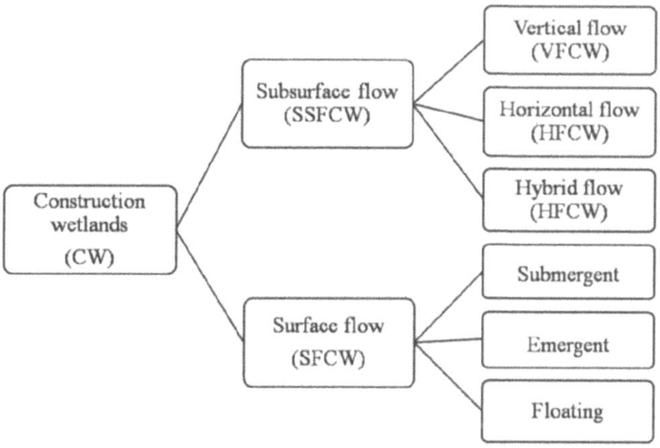

FIGURE 4.1 Classification of constructed wetlands.

macrophytes: submergent, emergent and floating types (Colares, 2019; Horn et al, 2014). SSFCWs can be sub-classified in to three types based on the direction of flow of wastewater: vertical flow CWs (VFCWs), horizontal flow CWs (HFCWs) and hybrid flow CW (HFCW) (del Castillo et al, 2022).

SFCWs are simple treatment units wherein wastewater is exposed the atmosphere and the degree of oxygenation is thus improved in these units while treating the wastewater (Figure 4.2A). The mean removal efficiencies and design parameters for SFCW from the literature are mentioned in Table 4.2.

Another kind of CW is SSFCWs that are further divided into three kinds based on flow pattern of wastewater. As can be seen from Figure 4.2, CWs comprise five components. They are basin or tank; substrate that is filler media made up of a combination of gravel or sand or brick above which plants are planted, vegetation, impermeable liners around the basin and inlet/outlet arrangements. In the case of HFCW (Figure 4.2B), the influent will enter into the basin from inlet and flow horizontally through the substrate and reaches the outlet after passing through the surface of the bed.

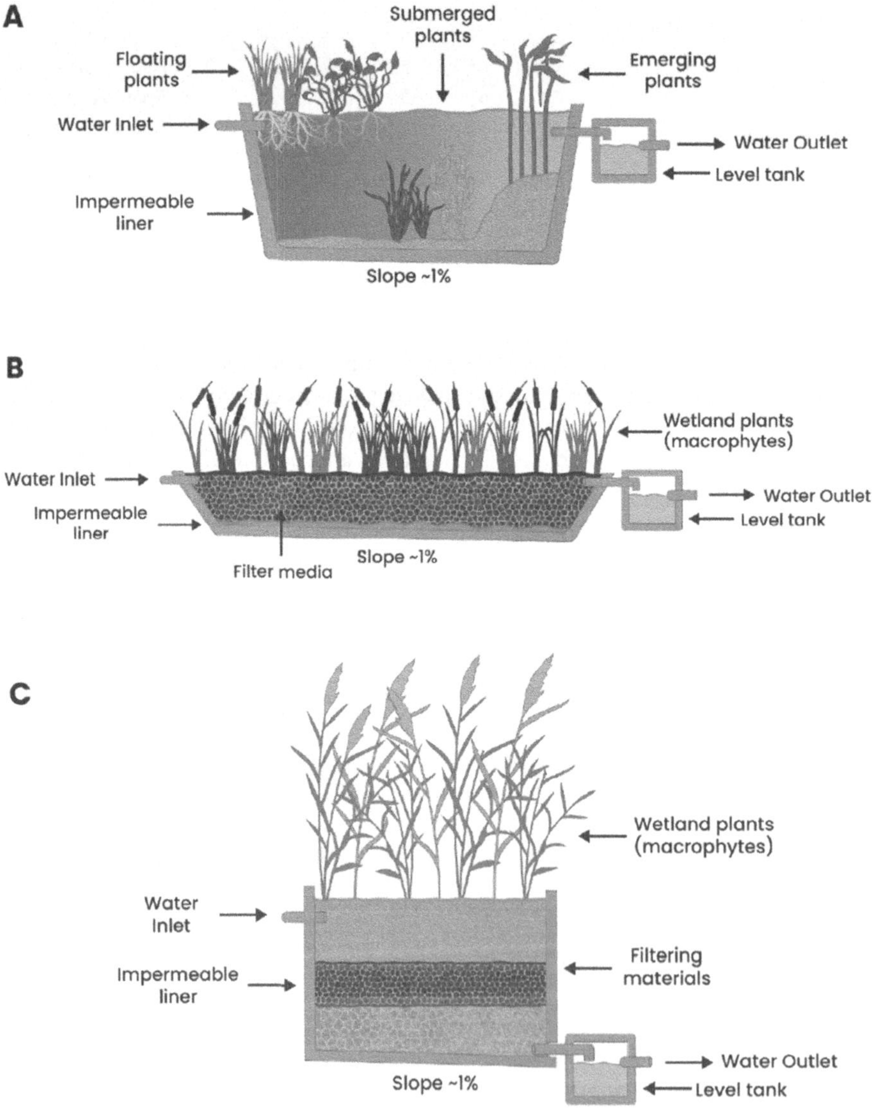

FIGURE 4.2 (A) Surface flow constructed wetland (SFCW), (B) horizontal flow constructed wetland (HFCW), and (C) vertical flow constructed wetland (VFCW). (Source: del Castillo et al, 2022.)

TABLE 4.2
Mean Removal Efficiencies and Design Parameters of Constructed Wetlands (Capodaglio et al, 2017; del Castillo et al, 2022; Philip et al, 2019; Thalla et al, 2019)

	Mean Removal Efficiencies (%)		
Parameter	**SFCW**	**HFCW**	**VFCW**
BOD[a]	45 ± 23	62 ± 20	84 ± 15
COD	48 ± 33	61 ± 24	78 ± 15
TSS	60 ± 23	66 ± 22	83 ± 19
NH_4^+-N	56 ± 20	38 ± 16	83 ± 12
TN	38 ± 16	43 ± 18	70 ± 19
TP	34 ± 16	33 ± 23	61 ± 26
HRT (days)	5–14	2–7	24 h[a]
Specific area (m^2/p.e)	5–10	5–10	1–2
Max. BOD loading rate (g/m^2/day)	8	7.5	4–6
Water depth (m)	0.5–1.0	0.1–1.0	0.4[a]
Hydraulic loading rate (mm/day)	7–60	2–30	40–80
Organic loading rate (g/m^2/day)	3–6	2.2[a]	3.5[a]
Length/width ratio	2:1 to 10:1	0.25:1 to 5:1	1.2:1[a]
Prevailing condition	Aerobic at surface and anaerobic at bottom	Anaerobic	Aerobic
Mosquito control	Required	Not required	Not required

[a] BOD – Biological oxygen demand, COD – chemical oxygen demand, TSS – total suspended solids, TN – total nitrogen, TP – total phosphorous, HRT – hydraulic retention time, SFCW – surface flow constructed wetland, HFCW – horizontal flow constructed wetland, VFCW – vertical flow constructed wetland.

SSFCWs are basically anaerobic treatment systems. In a HFCW, anaerobic and anoxic microenvironments exist predominantly with a possibility of aerobic condition at plant root zone (Vymazal, 2010). HFCWs are more common than VFCWs (Figure 4.2C) due to the ease in operation and construction. However, VFCWs are more efficient in terms of total suspended solids (TSS) removal and organics degradation than HFCW as can be noted from Table 4.3. HFCWs are a combination of more than one CW units that can combine the merits of aerobic and anaerobic conditions for achieving overall effluent quality as seen in Table 4.3 (Vymazal, 2013). The mechanisms of removal for various pollutants and nutrients in a CW are given in Figure 4.3.

TABLE 4.3
Design Factors and Performance of Lagoon Systems

Parameter	Facultative Lagoon	Anaerobic Lagoon	Aerobic Lagoon	Aerated Lagoon
BOD[a] (%)	75–95	50–80	NA	75–95
TSS (%)	90	NA	NA	90
Nitrogen (%)	60	NA	NA	10–20
Phosphorous (%)	50	NA	NA	15–20
Detention time (days)	20–180	1–50	3–50	10–30
Depth (m)	1.2–2.4	2.4–6.0	NA	1.8–6
Organic loading rate	15 – 80 kg/ha/d	0.04 – 0.3 kg/m^3/d	NA	NA
Surface area	NA	0.2–0.8 ha	NA	NA

[a] BOD – Biological oxygen demand, TSS – total suspended solids.

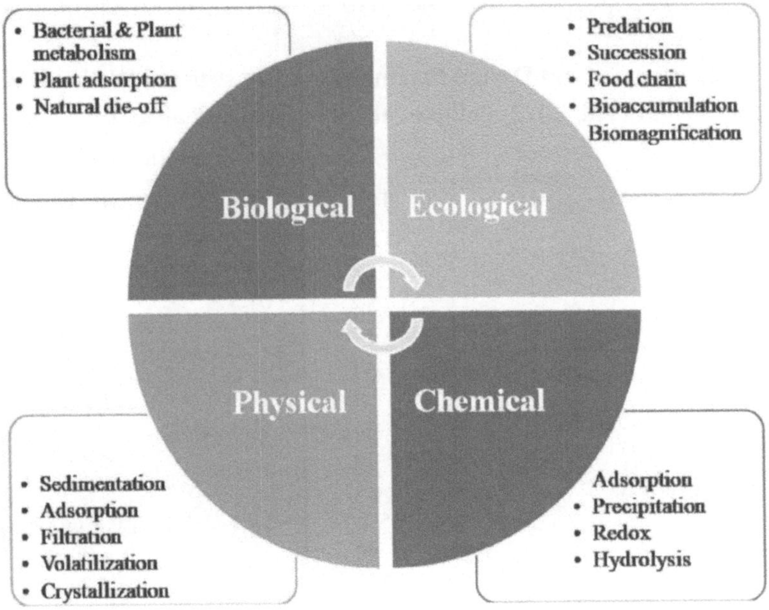

FIGURE 4.3 Mechanisms of pollutant and nutrient removal in a constructed wetland.

CWs are cost-effective and hence are highly recommended as decentralized treatment process in developing countries.

4.3.2 Lagoons (Facultative, Aerated and Anaerobic)

Lagoons are pond-like water bodies or water-basins used for the treatment of wastewater. They are similar to waste stabilization ponds (US EPA, 2002a). Lagoon systems are simple and economical, natural and energy-efficient. Lagoons are lined with clay or other materials to prevent the leakage. The treatment of wastewater in lagoons is achieved through physical, chemical and biological processes naturally. However, to achieve increased efficiency, sometimes aeration devices are installed (NIUA, 2019). Lagoon systems are classified into aerobic, anaerobic and facultative lagoons based on the type of treatment process. Lagoon systems have site-specific designs, and parameters considered are type of soil, available land area, climatic and weather conditions (Peavy et al, 1985).

Anaerobic lagoons are deep impoundments that can treat high strength organic wastewaters by anaerobic digestion in the absence of dissolved oxygen. The effluents of lagoons are suitable for the agricultural and urban water reuse (Bernal et al, 2021). Raw influent enters an anaerobic lagoon from the bottom of the pond and gets mixed with the sludge containing active biomass. The effluent of this type of lagoons cannot be discharged into the rivers and require further treatment. Surface layers of an anaerobic lagoon may retain grease and scum that are the reason for bad odors, and hence in few cases, aeration is done for the odor control. The anaerobic bacterium is responsible for the conversion of organic matter in to carbon dioxide and methane. This is achieved in two step process. First is the acid formation phase where complex organic compounds are converted into simpler compounds by acid formers removing certain quantities of BOD and COD. The second phase is methane formation with an intermediate step of acetogenesis. In this step, bacteria convert short chain organic acids to acetate, hydrogen and carbon dioxide gases. After this methane-forming bacteria become active in conversion of acetate, hydrogen and carbon dioxide into methane stabilizing the organics, and hence this process is called methanogenesis (Metcalf et al, 1991). Anaerobic lagoons are an attractive option because of the low cost and ability to handle high strength wastewaters originated from either industrial or municipal sources (US EPA, 2002b).

In an aerobic lagoon, the degradation occurs by aerobic process. No artificial aeration is provided in this case. Aerobic process happens due to the dissolved oxygen and also algal respiration. Algae utilize carbon dioxide from water and undergo photosynthesis under sunlight to produce oxygen. Then the released oxygen by algae along with reoxygenation due to atmospheric exchange helps aerobic bacteria to carry out aerobic degradation in aerobic lagoons. The biggest advantages with aerobic lagoons are that the effluents are rich in nutrients and devoid of pathogens and hence are more suitable for agricultural purposes.

In a facultative lagoon, aerobic decomposition process happens in the shallow surface layers, anoxic condition and facultative decomposition in intermittent layer and anaerobic decomposition in bottom layers of the lagoon. Design factors and performance of lagoons are tabulated in Table 4.3.

4.3.3 Sequential Batch Reactors

Sequential batch reactor (SBR) is an aerobic suspended growth process working in batch mode similar to an activated sludge process. SBRs can achieve equalization, aeration, clarification and biological treatment in a single tank. The performance of SBRs depends on detention time since all the mechanisms occur in a single tank. The major advantages of SBRs are as follows: less foot print, strong impact resistance, suitable for the treatment of toxic and high strength wastewaters, ease in maintenance and operation (Sadaf et al, 2022). SBRs can also achieve nitrification and desulfurization with prolonged retention times which in turn protect local ground and surface waters when disposed of (Massoud et al, 2009; Sekine et al, 2020). The operation cycle of SBR consists of five basic steps: fill, react, settle, draw or decant and idle (Ghangrekar & Behera, 2014). The essential elements of SBR are reactor tank, waste sludge draw-off mechanism, aerators, effluent decanter and process control system. The efficiency of SBR based on literature is BOD > 92%, COD > 94%, TSS > 85%, TN > 80%, TP > 60%, and further improvisation of the SBR can result in achieving enhanced effluent quality (Ding et al, 2011; Kim et al, 2008; Lin and Cheng, 2001; Sadaf et al, 2022; Sekine et al, 2020; Surampalli et al, 2000).

4.3.4 Other Methods

A large number of aerobic and anaerobic treatment systems are being used as decentralized wastewater treatment process.

4.3.4.1 Membrane Bioreactors (MBRs)

MBRs combine biological process with membrane filtration. The biological process can be aerobic or anaerobic referred to as aerobic MBRs (AeMBR) and anaerobic MBRs (AnMBR), respectively (Cashman et al, 2018). The membranes can be ultrafiltration or microfiltration membranes with a pore size < 1μm (Capodaglio et al, 2017; Varma et al, 2022). AeMBRs are compact units able to treat wastewaters from domestic, industrial sources and even pharmaceuticals effectively to generate good quality effluents (Cashman et al, 2018; Cecconet et al, 2019; Hazra & Durso, 2022; Sharma et al, 2022; Singh et al, 2015). The efficiency of AeMBR reaches 100% in TSS removal, > 98% BOD, up to 99% COD, 81–92% TN, 93% TP and > 97% NH_3-N with aeration demand of 0.34 to 2.3 $m^3/m^2/h$ for flat sheet membranes and energy demand of 2.3 kWh/m^3 for treating a specific black water (Capodaglio et al, 2017; DeCarolis & Adham, 2007; del Castillo et al, 2022). The major drawback in both the MBRs is membrane fouling and to a greater extent in the case of AnMBR (Lin et al, 2011). Higher temperatures are more favorable for anaerobic process; however, AnMBRs exhibited efficiency even at lower temperatures. AnMBRs at mesophilic and psychrophilic temperatures exhibited net energy benefits than AeMBRs making them a sustainable decentralized treatment choice (Cashman et al, 2018).

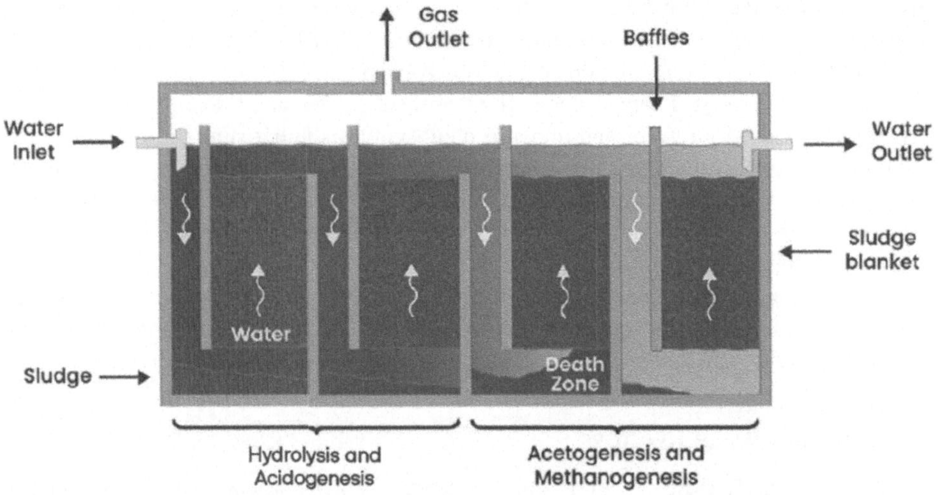

FIGURE 4.4 Anaerobic bioreactor (ABR). (Source: del Castillo et al, 2022.)

4.3.4.2 Anaerobic Baffled Reactor (ABR)

ABRs are anaerobic type consisting of fixed-bed reactors with baffle walls in which wastewater is allowed to flow in upward direction (Figure 4.4). Efficient anaerobic decomposition takes place in the reactors because of mixing of fresh wastewater with sludge containing activated biomass (Singh et al, 2019). An ABR consists of several chambers in which solids are settled first and then undergo anaerobic degradation. The activated sludge eventually settles down at the bottom of the reactor, and when the wastewater flows over the settled activated sludge, contaminants in the influent are consumed by the anaerobic bacteria and undergo bacterial metabolism (Varma et al, 2022). The baffles in the reactor resist the shock loads and increase the contact time resulting in progressive degradation in successive chambers. The ABRs are effective in treating wastewaters to achieve good quality effluents. Removal efficiencies of BOD are up to 90%, COD up to 85%, TSS up to 90%, TN up to 63% and TP up to 37% (del Castillo et al, 2022).

The other advantages of ABR lie in the fact that they are cost-effective, low energy requirements, long service life, high pollutant and BOD removal efficiency, low sludge production and less space foot print. The design components of ABR are the hydraulic retention time ranges between 48 and 72 h, upflow velocity of influent below 0.6 m/h, the number of chambers from three to six, sludge cleaning interval one to two years (Tilley et al, 2014; Varma et al, 2022).

4.3.4.3 Anaerobic Filters (AF)

AFs consist of fixed-bed biological reactors with a series of filtration chambers. They are simple units with advantages similar to ABR. The direction of flow of influent is also upward. The BOD removal is about 90%, whereas nitrogen removal is limited. Filter clogging is the biggest problem with these units. The influent needs pretreatment, and AFs are majorly used as polishing step following other secondary treatment units such as UASB or septic tanks (Alrajoula et al, 2009; De Oliveira Cruz et al, 2019; Dorji et al, 2021).

4.3.5 Combined Treatment

Combined treatment integrates various combinations of treatment options such as combination of aerobic, anaerobic and some natural systems (del Castillo et al, 2022; Singh et al, 2015; Subramaian et al, 2020), or hybrid systems coupling advanced oxidation processes or electrochemical methods with nature-based systems (Deng et al, 2022; Capodaglio et al, 2021; Colares et al, 20c19).

TABLE 4.4
Summary of a Few Selected Papers to Showcase the Performance of Combined Processes

	Removal Efficiencies (%)						
Process/Technique	**BOD**	**COD**	**TSS**	**NH_4^+-N**	**TN**	**TP**	**References**
UASB[a] + CW	93.3	88.5	86.5	46.1	52.9	66.3	del Castillo et al (2022)
ABR + CW	87.3	78.2	88.2	59.2	73.3	15.3	
UAF + CW	95.2	91.3	92.8	–	69.8	61.7	
Bioreactor septic tank + CW	99	99	92	97	95	100	Saeed et al (2021)
Anaerobic reactor + CW + ozonation	84.4	81.6	99	94.5	91	19.6	Colares et al (2019)
Septic tank + aerobic filter	90	30	72	–	–	–	Mesquita et al (2021)
Slow sand biofiltration + anaerobic sludge bioreactor + aeration + ozone-based disinfection	98	96	98	–	66	73	Subramanian et al (2020)
UASB + trickling filter	82	79	83	40	–	–	Vieira et al (2013)
Septic tank + anaerobic filter + intermittent sand filter	72	77	88	–	–	–	De Oliviera Cruz et al (2019)
Anaerobic reactor + VFCW + floating CW + HFCW	53	67	–	–	94	–	Pereira da silva et al (2021)

[a] UASB – Upflow anaerobic sludge blanket, CW – constructed wetland, ABR – anaerobic bio reactor, UAF – upflow anaerobic filter, VFCW – vertical flow constructed wetland, HFCW – horizontal flow constructed wetland, BOD – biological oxygen demand, COD – chemical oxygen demand, TSS – total suspended solids, TN – total nitrogen, TP – total phosphorous, NH_4^+-N – ammonia nitrogen.

The objective of combining different processes is to mitigate the limitations of a particular process and enhance the overall effluent quality to suit the standards or have reuse potential (Capodaglio et al, 2021; Castellar et al, 2022; Pereira da Silva et al, 2021; Saeed et al, 2021).

Table 4.4 summarizes few selected papers to showcase the performance of combined processes as an example. The removal efficiencies obtained in the individual studies through hybrid procedures are high compared to the individual treatment techniques. In particular, removal efficiencies of BOD, COD and TSS are high in the reported studies, whereas removal of phosphorous and nitrogen is difficult in independent systems, which was accomplished by hybrid processes with better removal efficiencies. Combinations of various treatment options depend on the specific effluent quality requirement, local effluent standards, financial investment, space availability and influent wastewater characteristics. No established design parameters are available for choice of combinations or combinatorial processes, and hence there are no standards or standard procedures.

4.4 SPECIFIC ON-SITE TECHNOLOGIES

On-site treatment of wastewater is an attribute of decentralized treatment that is located at either individual households or a cluster of dwellings (Capodaglio & Callegari, 2016; Chirisa et al, 2017; Estevez et al, 2022). On-site treatment of wastewater offers a sustainable solution by limiting the freshwater utilization through water reuse and immobilization of pollutants cross-boundaries (Li et al, 2018; Philip et al, 2019; Starkl et al, 2018; Zaharia, 2017). The reuse of energy and nutrient is also possible through decentralized on-site treatment systems (Libralato et al, 2012). The advantage of on-site treatment systems is the flexibility in adopting for either individual households or for a cluster of houses. The merits of this type of systems are usually low investment and maintenance costs and compact systems, wherein nuisance of odors and insects are minimal and zero power requirement in few cases. The biggest challenges with such systems are standardization of the systems. Undesirable environmental burden will increase as a result of improper maintenance. The efficiency of the systems in achieving the discharge standards is also not consistent, due to which often the effluents contain high COD and pathogens. A variety of on-site wastewater treatment systems are available; some of which are old and popular particularly in developing countries. The water treated through decentralized on-site treatment techniques was mostly recycled within the near vicinity of treatment. Few other researchers have further polished the treated effluents either separately or in conjunction with the treatment scheme for improving the quality of the treated effluent (Chen & Fukushi, 2016; Corales et al, 2019; Otter et al, 2020).

Septic tanks are preferable primary treatment option that is used either alone for the treatment of wastewater (Varma et al, 2022) or in alliance with other technologies like CWs or ABRs (Saeed et al, 2021). A conventional septic tank was modified in to a bioreactor septic tank with seven compartments consisting of settling chambers, ABR tanks, bio walls and effluent collection chamber to accomplish high quality effluent by increasing contact between wastewater and biomass (Saeed et al, 2021). Septic tanks with a combination of seepage pit or anaerobic filter were used in Brazil for on-site or cluster of 100 houses and reported that on-site systems are economical than cluster or central systems (Tonetti et al, 2022). Another on-site decentralized treatment module for the treatment of hand wash and kitchen wash waters from a school in Tamil Nadu, India, employed anaerobic biofilters and a series of anaerobic bioreactor and biofiltration tank respectively, both succeeded by aeration and ozonation achieving an overall removal efficiency of BOD, COD and TSS >95%. Even nitrate (98%) and total phosphorous (66%) removals are high making the on-site decentralized modules cost-effective and less cumbersome (Subramanian et al, 2020). Wastewaters from urban centers in Bhutan were treated with decentralized systems on-site using UASB filled with PET bottle pieces as biofilter media subsequently achieving high COD removal rate meeting Bhutan standards (Dorji et al, 2021). Development of technologies that can integrate or use any reclaimed or byproduct materials adds value to the sustainable goals. A permeable reactor with lateral flow design and down-flow design was employed in a rural sector in Canada and Brazil, respectively (Suhogusoff et al, 2019). The objective of this system is protecting groundwater from pathogens and nutrients such as nitrates and phosphates while achieving disposal standards. Ten villages with total of 146 on-site decentralized facilities surrounding Erhai Lake in China were analyzed for their performance (Yang et al, 2021). The sets of treatment technologies were anaerobic-anoxic-oxic membrane bioreactor (A_2O/-MBR), facultative MBR (FMBR) and soil purification tank (SPT). The average removal efficiencies of these technologies were not meeting the discharge standards especially in the removal of phosphorous, and the serviceability was reported to be low. The suggestion to improve the effluent quality is to pretreat the influent, and combining with natural treatment systems such as ponds or CWs (Yang et al, 2021). The selection and working efficiencies of on-site treatment depend on site conditions, wastewater characteristics, geographical location, climatic conditions, disposal methods, effluent and disposal standards, viability of a treatment option, availability of space, finance and expertise.

4.5 DISPOSAL METHODS

Disposal of the treated effluents or septage is an important component of DWWTS. Disposal literally means ultimately returning the treated effluents or byproducts back in to the environment, either to convert in to useful resources or to reduce the risk of the materials. The various methods in use for the disposal of liquid and solid end-products of after the treatment are shown in Figure 4.5.

Disposal methods can be basically two ways depending upon the quality of the wastewater or septage (Figure 4.5). The direct disposal methods include evaporation or evapotranspiration that occurs naturally in environment. Otherwise the treated effluents are discharged in the receiving water bodies or for land applications such as sewage farming. The stabilized decomposed sludge portion of septage is drawn out and used directly as manure or converted to cakes or pellets and then is used. Given its simplicity, stability and cost-effective nature, disposal of effluents for subsurface

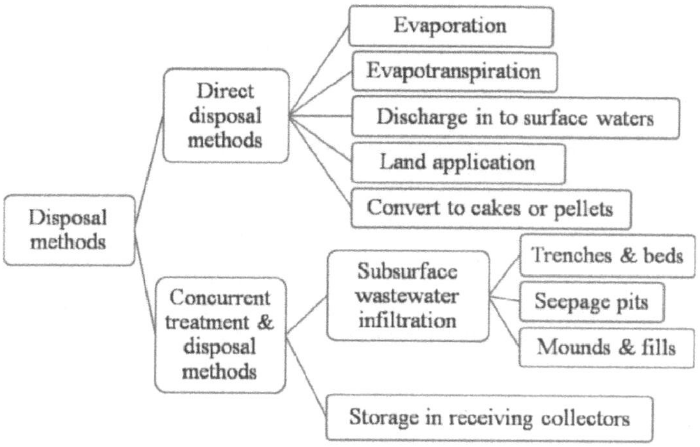

FIGURE 4.5 Disposal options in decentralized treatment systems.

soil absorption is the best option when the conditions such as soil and ground water characteristics and geological features are favorable. Seepage pits are a better alternative where land availability is limited and water table is low. The choice of the method obviously depends on the site conditions because in a decentralized treatment system, disposal is not far away from the treatment system.

4.6 RECYCLING SYSTEMS

Decentralized treatment systems are considered sustainable since they offer benefits of waste reduction at source, treatment consisting low energy requirement, safe disposal methods and facility to recycle or reuse the wastewater at site meeting the sustainable development goal of "ensure availability and sustainable management of water and sanitation for all" (Estevez et al, 2022; Liu et al, 2022; Prateep Na Talang et al, 2022). The term "reuse" meaning using again and again is old paradigm, in the context of water resources management, reuse of water is the recycling potential of treated wastewater for potable and other beneficial purposes (Mo & Zhang, 2013; Tortajada, 2020). Decentralized treatment systems are highly compatible with the local water use and reuse requirements (Capodaglio, 2017). The possible potential reuse/recycle of decentralized wastewater system treated effluents (Capodaglio, 2017; CPHEEO, 2021; Tortajada, 2020) is represented in Table 4.5.

4.7 PROBLEMS ASSOCIATED WITH TREATMENT METHODS

In the process of decentralized wastewater management, biggest challenge is choosing an appropriate technology with reduced problems and no failures. The appropriateness of a treatment method does not depend only on investment cost and general design considerations like site conditions, influent characteristics and required performance of the method (Massoud et al, 2009). There are certain failures reported in the literature (Anh et al, 2002; Wang & Xia, 2005) from which issues concerning decentralized wastewater treatment methods have been construed and studied. Figure 4.6 depicts the issues and corresponding aspects or characteristics for any decentralized treatment method.

The major issue of any treatment method is undoubtedly its performance in terms of efficiency, and cost. However, they alone do not define the success of the method; as shown in Figure 4.6, various socio-cultural, economic, institutional, environmental and technological issues may arise individually or in group. A strategic plan needs to be made before they are on the countries being developed or developing (Massoud et al, 2009).

TABLE 4.5
Possible Reuse/Recycle of Decentralized Wastewater Treatment Effluent

Reclamation Purpose	Application	Description/Remarks
Agriculture (irrigation)	Irrigation of cash and food crops	Irrigation is popular option for reuse of treated wastewater. Agricultural utility is more prominent in rural scenarios. The quality of effluents should meet standards and free from pathogens, groundwater should not be affected and soil properties should not be deteriorated.
	Horticulture	TDS should be same as groundwater TDS limits. Alkalinity should be controlled to preserve soil characteristics and plant health.
	Farm forestry	Farm forestry itself has environmental benefits and hence is a preferable reuse choice. However, it depends on climatic conditions and the quantity of treated effluents sufficient enough for farm forestry.
	Green houses	Improve nutrient removal and recovery through carbon dioxide sequestration consequently increasing the plant growth.
	Aqua culture	Used for fish hatcheries and fish ponds particularly popular in countries like Vietnam, India and Bangladesh.
	Landscaping and surface storage	Reclaimed water is used for urban farming, urban forestry or landscaping, irrigation of parks, green zones, golf courses, roadside greenery, cemeteries etc. Surface storage can be done for creating aesthetic value for real estate purposes.
Non-irrigation purposes	Urban uses	Street washing, fire-protection, air conditioning, vehicle washing, commercial laundry, ground cooling to mitigate heat island effect.
Industrial applications	Cooling, boiler feed water, process water, concrete making, soil compaction, dust control etc.	Every industry has specific standards for water to be used as process water and hence guidelines need to be followed. Use as cooling water is popular among others for industrial purposes.
Environment and recreation	Wild life habitats, augmentation of river flow, wetlands, silviculture, artificial lakes and ponds	These uses help restore environmental ecology and balance biodiversity.
		Creation of lakes and ponds for recreational purposes that does not depend on seasonal water inputs for maintenance.
Domestic and commercial	Toilet flushing	Treated gray water can be used for toilet flushing.
	Cleaning	Cleaning of outdoor surfaces and gardening.
Indirect recharge of impoundments (indirect potable use)	Groundwater recharge	Groundwater levels can be enhanced.
	Seawater intrusion barrier	Intrusion of saltwater can be controlled from entering in to freshwater aquifer in coastal areas.
	Soil aquifer treatment	A means to conserve freshwater being exploited for industrial and agricultural use.
	Surface water augmentation	The treated effluents are discharged in to perennial rivers for increasing the water quantity available for treatment plants. Releasing in to dry river courses helps in aquifer recharge.
	Surface water reservoirs	When surface water reservoirs receive inflow from rivers, they become impoundments and a source for public water supplies.
Direct potable purpose	Direct closed loop	Direct potable reuse of treated wastewater after advanced treatment.
	Discharge in to reservoirs	Since the perception of people and acceptance of toilet to tap concept is difficult, most of the cases, the treated effluents are discharged in to surface reservoirs and becomes source water for public water works.

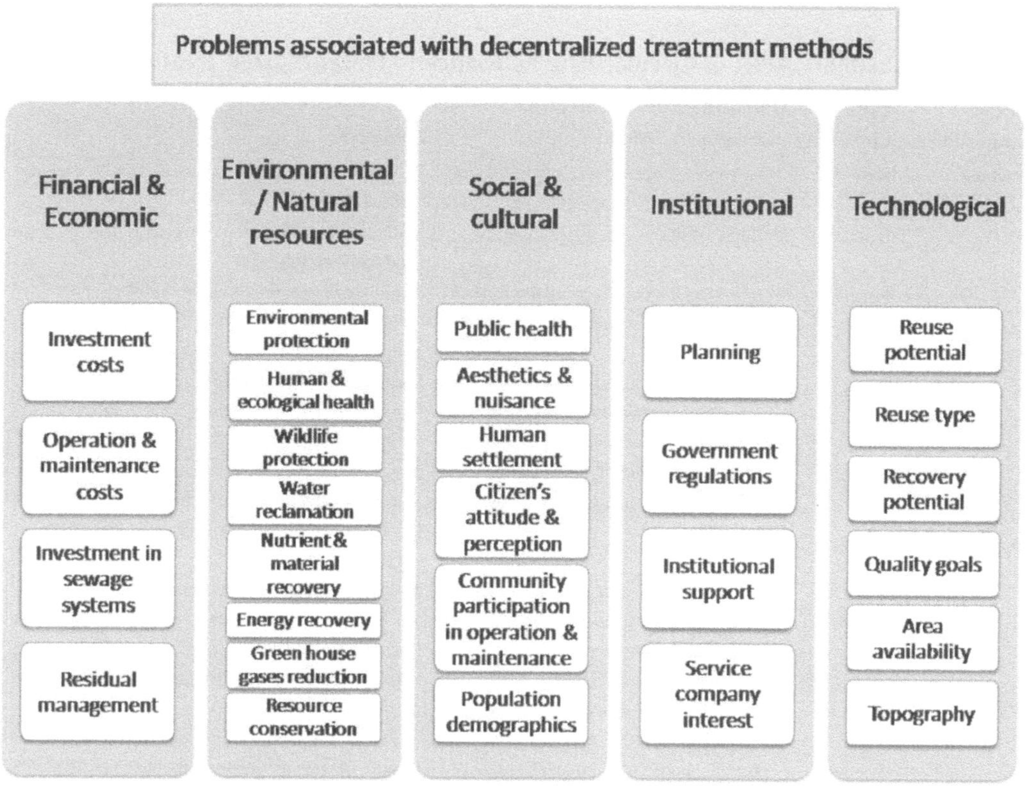

FIGURE 4.6 Issues and corresponding aspects of decentralized treatment methods.

4.8 ENVIRONMENTAL STANDARDS FOR DISPOSAL OF SEPTAGE

Septage is the term used for the excrement or other materials present in or removed from septic tank. Septage usually has high BOD and COD concentrations, total solids, nitrogen and pathogens. Improper treatment and disposal cause public health and environmental hazards. In India, there is no specific standard for the disposal of septage in particular. Nevertheless adequate legislative frameworks at national-, state- and city-level to protect water and environment are available. Public health and sanitation are constitutional responsibility. Environmental (Protection) Act, 1986 and water (Prevention and Control of Pollution) Act, 1974 provide structure for the control of septage discharge. Solid Waste Management (SWM) Rules, 2016, under Environment (Protection) Act, give framework for final safe disposal of post-processed residual fecal sludge and septage for the prevention of ground water, surface water and ambient air contamination and also for the disposal into landfills or use as compost (Rohilla et al, 2017).

4.9 MOVING TOWARD CIRCULARITY AND SUSTAINABLE SANITATION

Circular economy is a new paradigm that deals with global challenges such as climate change, loss of biodiversity, minimization of waste and pollution. Sustainable sanitation focuses on the on-site treatment of wastewater, recycling on-site or in near vicinity, reuse of the local resources (Liu et al, 2022). The concept of circularity is aimed to restore, regenerate, rely on system-wide innovation with in closed loop, using and reusing products to eliminate waste and minimize negative impacts. Table 4.6 showcases the popular 3Rs of sustainability and 10Rs of circular economy and their relevance to decentralized treatment along with some comments about their

TABLE 4.6
3Rs of Sustainability, 10Rs of Circularity in Relevance to Decentralized Treatment

3Rs of Sustainability	10Rs of Circularity	Relevance to Decentralized Treatment System	Comments
Reduce	Refuse	Prevention of raw material exploitation by using the local raw materials	Prevention from deterioration, and environmental conservation
	Reduce	Decreasing the use of raw materials	Reduction in production of waste, reduce pollution and exploitation
	Renew	Redesign the treatment method to suit circularity or sustainability goals	Eliminate waste, for example installation of treatments in green houses for carbon dioxide sequestration and simultaneous growth of plants
Reuse	Reuse	Increasing the life span of the treatment system and continuous use of the byproducts	Minimize negative impacts, reclaim the treated wastewaters for non-potable and potable purposes
	Repair	Maintenance and repair of the system to avoid for going to the new one	Proper maintenance of decentralized treatment is required such as sludge cleaning, removal of membrane clogging for overall increase in the performance
	Refurbish	Similar to renewal of product or making a new product from the materials	Making of cakes or pellets of manure instead of using directly as manure to add commercial value
	Repurpose	Reuse of the product for other purpose	Recirculation of sludge, or making building materials from treated sludge
Recycle	Recycle	Recycling the effluents or treatment byproducts	Artificial recharge of impoundments and rivers catering to public water works
	Recover	Energy, nutrient and material recovery	Biogas, electricity generation, microbial fuel cells, nutrient recovery can be done for recirculation within the system or impart commercial value
Other Rs	Restoration	Avoid large excavation and huge installations, pipelines etc.	Decentralized systems help in the restoration of the ecology and environmental quality because of the lesser manmade intervention boundaries
	Regeneration	Limiting of pollution and maximizing benefits	Limiting of pollution and exploitation, increasing the potential use of benefits help in regeneration of earth's ecology, biodiversity, and public health

connectivity (2030 Water Resources Group and Council of Energy (Environment and Water), 2016; Cramer, 2017). The entries in the table are in the order of priority from most preferred to least preferred (top to bottom). A decentralized treatment system has potential to recovery of energy, integrate and implement eco-efficient solutions to decrease costs (Lopes et al, 2020), tackle with global problems such as environmental protection, climate change and water scarcity, resource recovery like nutrients and energy (del Castillo et al, 2022; Estevez et al, 2022). Recently, reclaimed gray water has shown potential to overcome quality and risk barriers through decentralized treatment making it in line with circularity and sustainability concepts (Sauve et al, 2021; Vuppaladadiyam et al, 2019).

4.10 CONCLUSIONS AND PERSPECTIVES

This chapter discusses various decentralized wastewater treatment methods and its efficiency in treatment and also highlights the pros and cons of different decentralized treatment methods in terms of various physico-chemical and biological parameters. The first and foremost goal is to use optimum/minimal quantity of water (refuse/reduce) in all sectors to minimize the wastewater production. After reduce, 3Rs (reuse, recycle and recovery) are in the second option prepared for wastewater management. The third option is to treat the polluted water. Most of the primary treatment consists of settlers, sedimentation tanks, septic tanks. Secondary treatment consists of several aerobic and anaerobic processes. Among the secondary treatment processes, SBR shows maximum removal efficiency and results in achieving enhanced effluent quality. The combination of bioreactor septic tank and CWs shows maximum removal efficiency. The disposal is the final/least preferred option in wastewater management. In developed countries especially for densely populated areas in developing and non-developed countries, effective sanitation and wastewater management are becoming increasingly difficult. In many developing nations, centralized sewage and wastewater treatment facilities serve only a section of big cities, and on-site sanitation is frequently insufficient in heavily populated areas. Many developed countries have chosen to use decentralized wastewater treatment because of reduced infrastructural requirements. However, achieving compliance of regulatory standards using a decentralized strategy can be difficult and costly.

The decentralized wastewater treatment should be economic and cost-effective. An effective decentralized wastewater treatment provides secure water quality by advanced treatment methods toward environmental protection thereby improving the public health. The design of decentralized wastewater treatment should enhance resource recovery, such as water reclamation for agricultural and irrigation activities, energy recovery and reduction of global warming. More holistic approach is required integrated with engineering and site-specific knowledge along with people participation and awareness for successful implementation of decentralized treatment methods under the umbrella of circular economy concept.

REFERENCES

2030 Water Resources Group and Council of Energy (Environment and Water). 2016. Circular Economy Pathways for Municipal Wastewater Management in India: A Practitioner's Guide. https://www.2030wrg.org/wp-content/uploads/2017/01/Circular-Economy-Pathways-India.pdf.

Al Momani, F., 2015. Mathematical model for predicting sludge accumulation in septic tanks based on operational parameters: Effect of hydraulic mixing. Water and Environment Journal, 29(4):499–506.

Alrajoula, M., Halalsheh, M., & Fayyad, M. 2009. Anaerobic filter for polishing effluent of UASB reactor treating strong sewage at 23°C. Water Science and Technology, 59:1975–1981.

Anh, N. V., Ha, T. D., Nhue, T. H., Heinss, U., Morel, A., Moura, M., & Schertenleib, R. 2002. Decentralized wastewater treatment-new concept and technologies for Vietnamese conditions. In 5th Specialised conference on small water and wastewater treatment systems, Istanbul, Turkey (September):24–26.

Arora, J., Yadav, A., & Saroj, D. 2016. Potential of decentralised wastewater treatment systems applicable to India. Current World Environment, 11:338.

Avsar, Y., Tarabeah, H., Kimchie, S., Naamneh, H., & Ozturk, I. 2008. Rehabilitation of an available facultative pond unit using a trickling biofilter. Environmental Engineering Science, 25:106–113.

Bajpai, M., Katoch, S. S., & Chaturvedi, N. K. 2019. Comparative study on decentralized treatment technologies for sewage and graywater reuse—a review. Water Science and Technology, 80:2091–2106.

Bernal, D., Restrepo, I., & Grueso-Casquete, S. 2021. Key criteria for considering decentralization in municipal wastewater management. Heliyon, 7:e06375.

Biswas, A., Tewari, S., & Parida, S. 2021. Decentralized management of segregated organic waste. Centre for science and environment, New Delhi. https://www.cseindia.org/content/downloadreports/10877.

Capodaglio, A. G. 2017. Integrated, decentralized wastewater management for resource recovery in rural and peri-urban areas. Resources, 6:22.

Capodaglio, A.G., Bolognesi, S., & Cecconet, D., 2021. Sustainable, decentralized sanitation and reuse with hybrid nature-based systems. Water, 13(11):1583.

Capodaglio, A. G., & Callegari, A. 2016. Domestic wastewater treatment with a decentralized, simple technology biomass concentrator reactor. Journal of Water, Sanitation and Hygiene for Development, 6:507–510.

Cashman, S., Ma, X., Mosley, J., Garland, J., Crone, B., & Xue, X. 2018. Energy and greenhouse gas life cycle assessment and cost analysis of aerobic and anaerobic membrane bioreactor systems: influence of scale, population density, climate, and methane recovery. Bioresource Technology, 254:56–66.

Castellar, J.A., Torrens, A., Buttiglieri, G., Monclus, H., Arias, C.A., Carvalho, P.N., Galvao, A. and Comas, J., 2022. Nature-based solutions coupled with advanced technologies: An opportunity for decentralized water reuse in cities. Journal of Cleaner Production, 340:130660.

Cecconet, D., Callegari, A., Hlavínek, P., & Capodaglio, A. G. 2019. Membrane bioreactors for sustainable, fit-for-purpose greywater treatment: a critical review. Clean Technologies and Environmental Policy, 21:745–762.

Chen, X., & Fukushi, K. 2016. Development of an innovative decentralized treatment system for the reclamation and reuse of strong wastewater from rural community: effects of elevated CO_2 concentrations. Journal of Environmental Management, 180:401–408.

Chirisa, I., Bandauko, E., Matamanda, A., & Mandisvika, G. 2017. Decentralized domestic wastewater systems in developing countries: the case study of Harare (Zimbabwe). Applied Water Science, 7:1069–1078.

Colares, G. S., de Souza Celente, G., da Silva, F. P., de Loreto, A. C., Lutterbeck, C. A., Kist, L. T., & Machado, Ê. L. 2019. Combined system for the treatment and reuse of urban wastewater: the efficiency of anaerobic reactors+ hybrid constructed wetlands + ozonation. Water Science and Technology, 80:254–264.

CPHEEO. 2021. Compendium of recycle and reuse of wastewater in 54 million plus cities. Ministry of Housing and Urban Affairs, Government of India. https://mohua.gov.in/pdf/6156db8e5ca2fCompendium_of_Recycle_and_Reuse_of_Wastewater_in_54_million_plus_cities.pdf.

Cramer, J. 2017. The raw materials transition in the Amsterdam metropolitan area: added value for the economy, well-being, and the environment. Environment: Science and Policy for Sustainable Development, 59:14–21.

De Oliveira Cruz, L. M., Gomes, B. G. L. A., Tonetti, A. L., & Figueiredo, I. C. S. 2019. Using coconut husks in a full-scale decentralized wastewater treatment system: the influence of an anaerobic filter on maintenance and operational conditions of a sand filter. Ecological Engineering, 127:454–459.

DeCarolis, J. F. Jr., & Adham, S. 2007. Performance investigation of membrane bioreactor systems during municipal wastewater reclamation. Water Environment Research, 79:2536–2550.

Del Nery, V., Damianovic, M. H. Z., Pozzi, E., De Nardi, I. R., Caldas, V. E. A., & Pires, E. C. 2013. Long-term performance and operational strategies of a poultry slaughterhouse waste stabilization pond system in a tropical climate. Resources, Conservation and Recycling, 71:7–14.

Deng, H., Ren, H., Fan, J., Zhao, K., Hu, C., & Qu, J. 2022. Membrane fouling mitigation by coagulation and electrostatic repulsion using an electro-AnMBR in kitchen wastewater treatment. Water Research, 222:118883.

Ding, D., Feng, C., Jin, Y., Hao, C., Zhao, Y., & Suemura, T. 2011. Domestic sewage treatment in a sequencing batch biofilm reactor (SBBR) with an intelligent controlling system. Desalination, 276:260–265.

Dorji, U., Tenzin, U., Dorji, P., Pathak, N., Johir, M. A., Volpin, F., & Phuntsho, S. 2021. Exploring shredded waste PET bottles as a biofilter media for improved on-site sanitation. Process Safety and Environmental Protection, 148:370–381.

del Castillo, A. F., Garibay, M. V., Senés-Guerrero, C., Orozco-Nunnelly, D. A., de Anda, J., & Gradilla-Hernández, M. S. 2022. A review of the sustainability of anaerobic reactors combined with constructed wetlands for decentralized wastewater treatment. Journal of Cleaner Production, 371:133428.

Estévez, S., Feijoo, G. and Moreira, M.T., 2022. Environmental synergies in decentralized wastewater treatment at a hotel resort. Journal of Environmental Management, *317*, 115392.

Ghangrekar, M., & Behera, M. M. 2014. Suspended Growth Treatment Processes. Chapter 3.5 in the book Comprehensive Water Quality and Purification, Elsevier Publishers. Amsterdam 74–89.

Gray, N. F. 1995. The influence of sludge accumulation rate on septic tank design. Environmental Technology, 16:795–800.

Hayati, H., Doosti, M., & Sayadi, M. 2013. Performance evaluation of waste stabilization pond in Birjand, Iran for the treatment of municipal sewage. Proceedings of the International Academy of Ecology and Environmental Sciences, 3:52.

Hazra, M., & Durso, L. M. 2022. Performance efficiency of conventional treatment plants and constructed wetlands towards reduction of antibiotic resistance. Antibiotics, 11:114.

Horn, T. B., Zerwes, F. V., Kist, L. T., & Machado, Ê. L. 2014. Constructed wetland and photocatalytic ozonation for university sewage treatment. Ecological Engineering, 63:134–141.

Huang, M. H., Li, Y. M., & Gu, G. W. 2010. Chemical composition of organic matters in domestic wastewater. Desalination, 262:36–42.

Juan-Garcia, P., Butler, D., Comas, J., Darch, G., Sweetapple, C., Thornton, A., & Corominas, L. 2017. Resilience theory incorporated into urban wastewater systems management. State of the art. Water Research, 115:149–161.

Kim, D.S., Jung, N.S. and Park, Y.S., 2008. Characteristics of nitrogen and phosphorus removal in SBR and SBBR with different ammonium loading rates. Korean Journal of Chemical Engineering, 25:793–800.

Kohler, L. E., Silverstein, J., & Rajagopalan, B. 2020.Resilience of on-site wastewater treatment systems after extreme storm event. Journal of Sustainable Water in the Built Environment, 6(2):1–9.

Koottatep, T., Connelly, S., Pussayanavin, T., Khamyai, S., Sangchun, W., Sloan, W., & Polprasert, C. 2020. 'Solar septic tank': evaluation of innovative decentralized treatment of blackwater in developing countries. Journal of Water, Sanitation and Hygiene for Development, 10:828–840.

Langergraber, G., Haberl, R., Laber, J., & Pressl, A. 2003. Evaluation of substrate clogging processes in vertical flow constructed wetlands. Water Science and Technology, 48:25–34.

Li, H., Liu, F., Luo, P., Xie, G., Xiao, R., Hu, W., ... & Wu, J. 2018. Performance of integrated ecological treatment system for decentralized rural wastewater and significance of plant harvest management. Ecological Engineering, 124:69–76.

Libralato, G., Ghirardini, A. V., & Avezzù, F. 2012. To centralise or to decentralise: an overview of the most recent trends in wastewater treatment management. Journal of Environmental Management, 94:61–68.

Lin, H., Chen, J., Wang, F., Ding, L., & Hong, H. 2011. Feasibility evaluation of submerged anaerobic membrane bioreactor for municipal secondary wastewater treatment. Desalination, 280:120–126.

Liu, Y., He, L. F., Deng, Y. Y., Zhang, Q., Jiang, G. M., & Liu, H. 2022. Recent progress on the recovery of valuable resources from source-separated urine on-site using electrochemical technologies: a review. Chemical Engineering Journal, 442:136200.

Lopes, T. A., Queiroz, L. M., Torres, E. A., & Kiperstok, A. 2020. Low complexity wastewater treatment process in developing countries: a LCA approach to evaluate environmental gains. Science of The Total Environment, 720:137593.

Lossing, H., Champagne, P., & McLellan, P. J. 2010. Examination of sludge accumulation rates and sludge characteristics for a decentralized community wastewater treatment systems with individual primary clarifier tanks located in Wardsville (Ontario, Canada). Water Science and Technology, 62:2944–2952.

Mac Mahon, J., Knappe, J. and Gill, L.W., 2022. Sludge accumulation rates in septic tanks used as part of the on-site treatment of domestic wastewater in a northern maritime temperate climate. Journal of Environmental Management, 304:114199.

Martins, P. H. N., Cecchin, D., Azevedo, A. R. G., Carmo, D. F., Danagemma, R. A., Waite, R. M. M., Rodrigues, N. F., Sousa, F. A., Amaral, P. I. S., Huther, C. M., Pereira, C. R., & da Cruz, V. M. F. 2021. Technical and economic pre-feasibility study for the construction of septic tank-filter-sinkhole with alternative material. Agronomy Research, 19:1586–1601.

Massoud, M. A., Tarhini, A., & Nasr, J. A. 2009. Decentralized approaches to wastewater treatment and management: applicability in developing countries. Journal of Environmental Management, 90:652–659.

Moussavi, G., Kazembeigi, F. and Farzadkia, M., 2010. Performance of a pilot scale up-flow septic tank for on-site decentralized treatment of residential wastewater. Process Safety and Environmental Protection, 88(1):47–52.

Lin, S.H. and Cheng, K.W., 2001. A new sequencing batch reactor for treatment of municipal sewage wastewater for agricultural reuse. Desalination, 133(1):41–51.

Mesquita, R. T. C., Pereira Rosa, A., de Oliveira Santos, T. F., Carraro Borges, A., Calijuri, M. L., & de Paula Souza, F. M. 2021. Decentralized management of sewage using septic tanks and anaerobic filters and its potential to comply with required standards in a developing country: a case study in Brazil. Environmental Science and Pollution Research, 28:50001–50016.

Metcalf, L., Eddy, H. P., & Tchobanoglous, G. 1991. Wastewater Engineering: Treatment, Disposal, and Reuse (Vol. 4). McGraw-Hill, New York.

Mo, W., & Zhang, Q. 2013. Energy–nutrients–water nexus: integrated resource recovery in municipal wastewater treatment plants. Journal of Environmental Management, 127:255–267.

Nasr, F. A., & Mikhaeil, B. 2013. Treatment of domestic wastewater using conventional and baffled septic tanks. Environmental Technology, 34:2337–2343.

NIUA. 2016. Handbook on decentralized waste water treatment module, Delhi, India. https://smartnet.niua.org/sites/default/files/resources/booklet_on_technology_options_dewats.pdf.

NIUA. 2019. Decentralized wastewater management: complementing centralized solutions for achieving 100% sanitation coverage – learning notes. A hand book by National Institute of Urban Affairs, New Delhi. http://urbanrivers.niua.org/sites/default/files/202007/Participant%20Handbook_Advance%20Training.pdf.

Otter, P., Hertel, S., Ansari, J., Lara, E., Cano, R., Arias, C., Gregersen, P., Grischek, T., Benz, F., Goldmaier, A., & Alvarez, J. A. 2020. Disinfection of decentralized wastewater reuse in rural areas through wetlands and solar driven onsite chlorination. Science of the Total Environment, 721:137595.

Papadopoulos, F. H., & Tsihrintzis, V. A. 2011. Assessment of a full-scale duckweed pond system for septage treatment. Environmental Technology, 32:795–804.

Peavy, H. S., Rowe, D. R., & Tchobanoglous, G., 1985. Environmental Engineering. McGraw-Hill, Inc., New York.

Pereira da Silva, F., Lutterbeck, C. A., Colares, G. S., Oliveira, G. A., Rodrigues, L. R., Osbel, N. D., Rodriguez, A. L., Lopz, D. A. R., Gehlen, G., & Machado, E. L. 2021. Treatment of university campus wastewaters by anaerobic reactor and multi-stage constructed wetlands. Journal of Water Process Engineering, 42:102119.

Philip, L., Ramprasad, C., & Krithika, D. 2019. Sustainable wastewater management through decentralized systems: case studies. In Water Scarcity and Ways to Reduce the Impact. Springer, Cham.15–45.

Prateep Na Talang, R., Sirivithayapakorn, S., & Polruang, S. 2022. Life cycle impact assessment and life cycle cost assessment for centralized and decentralized wastewater treatment plants in Thailand. Scientific Reports, 12:1–13.

Rodriguez, C., Van Buynder, P., Lugg, R., Blair, P., Devine, B., Cook, A. and Weinstein, P., 2009. Indirect potable reuse: a sustainable water supply alternative. International journal of environmental research and public health, 6(3):1174–1209.

Rohilla, S. K., Luthra, B., Bhatnagar, A., Matto, M., & Bhonde, U. 2017. Septage management: A practioner's Guide, Centre for Science and Environment, New Delhi. https://nmcg.nic.in/writereaddata/fileupload/5_Septage%20Practitioner's%20Guide-EnglishVersion.pdf

Sadaf, S., Singh, A.K., Iqbal, J., Kumar, R.N., Sulejmanović, J., Habila, M.A., Américo-Pinheiro, J.H.P. and Sher, F., 2022. Advancements of sequencing batch biofilm reactor for slaughterhouse wastewater assisted with response surface methodology. Chemosphere, 307:135952.

Saeed, T., Afrin, R., Al-Muyeed, A., Miah, M. J., & Jahan, H. 2021. Bioreactor septic tank for on-site wastewater treatment: floating constructed wetland integration. Journal of Environmental Chemical Engineering, 9:105606.

Sauvé, S., Lamontagne, S., Dupras, J., & Stahel, W., 2021. Circular economy of water: Tackling quantity, quality and footprint of water. Environmental Development, 39:100651.

Sekine, M., Akizuki, S., Kishi, M., Kurosawa, N., & Toda, T. 2020. Simultaneous biological nitrification and desulfurization treatment of ammonium and sulfide-rich wastewater: effectiveness of a sequential batch operation. Chemosphere, 244:125381.

Sharma, M., Yadav, A., Dubey, K. K., Tipple, J., & Das, D. B. 2022. Decentralized systems for the treatment of antimicrobial compounds released from hospital aquatic wastes. Science of The Total Environment, 840:156569.

Siegrist, R. L. 2017. Decentralized Water Reclamation Engineering. A Curriculum Workbook. Springer International Publishing, New York.

Singh, N. K., Kazmi, A. A., & Starkl, M. 2015. A review on full-scale decentralized wastewater treatment systems: techno-economical approach. Water Science and Technology, 71:468–478.

Singh, A., Sawant, M., Kamble, S. J., Herlekar, M., Starkl, M., Aymerich, E., & Kazmi, A. 2019. Performance evaluation of a decentralized wastewater treatment system in India. Environmental Science and Pollution Research, 26:21172–21188.

Starkl, M., Amerasinghe, P., Essl, L., Jampani, M., Kumar, D., & Asolekar, S. R. 2013. Potential of natural treatment technologies for wastewater management in India. Journal of Water, Sanitation and Hygiene for Development, 3:500–511.

Starkl, M., Anthony, J., Aymerich, E., Brunner, N., Chubilleau, C., Das, S., & Singh, A. 2018. Interpreting best available technologies more flexibly: a policy perspective for municipal wastewater management in India and other developing countries. Environmental Impact Assessment Review, 71:132–141.

Subramanian, P. G., Raj, A. V., Jamwal, P., Connelly, S., Yeluripati, J., Richards, S., Ellis, R., & Rao, L. 2020. Decentralized treatment and recycling of greywater from a school in rural India. Journal of Water Process Engineering, 38:101695.

Suhogusoff, A. V., Hirata, R., Ferrari, L. C. K., Robertson, W. D., Stimson, J., Forbes, D., & Blowes, D. W. 2019. Field performance of two on-site wastewater treatment systems using reactive media layers for nutrient and pathogen removal. Journal of Water Process Engineering, 32:100905.

Surampalli, R. Y., Tyagi, R. D., & Scheible, O. K. 2000. Sequencing batch reactors – technology and performance evaluation. Journal of Environmental Systems, 28:25–43.

Thalla, A. K., Devatha, C. P., Anagh, K., & Sony, E. 2019. Performance evaluation of horizontal and vertical flow constructed wetlands as tertiary treatment option for secondary effluents. Applied Water Science, 9:1–9.

Tilley, E., Ulrich, L., Luthi, C., Reymond, P., & Zurbrugg, C. 2014. Compendium of Sanitation Systems and Technologies 2nd revised edition. Swiss Federal Institute of Aquatic Science and Technology (Eawag), Dubendorf, Switzerland.
Tonetti, A. L., Figueiredo, I. C. S., Madrid, F. J. P. L., Magalhães, T. M., & Miyazaki, C. K. 2022. Cost confrontation study for decentralized wastewater treatment: when to adopt a cluster or onsite system? International Journal of Environmental Science and Technology, 19:3529–3538.
Tortajada, C. 2020. Contributions of recycled wastewater to clean water and sanitation sustainable development goals. NPJ Clean Water, 3:1–6.
US EPA. 2002a. Wastewater technology fact sheet: Anaerobic lagoons. https://www3.epa.gov/npdes/pubs/alagoons.pdf
US EPA. 2002b. Wastewater technology fact sheet: Aerated, parial mix lagoons. https://www3.epa.gov/npdes/pubs/apartlag.pdf
Varma, V. G., Jha, S., Raju, L. H. K., Kishore, R. L., & Ranjith, V. 2022. A review on decentralized wastewater treatment systems in India. Chemosphere, 300:134462.
Vieira, P. C., Sperling, M. V., Nogueira, L. C. M., & Assis, B. F. S. 2013. Performance evaluation of a novel open trickling filter for the post-treatment of anaerobic effluents from small communities. Water Science & Technology, 67:2746–2752.
Vuppaladadiyam, A. K., Merayo, N., Prinsen, P., Luque, R., Blanco, A., & Zhao, M. 2019. A review on greywater reuse: quality, risks, barriers and global scenarios. Reviews in Environmental Science and Bio/Technology, 18:77–99.
Vymazal, J. 2010. Constructed wetlands for wastewater treatment. Water, 2:530–549.
Vymazal, J. 2013. The use of hybrid constructed wetlands for wastewater treatment with special attention to nitrogen removal: a review of a recent development. Water Research, 47:4795–4811.
Wang, G. S., & Xia, J.. 2005.sustainable water resources exploitation and management in Beijing. In sustainable water management solutions for large cities; Savic, D., Marino, M.A., Savenije, H.H.G., Bertoni, J.C., Eds.; IAHS Publication 293; IAHS Press, Centre for Ecology and Hydrology, Wallingford.
Yang, F., Zhang, H., Zhang, X., Zhang, Y., Li, J., Jin, F., & Zhou, B. 2021. Performance analysis and evaluation of the 146 rural decentralized wastewater treatment facilities surrounding the Erhai Lake. Journal of Cleaner Production, 315:128159.
Zaharia, C. 2017. Decentralized wastewater treatment systems: efficiency and its estimated impact against onsite natural water pollution status. A Romanian case study. Process Safety and Environmental Protection, 108:74–88.
Zhang, D. Q., Jinadasa, K. B. S. N., Gersberg, R. M., Liu, Y., Tan, S. K., & Ng, W. J. 2015. Application of constructed wetlands for wastewater treatment in tropical and subtropical regions (2000–2013). Journal of Environmental Sciences, 30:30–46.

5 Electrochemistry Applications in Decentralized Wastewater Treatment

Hui Guo and Younggy Kim
McMaster University, Hamilton, Canada

5.1 INTRODUCTION

Electrochemical systems that contain two electrodes separated by electrolyte and connected via an external electronic conductor (Newman and Thomas-Alyea, 2012) have a strong potential to be used in decentralized wastewater treatment. Compared with conventional centralized wastewater treatment technologies, electrochemical systems present higher treatment efficiency, lower energy consumption, and less biosolids generation. This chapter will discuss the application of electrochemical systems, including bioelectrochemical systems (BESs) and electrodialysis (ED), in decentralized wastewater treatment.

BESs are designed to use microorganisms as biocatalysts (electrode reactions) and recover energy from wastewater (ElMekawy et al., 2015; Hamelers et al., 2010; Rabaey et al., 2010; Wang et al., 2015). In BESs, electrons produced by the microorganisms from substrates (e.g., organic matter) are transferred to the anode, generating electric current (Figure 5.1) (Logan et al., 2006). Depending on the cathodic reactions, BESs can be categorized as microbial fuel cells (MFCs) where electricity is generated (Figure 5.1a) or microbial electrolysis cells (MECs) in which hydrogen gas is produced with an addition of a small voltage (Figure 5.1b) (Liu et al., 2005; Logan et al., 2006, 2008).

BESs are suitable for decentralized wastewater treatment because of the efficient treatability, easily scalable structure, and low energy consumption. For instance, a septic tank with the MEC electrode reactions showed enhanced biogas production, improved phosphorus removal efficiency, and reduced sulfide hydrogen generation (Zamalloa et al., 2013). BESs can also be used for nitrogen removal in constructed wetlands, implying potential applications in wastewater treatment for nutrient control (Gupta et al., 2021; Jingyu et al., 2020; Yadav et al., 2012). Therefore, in this chapter, we focused on various beneficial aspects of BESs in wastewater treatment.

ED is an electrochemical process to separate ions under electric current (Strathmann, 2004; Xu and Huang, 2008). A conventional ED consists of electrodes, cation exchange membranes (CEMs), and anion exchange membranes (AEMs) (Figure 5.2a). Negative charges (i.e., anions) of the feed solution migrate toward the anode and pass through the AEM but are blocked by CEM. Oppositely, positive charges (i.e., cations) are attracted by the cathode and can pass through the CEM but are blocked by AEM. The ionic separation processes contribute to the removal of salt compounds in the feed solution and achieve the recovery of resources in the concentrate cells (Figure 5.2a). ED has been widely applied in desalination and nutrient recovery from water and wastewater (Mondor et al., 2008; Strathmann, 2010; Zhang et al., 2009).

Bipolar membrane electrodialysis (BMED) has gained more attention in the last two decades. A typical BMED consists of electrodes, CEMs, AEMs, and bipolar membranes (BPMs), which are manufactured by chemically attaching a CEM and an AEM (Simons, 1993) (Figure 5.2b). At the interface of BPMs (the interface between the CEM and AEM), water is directly dissociated into H^+ and OH^- (Figure 5.2b). The H^+ can permeate through the cation exchange side of the bipolar membrane and form the acid (HZ) with the anions from the adjacent cell (Figure 5.2b). Similarly,

DOI: 10.1201/9781003381419-7

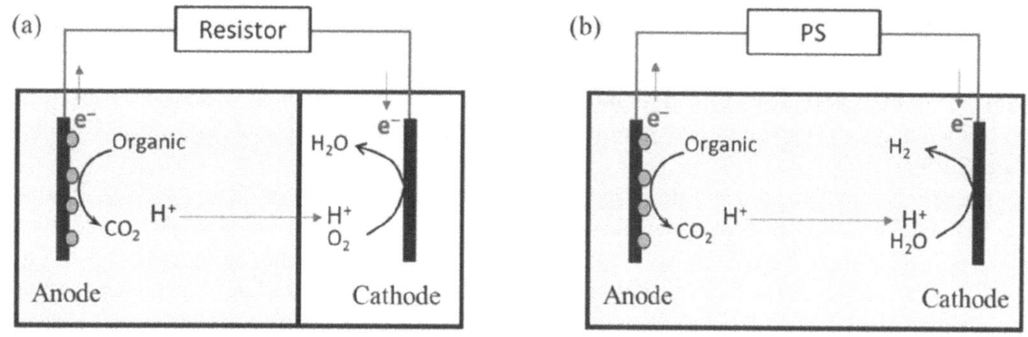

FIGURE 5.1 The schematic diagram of (a) MFCs and (b) MECs (PS: power supply).

The OH^- can pass through the anion exchange side of the bipolar membrane and form the base (MOH) with the cations from the feed cells. Therefore, BMED has been applied in many areas such as the production of mineral acids and bases from corresponding salts (Huang and Xu, 2006; Lei et al., 2020; Pourcelly, 2002), the recovery of organic acid (Ferrer et al., 2006; Koter and Warszawski, 2000; Trivedi et al., 1997), the regeneration of acid or base (Huanget al., 2006), and the treatment of wastewater (Zhang et al., 2011). Recently, BMED has emerged as an efficient technology for nutrient recovery from water and wastewater systems (Guo et al., 2021; Li et al., 2016; Shi et al., 2018; van Linden et al., 2020). In the BMED system, there is no gas or by-product generated (Huang and Xu, 2006) during the water dissociation. This contributes to the high energy utilization of the BMED systems. In addition, only one pair of electrodes is needed for BMED, while the reactor can have many bipolar membranes, which results in some benefits, such as space-saving, operation convenience, and low costs (Huang and Xu, 2006). In this chapter, we will review the application of ED in nutrient recovery and discuss the potential challenges of ED development.

5.2 BIOELECTROCHEMICAL SYSTEMS FOR DECENTRALIZED WASTEWATER TREATMENT

5.2.1 Rapid Organic Removal and Low Sludge Generation

High energy demand and expensive sludge management are two main challenges in large-scale conventional wastewater treatment plants (Burton Tchobanoglus et al., 2003; Metcalf et al., 2014). In previous studies, BES with different configurations, operation conditions, and electrode materials

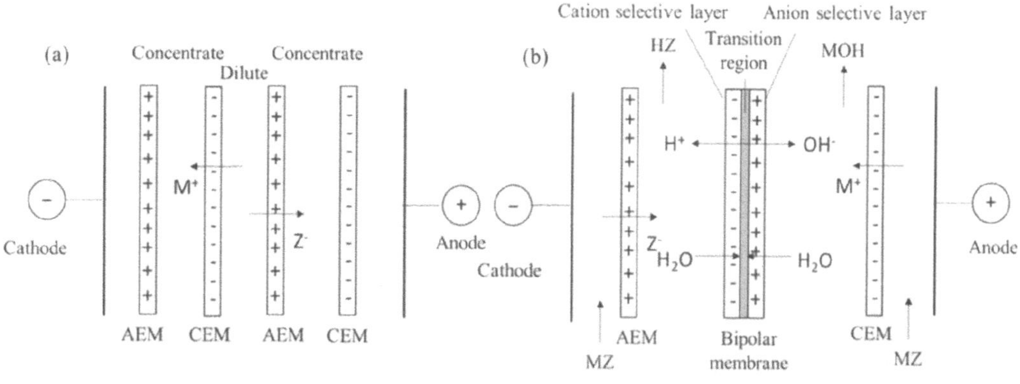

FIGURE 5.2 (a) Scheme of electrodialysis stack; (b) scheme of bipolar membrane electrodialysis.

were examined to treat synthetic/real wastewater. The COD removal efficiency ranged from 40 to 100% with hydrogen recovery efficiency up to 90% were observed (Chae et al., 2008; Lalaurette et al., 2009; Mook et al., 2013; Selembo et al., 2009; Tartakovsky et al., 2011). Among these studies, MECs with the stacked electrode design demonstrated excellent municipal wastewater treatability with low energy consumption and minimal biosolids production (Guo and Kim, 2018; Guo and Kim, 2019).

The MECs with stacked electrode design were built with activated carbon cloth (ACC100, Evertech Envisafe Ecology, Taiwan) as the bioanode material and stainless-steel mesh (304 stainless steel, 200 × 200 mesh, McMaster Carr, USA) as cathode without any precious metal catalyst (Guo and Kim, 2018; Guo and Kim, 2019). Each electrode pair consists of two rubber gaskets, one plastic mesh (to avoid potential electric short circuits among the electrodes), one piece of carbon cloth (bioanode), and one piece of stainless-steel mesh (cathode). A very short distance (2.8 mm) was maintained between the bioanode and cathode as well as between the electrode pairs to reduce the internal resistance in the MECs (Cotterill et al., 2017; Hu et al., 2009; Jwa et al., 2019). This stacked MEC is easy to scale up by repeating additional electrode pairs. For instance, MECs with one, five, and ten electrode pairs, the surface area of which is 5.42 m^2 for each electrode (Figure 5.3), were built to treat primary sludge collected from the local municipal wastewater treatment plant (Guo and Kim, 2018; Guo and Kim, 2019). In the study, an MEC with ten electrode pairs and another MEC with five electrode pairs are arranged in series to demonstrate highly rapid treatment of real wastewater (hydrologic retention time of 0.63 h) for $90 \pm 2\%$ carbonaceous biochemical oxygen demand (CBOD) removal (from 192 to 19 mg/L as CBOD). The concentration of CBOD in the effluent meets the standards of wastewater system effluent regulations in Canada (25 mg/L, Government of Canada, (2012)).

The electric current density, which represents the rate of electrode reactions in the MEC, was governed by the organic loading rate (OLR) rather than the substrate concentration, conductivity of solutions, and other operating conditions (Asztalos and Kim, 2015; Guo and Kim, 2019). For instance, the low OLR (smaller than 2 kg-COD/m^3/d) for the stacked MECs resulted in low electric current densities (<40 A/m^3) (Guo and Kim, 2019). However, a significant increase was observed in the electric current densities (up to 190 A/m^3) when the OLR increased to 8–16 kg-COD/m^3/d. Considering the typical OLR (2 kg-COD/m^3/d) in conventional activated sludge systems (Burton et al., 2013), the higher OLR in the stacked MEC system indicated that the MECs can treat municipal wastewater more efficiently.

Nearly complete biosolids (>94%), including total suspended solids (TSS) and volatile suspended solids (VSS), were removed in the MEC system. The low biosolids concentration ranged from 2.0 to 10.5 mg-TSS/L or 1.6 to 6.4 mg-VSS/L in the effluent implied that the MEC effluent can be disposed even without further clarification. The enhanced reduction of biosolids is beneficial to minimizing the cost of sludge treatment/disposal and simplifying the operation of decentralized wastewater treatment.

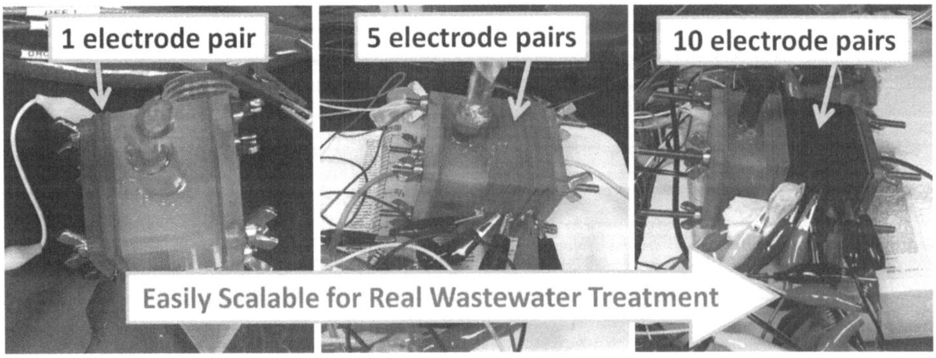

FIGURE 5.3 Easily scalable MECs for wastewater treatment.

The stacked electrode MECs are energy-efficient since the energy consumption was only 0.40 kWh/kg-COD removed or 0.058 kWh/m^3 wastewater treated in the MECs (Guo and Kim, 2019). The capacities of the designed MECs could be in a large range (e.g., several to dozens of tons per day) since the MECs with multiple electrodes (e.g., 5–20) can be used to receive the wastewater in parallel and operate independently without further scale-up. However, there is a challenge in energy recovery for stacked MEC with short hydraulic retention time. Hydrogen gas can be rapidly converted into methane via hydrogenotrophic methanogens when MECs are used to treat real wastewater. Methane is highly dissolved in the wastewater and flows out of the reactor with water streams. Therefore, the recovery of energy (i.e., hydrogen or methane) is a main challenge in the wide application of stacked MEC systems.

5.2.2 Bioelectrochemical Anaerobic Digestion

Anaerobic digestion (AD) is widely used in centralized wastewater treatment processes to produce bioenergy and manage biosolids (Kelleher et al., 2002; Li et al., 2019). However, the challenges, such as poor operational stability and high operation cost, have limited AD applications in small and decentralized wastewater treatment facilities (Gude, 2018; Park et al., 2020; Wang, 2014). Bioelectrochemical AD (BEAD) that combines the BES and AD was proposed to enhance organic removal and biogas production (Park et al., 2020). In previous studies, the application of BEAD was beneficial to energy production, treatment efficiency, and operation cost (Guo et al., 2013; Lin et al., 2019; Pham et al., 2006; Yu et al., 2018).

In BEAD, the organic substrates (e.g., acetate, butyrate, propionate) are oxidized and generate protons, electrons, and CO_2 (Figure 5.4). Methane can be produced by the cathodic reactions through three mechanisms (Blasco-Gómez et al., 2017; Fudge et al., 2021; Nevin et al., 2011; Park et al., 2020; Villano et al., 2010): (1) direct bioelectrochemical methanogenesis, in which the electroactive biofilm reduces CO_2 into methane by utilizing the electrons provided at the cathode (Figure 5.4a);

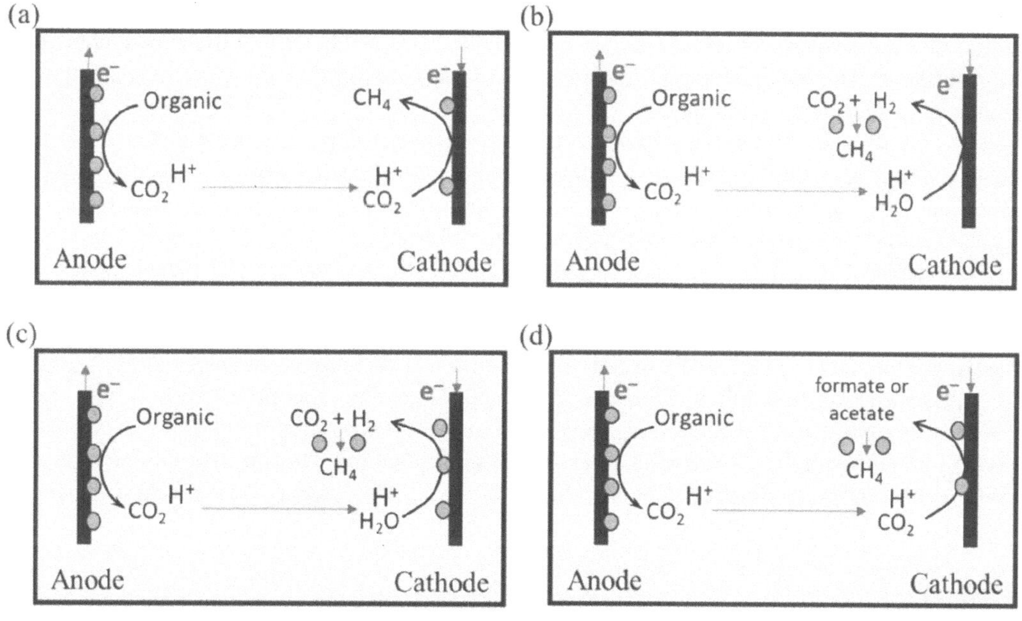

FIGURE 5.4 Mechanisms of methane production in BEADs. (a) Direct bioelectrochemical methanogenesis; (b) hydrogenotrophic methanogenesis with electrochemical hydrogen production; (c) hydrogenotrophic methanogenesis with bioelectrochemical hydrogen production; (d) carbon compounds-mediated methanogenesis.

(2) hydrogenotrophic methanogenesis, where hydrogen is produced either electrochemically (Figure 5.4b) or bioelectrochemically (Figure 5.4c) and then reacts with CO_2 to produce methane; and (3) carbon compound-mediated methanogenesis, where the electroactive biofilm reduces CO_2 into formate or acetate, and then other microorganisms generate methane by using the formate or acetate (Figure 5.4d). The direct bioelectrochemical methanogenesis and carbon compound-mediated methanogenesis do not need an external voltage to drive the electron transfer reaction since the overall reactions are thermodynamically spontaneous (Park et al., 2020). However, the production of hydrogen in BES requires a small voltage (>0.13 V) to initiate methane production (Logan et al., 2008). Therefore, mechanisms of (1) and (3) can be achieved in MFCs, while MECs can produce methane via all three mechanisms.

Compared to conventional AD, BEAD is capable of maintaining the process stability under high OLR conditions. This high capacity of organic removal is explained by the enhanced hydrogenotrophic methanogenesis in BEAD (i.e., MEC-AD). In conventional AD, around 67% of methane is produced by acetoclastic methanogens, while hydrogenotrophic methanogenesis only contributes to 33% of methane production (Batstone et al., 2002; Kotsyurbenko et al., 2004). However, in BEAD, microbial community analysis on the cathode indicated that hydrogenotrophic methanogenesis was more favored than acetoclastic methanogenesis (Cai et al., 2016; Saheb-Alam et al., 2018; Villano et al., 2010), especially when a small voltage (0.3–0.6 V) was provided (Li et al., 2016; Park et al., 2019). The enhanced hydrogenotrophic methanogenesis can convert protons that are released from organic matter to hydrogen gas rapidly, resulting in a stable pH in the BEAD. The pH drop in AD is the main challenge to maintaining stable treatment efficiency (Lyberatos and Skiadas, 1999). In addition, hydrogenotrophic methanogens are more robust to the variation in environmental conditions (Zakaria and Dhar, 2019). In previous studies, the methane production rate in BEAD was 1.7–3.2 times higher than it was in AD, while the OLR in BEAD was increased by 30–180% (Guo et al., 2017; Hirano and Matsumoto, 2018; Park et al., 2020; Song et al., 2016). The high tolerance of OLR in BEAD contributes to land saving and helps to reduce the operation time of biosolids treatment.

There are many factors, such as voltage supply, feed substrates, electrode materials, inoculation, and system structures, which govern the performance of BEAD (Fudge et al., 2021; Park et al., 2020). Table 5.1 summarizes the BEAD design with the methane yield in previous studies. Among these studies, the volume of BEAD ranged from 0.02 to 20 L, while OLR is in the range of 0.24 to 10 g COD/L·d. The operational temperature varied from 10 to 35°C. Up to 700 mL methane was produced for 1 L substrate per day.

BEAD is a suitable technology for decentralized wastewater, especially for high OLR with food waste and animal manure. The active electrochemical and bioelectrochemical reactions in BEAD contribute to the highly robust stability even under harsh operation conditions (e.g., high OLR, volatile fatty acids (VFAs) accumulation, and significant pH drop).

Therefore, BEAD can be used as a supplement for existing AD systems or independent wastewater treatment systems.

BES-integrated septic tanks were examined to increase the energy recovery in decentralized wastewater treatment (Gude, 2018). In addition, Yazdi et al. (2015) developed pluggable MFC stacks for septic wastewater treatment. A power density of 142 mW/m^2 was produced in 3 MFC units. Preliminary calculation indicated that a 6-W LED light can be powered for 4 h/day with a cost of US$25 for MFC operation. In another study, a 2.93-m^3 septic tank with similarly designed stacked MFCs generated power in the range of 109–131 mW/m^3 under various OLR conditions (Alzate-Gaviria et al., 2016).

5.2.3 Heavy Metal Removal

Heavy metals are generated from various industries, such as metal plating, fertilizer production, battery manufacturing, paint factories, and pesticide-producing establishments (Bagchi and Behera, 2020; Srivastava and Majumder, 2008). Most heavy metals are highly toxic, and thus they pose a

TABLE 5.1
Methane Yield and Operation Conditions in BEAD

Configuration	Electrode	Voltage	Substrate	Inoculation	T	OLR	Methane Yield	Reference
CSTR (0.8 L)	Carbon felt	0.5/1.0 V	AS Molasses	Anaerobic sludge	34°C	1.5–2.0 g COD/L·d	Up to 577 mL/L·d	De Vrieze et al. (2014)
Batch reactor (1.0 L)	Carbon fiber brush	0.8 V	FW	Anaerobic sludge	35–37°C	1.7 g COD/L·d	300–600 mL/g·d	Liu et al. (2019)
Batch reactor (20 L)	Graphite carbon mesh	0.3 V	FW	Anaerobic sludge (food)	35°C	2.0–10.0 g COD/L·d	400 mL/g COD	Park et al. (2019)
Batch reactor (0.3,0.18 L)	Ti/Ru alloy	0–2.3 V	AS	Anaerobic sludge	35°C	0.8 g COD/L·d	38–200 mL/g VSS	Xiao et al. (2018)
Batch reactor (0.02 L)	Granular activated carbon	0.9 V	Acetate	Anaerobic sludge	10°C	0.24 g COD/L·d	30 mg COD/g VSS	Liu et al. (2016)
Batch reactor (0.33 L)	Carbon fiber brush	0.5/0.7/ 1.0/1.5 V	Glucose	Anaerobic digestate	33–37°C	0.36 g COD/L·d	369–408 mL/g COD	Choi et al. (2017)
CSTR (12 L)	Graphite fiber fabric	0.3 V	PS + AS	BEAD sludge	25°C	2.0–5.2 g COD/L·d	Up to 700 mL/L·d	Feng et al. (2019)
Batch reactor (1 L)	Carbon brush Nickel foam	0.8 V	WAS	MEC effluent	Room	/	Up to 146 mL/L·d	Wang et al. (2020)

AS: activated sludge; PS: primary sludge; WAS: wasted activated sludge; FW: food waste; OLR: organic loading rate.

high risk to the ecosystem and human health. For instance, a high concentration of cadmium (Cd) in the soil causes growth inhibition, chlorosis, browning of root tips, and death of plants (Guo et al., 2008; Mohanpuria et al., 2007; Nagajyoti et al., 2010). Cadmium also affects enzyme activities and thus results in kidney damage and bone diseases (Manahan, 2017). Long-term exposure to uranium, arsenic, and hexavalent chromium can cause cancers (Achparaki et al., 2012). Considering the toxicity of heavy metals, the World Health Organization (WHO) and the United States Environmental Protection Agency (USEPA) posed strict limitations on heavy metal concentration in water systems (USEPA, 2009; WHO, 2011).

Heavy metal removal is another serious challenge in conventional wastewater treatment. Wastewater containing heavy metals is more suitable to treat on-site (i.e., decentralized treatment) since the sources of heavy metals are specific. BESs were proposed in various studies as a promising method for heavy metal removal with low energy consumption (Bagchi and Behera, 2020; Nancharaiah et al., 2015; Wang and Ren, 2014). In addition, BES also offers the possibility to recover heavy metals from wastewater.

Reduction, precipitation, and adsorption are the main mechanisms for heavy metal removal and/or recovery in BESs (Bagchi and Behera, 2020; Guo and Kim, 2020; Nancharaiah et al., 2015; Wang and Ren, 2014). Heavy metal ions can be reduced to metallic forms by accepting electrons directly from the cathode or indirectly via microorganisms (Figure 5.5a). The reduction potential represents the capability of heavy metal ions to gain electrons for reduction and Logan et al. (2006) reported that the typical cathode potential in MFCs ranges from 0.1 to 0.3 V. Therefore, MFCs can be applied to reduce heavy metals, the reduction potential of which is higher than 0.1–0.3 V (e.g., Co^{3+} to Co^{2+}:1.82 V vs. SHE). For heavy metals, the reduction potential of which is lower than 0.1 V, an additional applied voltage needs to be provided to BES (i.e., MECs). The cathode potential in MECs varies from −0.25 to −1.5 V depending on the applied voltage (Colantonio and Kim, 2016; Huang et al., 2014; Logan et al., 2006).

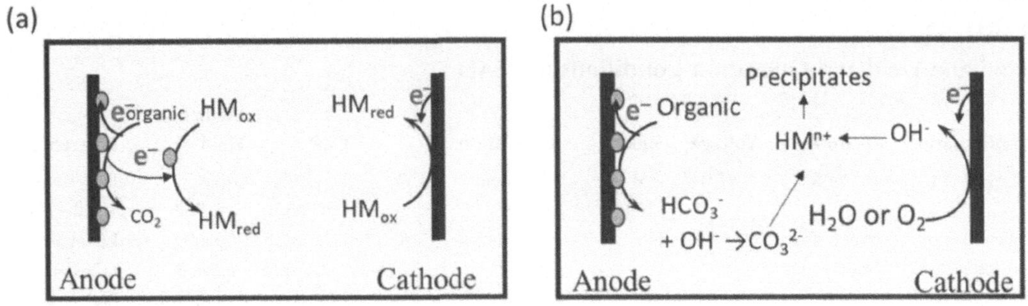

FIGURE 5.5 Mechanisms of heavy metal separation in BES. (a) reduction; (b) precipitation. (HM_{red}: reduced heavy metals; HM_{ox}: Oxidized heavy metals; HM^{n+}: heavy metal ions).

Precipitation also contributes to heavy metal removal in BES. The cathodic reactions (e.g., oxygen reduction in MFCs or hydrogen production in MECs) result in a sharp pH increase near the cathode. The heavy metal ions can react with OH^- or CO_3^{2-} to form precipitates (Figure 5.5b). The solubility product indicates the removal efficiency of heavy metal ions in BES. Some heavy metal ions such as CO^{2+} require a high pH to form precipitates (e.g., 10^{10} times difference in the concentration of CO^{2+} to form precipitates at pH 7 and 12, Guo and Kim, 2019). Other metal ions, such as Cu^{2+}, Hg^{2+}, and Cr^{3+}, are easy to form precipitates even at natural pH.

The removal efficiencies of heavy metal ions, including cadmium, cobalt, copper, nickel, mercury, gold, silver, vanadium, zinc, selenium, and chromium, were investigated in BES (Table 5.2). More than 90% of heavy metals were removed with different operation times (in the range of 4.5–480 h, Table 5.2). The high removal efficiency in BES indicated that BES is a suitable technology for heavy metal removal.

5.2.4 Challenges and Techno-Economic Assessment of BESs Application

There are still challenges for broad applications of BES in decentralized wastewater systems. First, the high capital costs for electrodes (especially precious metal-catalyzed cathodes), membranes, and current collectors. A purchase cost between £396 and £1149/m^3 for a MEC system in wastewater treatment plants was reported by Escapa et al. (2012). The cost increased to £1308 and £2344/m^3 for pilot systems (Cotterill et al., 2017; Heidrich et al., 2013). However, considering the beneficial aspects (e.g., energy/resource recovery), the overall cost of BES construction and operation is competitive when compared to traditional methods for decentralized wastewater treatment. For instance, Ge and He (2016) demonstrated that a 200-L MFC can be cost-effective compared to small-scale wastewater treatment systems.

Membrane fouling that negatively affects the performance of BES is another challenge for commercialization of BES. Although membrane-less BES is developed and demonstrated in practical applications, oxygen transfer from the cathode to the anode can seriously decrease the power generation and microbial activity in MFC (Kim et al., 2016).

Stable operation is also a challenge for BES applications for various types of wastewater. Variations in the characteristics of wastewater (e.g., low pH, high OLR, high free ammonia, high dissolved oxygen, and presence of toxic chemicals) can potentially fail BES operations. In addition, MEC systems with long-term operation may face the challenge of low hydrogen production due to the growth of methanogens. Toxic components, such as free ammonia and heavy metals, can inhibit the exoelectrogenic activity, thus resulting in the poor performance of BES.

TABLE 5.2
The Summary of Heavy Metal Separation in BES (Studies in the Last Ten Years)

Heavy Metals	Concentration	Reactor Type	Removal Efficiency	Mechanisms	Reference
Cobalt (Co^{2+} or Co^{3+})	Co^{2+} 0.42–2.12	MEC	72–99.2% in 6 h	Reduction and precipitation	Saad et al. (2020)
	Co^{2+} 0.68–1.01	MFC	98.47% in 48 h	Adsorption and precipitation	Huang et al. (2019)
	Co^{2+} 0.847	MEC	92% in 6 h	Reduction and anode biosorption	Jiang et al. (2014)
	$LiCoO_2$ (s) and Co^{2+} 0.34	MFC-MEC	88% in 6 h	Reduction and adsorption	Huang et al. (2014)
	$LiCoO_2$ (s)	MFC	62–70% in 48 h	Reduction	Huang et al. (2013)
	Co^{2+} 0.36	MFC	93% in 6 h	Reduction and precipitation	Huang et al. (2015)
Nickel (Ni^{2+})	0.85–17	MEC	33–99% in 19.8 h	Reduction and precipitation	Qin et al. (2012)
Gold (Au^{4+})	0–10.1	MFC	~100% in 12 h	Reduction and precipitation	Choi and Hu (2013)
Silver (Ag^{+})	0.46–1.84	MFC	>98% in 8 h	Reduction	Choi and Cui (2012)
	$AgNO_3$ or AgS_2O_3	MFC	95% in 35 h	Reduction	Tao et al. (2012)
	9.26	MFC	99.9% in 21 h	Reduction	Wang et al. (2013a,b)
	$[Ag(NH_3)_2]^+$ and $[Ag(S_2O_3)_2]^{3-}$ 9.35–10	MFC	81.8–99% in 48 h	Reduction	Ho and Babel (2019)
	$[Ag(NH_3)_2]^+$ 9.35–18.70	MFC	96% in 48 h	Reduction	Ho et al. (2018)
Vanadium (V^{5+})	3.9–7.8	MFC	Up to 100% in 168 h	Reduction and precipitation	Qiu et al. (2017)
	1.0–5.9	MFC	87% in 12 h	Reduction	Zhang et al. (2009)
Zinc (Zn^{2+})	1.54–12.3	MEC	60–99% in 23 h	Reduction and precipitation	Modin et al., 2017
	1.9	MFC	96% in 22 h	Reduction and precipitation	Lim et al. (2021)
	2.3	MEC	98.7% in 6 h	Reduction	Zhang et al. (2020)
	0.2–0.5	MFC	94–99% in 48 h	Precipitation and biosorption	Abourached et al. (2014)
Selenium (Se^{4+})	0.32–1.29	MFC	73.6% in 72 h	Reduction	Sravan et al. (2020)
Lead (Pb^{2+})	0.56	MEC	98.5% in 10 h	Reduction	Zhang et al. (2020)
Chromium (Cr^{6+})	0.58	MEC	72.65%	Reduction	Wang et al. (2018)
	0.18–0.37	MFC	>90% in 120 h	Reduction and precipitation and adsorption	Habibul et al. (2016)
	0.19–1.92	MFC	Up to 88.7% in 24 h	Reduction and precipitation	More and Gupta (2018)
	0.19	MFC and MEC	66.2%	Reduction and precipitation	Gajaraj et al. (2019)
	0.96	MFC	87.9 % in 24 h	Reduction	More and Gupta (2021)
	0–2.88	MFC	65.6% in 3 h	Chemical reduction, cathodic reduction	Wang et al. (2017)

(*Continued*)

TABLE 5.2 (*Continued*)
The Summary of Heavy Metal Separation in BES (Studies in the Last Ten Years)

Heavy Metals	Concentration	Reactor Type	Removal Efficiency	Mechanisms	Reference
Cadmium (Cd^{2+})	0.1	MEC	71–91% in 48 h	Reduction and precipitation	Colantonio and Kim (2016)
	0.45	MEC	39–47% in 4 h	Cathodic reduction, adsorption	Wang et al. (2016)
	10^{-4}–10^{-1} μM 0.02 mM	MEC	69% in 168 h	Cathodic reduction, anode biosorption	Colantonio et al. (2016)
Copper (Cu^{2+})	$Cu(NH_3)_4^{2+}$	MFC	84% in 8 h	Cathodic reduction	Zhang et al. (2012)
	Cu^{2+} 3.125	MFC	~100% in 576 h	Reduction and precipitation	Wang et al. (2020)

5.3 ELECTRODIALYSIS FOR DECENTRALIZED WASTEWATER TREATMENT

5.3.1 Nutrient Recovery

Accumulation of nutrient in natural water systems can result in eutrophication. In recent years, the growing population has increased the demand for food and water. On one hand, the limited resource of nutrient (especially phosphorus) poses a threat to the long-term development of food industries. On the other hand, the excessive use of fertilizer contributes to the increased concentration of nutrient in water systems. Therefore, efficient nutrient recovery from wastewater is necessary for sustainable development.

ED and BMED have been examined for ionic nutrient recovery from water and wastewater (Guo et al., 2021; Li et al., 2016; Shi et al., 2018; van Linden et al., 2020; Zhang et al., 2009, 2010; Mondor et al., 2008). Table 5.3 summarizes the configurations and performance of ED and BMED stacks reported in previous studies. Up to 21.3 g-N/L and 8.2 g-P/L can be recovered from wastewater by using ED or BMED systems (Table 5.3).

Compared to conventional ED systems, BMED can significantly improve nutrient recovery efficiency and produce high-purity products. In BMED, the anions and cations are separated and concentrated in acid chambers and base chambers, respectively (Figure 5.2). For instance, 78% of ammonium was recovered in the base cell, while 75% of phosphate was recovered in the acid cell (Shi et al., 2018). However, high energy consumption was observed in BMED systems since the water dissolution required a high applied voltage. Li et al. (2016) found that the energy consumption in a two-cell BMED (either BPMs only with CEMs or BPMs only with AEMs) is lower than that in a three-cell BMED. In addition, for some kinds of wastewater (e.g., dewatering centrate), acid (or anions) recovery is not meaningful since the organic anions are relatively low. Therefore, a two-cell BMED (BPMs with CEMs) was proposed to recover ammonia from dewatering centrate (Guo et al., 2021) (Figure 5.6). Highly concentrated and high-purity ammonium sulfate (i.e., 4.34 g-N/L in 90 min, 88.4% ammonia recovery from dewatering centrate) was produced by using the two-cell BMED. The final products can be used as fertilizers for agricultural industries.

The ED technologies are suitable for decentralized wastewater treatment. They are especially applicable for agricultural and food-processing wastewater with high nutrient concentration. The separated nutrient using ED or BMED stacks can be used as land fertilizers in the agricultural industry. ED and BMED can also be integrated or modified with specific treatment processes, to achieve higher recovery efficiency. For instance, Ward et al. (2018) induced struvite precipitation to recover phosphate from anaerobic sludge before feeding the sludge into ED. The removal of phosphate controls the membrane scaling problem in the ED system. De Paepe et al. (2018) built

TABLE 5.3
Summary of Previous ED Studies for Nutrient Recovery

Configuration	IEMs	Membrane Area	Feed	P Recovery	N Recovery	Energy Consumption	Reference
ED	30 pairs CEM + AEM	7.2 m^2	Treated centrate	Removed by struvite	7.1 g-N/L	4.9 kWh/kg-N	Ward et al. (2018)
BMED	7 pairs CEM + BPM	6.0 dm^2	Dewatering centrate	/	4.3 g-N/L	9.6 kWh/kg-N	Guo et al. (2021)
ED	2 CEM + AEM + MVA	1 dm^2	Synthetic wastewater	21.2 mg-P/L	40.6 mg-N/L	1.5 kWh/m^3	Liu et al. (2017)
BMED	BPM + CEM + AEM BPM + CEM BPM + AEM	5.4 dm^2 3.6 dm^2	Synthetic wastewater	8.2 g-P/L ~0 g-P/L 6.8 g-P/L	13.2 g-N/L 12.0 g-N/L 2.6 g-N/L	0.33 kWh	Shi et al. (2020)
ED	10 pairs CEM + AEM	13.0 dm^2	Secondary clarifier	/	1.9 g-N/L	4.3kWh/kgNO_3^-	Mohammadi et al. (2021a,b)
ED	10 pairs CEM + AEM	2.2 dm^2	Swine manure	/	21.3 g-N/L	18.0 kWh/kg-N	Ippersiel et al. (2012)
ED	10 pairs CEM + AEM	64.0 cm^2	Human urine	0.06 g-P/L	Nitrification	/	De paepe et al. (2018)
BMED	4 pairs BPM + CEM + AEM	7.1 cm^2	Synthetic sludge	1.55 g-P/L	/	29.3 kWh/kg H_3PO_4	Wang et al. (2013)
ED	2 pairs CEM + AEM	49.0 cm^2	Urine	0.61 g-P/L	14.1 g-N/L	/	Pronk et al. (2006)
ED	20 pairs CEM + AEM	80.0 cm^2	Urine	0.74 g-P/L	11.89 g-N/L	/	Wang et al. (2015)
BMED	10 pairs BPM + CEM + AEM	19.2 dm^2	Synthetic wastewater	/	4.5 g-N/L	5.3 kWh/kg-N	van Linden et al. (2020)
BMED	5 pairs BPM + CEM + AEM	2.0 dm^2	Synthetic pig manure	~2.4 mg-P/L	11.3 g-N/L	0.4 kWh	Shi et al. (2018)

ED: electrodialysis; BMED: bipolar membrane electrodialysis; CEM: cation exchange membrane; AEM: anion exchange membrane; BPM: bipolar membrane; MVA: monovalent anion exchange membrane.

a pilot-scale ED stack to enhance nutrient recovery by combining precipitation and nitrification. Mondor et al., (2008) integrated ED with reverse osmosis to produce a readily applicable nitrogen fertilizer from swine manure.

5.3.2 Fouling and Scaling Problems in ED Operation

Fouling is a phenomenon of undesirable molecules of substances depositing on surfaces of or inside the membrane (Dammak et al., 2021; Mikhaylin and Bazinet, 2016). The membrane fouling problem can limit broad applications of the ED technologies since the foulants cause high electrical resistance as well as shortened membrane lifetime. There are four classifications of ion exchange membrane (IEM) fouling/scaling in ED systems (Guo et al., 2012; Mikhaylin and Bazinet, 2016): (1) Particulate/colloidal fouling, which organic or inorganic particles/colloids physically blind the membrane surface. The sizes of the particulate and colloidal foulants vary from 0.001 to 2 μm in diameter. (2) Organic fouling occurs when organic substances, such as humic acid, proteins,

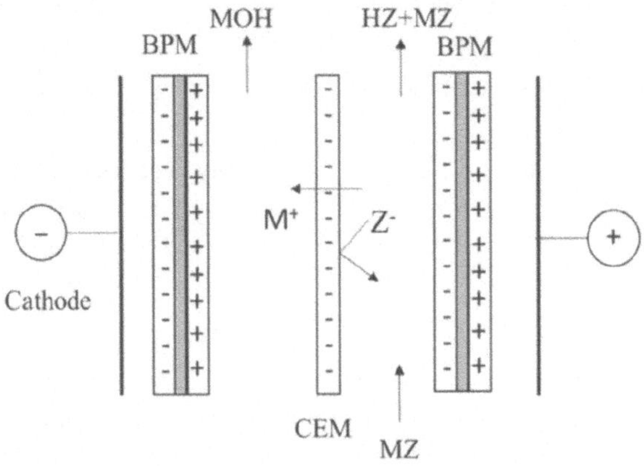

FIGURE 5.6 BMED with BPMs and CEMs.

carbohydrates, and oil, attach/stick to the membranes. (3) Scaling or inorganic fouling is triggered when salts precipitate onto the membrane surface or inside the membranes. (4) Biofouling happens when microorganisms or vegetative matter (e.g., algae) adhere and grow on the membranes.

There are many factors, such as the characteristics of feed solutions (e.g., pH, solids concentration, and concentration of divalent ions), membrane properties (e.g., hydrophobicity and wettability), and operation conditions (e.g., temperature and flow rate), which govern membrane fouling. For instance, the inorganic scaling problem with calcium and magnesium precipitants in ED reactors fed with municipal and food waste liquid digestate can increase electric resistance and even damage IEMs (Guo and Kim, 2020). A large amount of calcium carbonate scalants were found both with municipal and food waste liquid digestate, while the formation of struvite was limited by magnesium concentration for the food waste liquid digestate and phosphate concentration for the municipal liquid digestate, respectively. Tanaka et al. (2012) observed more severe organic fouling on AEMs with divinylbenzene and chloromethyl styrene base matrix than aliphatic organic substances when aromatic organic substances are fed. Bukhovets et al. (2010) found that IEMs were fouled by phenylalanine when the current densities were close to the limiting values.

Chemical cleaning is a common procedure in membrane application processes (Goode et al., 2013). The Clean-In-Place operations, which involve the use of acid or salt solutions, are widely used to remove membrane foulants (Guo and Kim, 2021; Wang et al., 2011). However, frequent chemical cleaning (e.g., use of sodium hypochlorite to control organic fouling and biofouling) can cause the degradation of ion exchange sites in IEMs (Garcia-Vasquez et al., 2014). Therefore, other methods (e.g., electrodialysis reversal (EDR), enzymatic cleaning, mechanical cleaning, and pre-treatment) are used to solve membrane fouling problems (Bdiri et al., 2019; Lee et al., 2012; Sui et al., 2008). Recent studies also examined new strategies to prevent and control membrane fouling. A pulsed electric field was found to reduce fouling and increase the current efficiency (Cifuentes-Araya et al., 2011; Lee et al., 2002). Modification of IEMs, including dip coating, adding additional layers, surface modification by chemicals, and electrodeposition (Li et al., 2018; Li et al., 2018; Vaselbehagh et al., 2014; Zhao et al., 2016, 2018), work efficiently to prevent membrane fouling. In addition, the application of two-cell BMED also helps to control membrane fouling (Guo et al., 2021).

The BMED with CEMs (i.e., two-cell BMED with CEMs) can also help control membrane scaling problems (Mohammadi et al., 2021a,b). In the BMED, cations (e.g., ammonium and calcium) are concentrated in the base cell, while anions (e.g., carbonate and phosphate) are retained in the feed (or acid) cell (Figure 5.6). Therefore, the formation of typical scalants such as calcium carbonate or struvite can be inhibited.

5.4 CONCLUSIONS AND PERSPECTIVES

This chapter summarizes the potential applications of electrochemical systems, including BES, ED, and BMED, in decentralized wastewater treatment. The compact modulation of BES can achieve rapid organic removal, complete biosolids reduction, stable sludge treatment in the downstream process (i.e., AD), and efficient heavy metal separation in small-scale wastewater treatment facilities. Compared to conventional wastewater treatment systems, BES is more stable under high OLR conditions. For instance, high OLR contributed to the high electric current density, which indicates the rapid electrode reactions for hydrogen gas production and organic removal in MECs. Hydrogen gas evolution at the cathode significantly enhances methane production in BEAD and benefits the process stability under high OLR conditions. In addition, modulated BES with stacked electrodes is an energy-efficient technology for small-scale wastewater treatment. The electric energy consumption in stacked electrode MECs (without considering the hydrogen production from MECs) was two to five times lower than the energy consumption in activated sludge systems. Also, MFC electrode stacks in septic can provide power for LED lights.

ED provides a suitable method for nutrient recovery from wastewater, especially for agricultural and food-processing wastewater with high nutrient concentrations. The concentrated streams from ED can be utilized as fertilizer sources. However, there are still challenges, such as the IEM fouling/scaling and low purity of the final products. Besides conventional membrane fouling control strategies, such as chemical cleaning, electrode reversal, and enzymatic cleaning, the application of BMED without AEMs can control membrane scaling efficiently. BMED without AEMs can also reduce energy consumption and improve the purity of fertilizers.

The treatment capacity of electrochemical systems widely ranges from several tons to thousands of tons of wastewater per day since the systems are easy to scale up. For instance, both BES and ED systems can be scaled up by repeating additional electrode and IEM pairs, respectively. They can also be used to receive influent wastewater in parallel and operate independently without further scale-up, allowing decentralized wastewater treatment systems with low maintenance cost and high treatment stability.

REFERENCES

Abourached, C., Catal, T., & Liu, H. (2014). Efficacy of single-chamber microbial fuel cells for removal of cadmium and zinc with simultaneous electricity production. *Water Research*, 51, 228–233.

Achparaki, M., Thessalonikeos, E., Tsoukali, H., Mastrogianni, O., Zaggelidou, E., Chatzinikolaou, F., Vasilliades, N., & Raikos, N. (2012). Heavy metals toxicity. *Aristotle University Medical Journal*, 39(1), 29–34.

Alzate-Gaviria, L., García-Rodríguez, O., Flota-Bañuelos, M., Del Rio Jorge-Rivera, F., Cámara-Chalé, G., & Domínguez-Maldonado, J. (2016). Stacked-MFC into a typical septic tank used in public housing. *Biofuels*, 7(2), 79–86.

Asztalos, J. R., & Kim, Y. (2015). Enhanced digestion of waste activated sludge using microbial electrolysis cells at ambient temperature. *Water Research*, 87, 503–512.

Bagchi, S., & Behera, M. (2020). Assessment of heavy metal removal in different bioelectrochemical systems: A review. *Journal of Hazardous, Toxic, and Radioactive Waste*, 24(3), 04020010.

Batstone, D. J., Keller, J., Angelidaki, I., Kalyuzhnyi, S. V., Pavlostathis, S. G., Rozzi, A., …, & Vavilin, V. A. (2002). The IWA anaerobic digestion model no 1 (ADM1). *Water Science and Technology*, 45(10), 65–73.

Bdiri, M., Bensghaier, A., Chaabane, L., Kozmai, A., Baklouti, L., & Larchet, C. (2019). Preliminary study on enzymatic-based cleaning of cation-exchange membranes used in electrodialysis system in red wine production. *Membranes*, 9(9), 114.

Blasco-Gómez, R., Batlle-Vilanova, P., Villano, M., Balaguer, M. D., Colprim, J., & Puig, S. (2017). On the edge of research and technological application: A critical review of electromethanogenesis. *International Journal of Molecular Sciences*, 18(4), 874.

Bukhovets, A., Eliseeva, T., & Oren, Y. (2010). Fouling of anion-exchange membranes in electrodialysis of aromatic amino acid solution. *Journal of Membrane Science*, 364(1–2), 339–343.

Cai, W., Liu, W., Yang, C., Wang, L., Liang, B., Thangavel, S., Guo, Z., & Wang, A. (2016). Biocathodic methanogenic community in an integrated anaerobic digestion and microbial electrolysis system for enhancement of methane production from waste sludge. *ACS Sustainable Chemistry & Engineering*, 4(9), 4913–4921.

Chae, K. J., Choi, M. J., Lee, J., Ajayi, F. F., & Kim, I. S. (2008). Biohydrogen production via biocatalyzed electrolysis in acetate-fed bioelectrochemical cells and microbial community analysis. *International Journal of Hydrogen Energy*, 33(19), 5184–5192.

Choi, C., & Cui, Y. (2012). Recovery of silver from wastewater coupled with power generation using a microbial fuel cell. *Bioresource Technology*, 107, 522–525.

Choi, C., & Hu, N. (2013). The modeling of gold recovery from tetrachloroaurate wastewater using a microbial fuel cell. *Bioresource Technology*, 133, 589–598.

Choi, K. S., Kondaveeti, S., & Min, B. (2017). Bioelectrochemical methane (CH4) production in anaerobic digestion at different supplemental voltages. *Bioresource Technology*, 245, 826–832.

Cifuentes-Araya, N., Pourcelly, G., & Bazinet, L. (2011). Impact of pulsed electric field on electrodialysis process performance and membrane fouling during consecutive demineralization of a model salt solution containing a high magnesium/calcium ratio. *Journal of Colloid and Interface Science*, 361(1), 79–89.

Colantonio, N., Guo, H., & Kim, Y. (2016). Effect of low cadmium concentration on the removal efficiency and mechanisms in microbial electrolysis cells. *ChemistrySelect*, 1(21), 6920–6924.

Colantonio, N., & Kim, Y. (2016). Cadmium (II) removal mechanisms in microbial electrolysis cells. *Journal of Hazardous Materials*, 311, 134–141.

Cotterill, S. E., Dolfing, J., Jones, C., Curtis, T. P., & Heidrich, E. S. (2017). Low temperature domestic wastewater treatment in a microbial electrolysis cell with 1 m^2 anodes: Towards system scale-up. *Fuel Cells*, 17(5), 584–592.

Dammak, L., Fouilloux, J., Bdiri, M., Larchet, C., Renard, E., Baklouti, L., ... & Pismenskaya, N. (2021). A review on ion-exchange membrane fouling during the electrodialysis process in the food industry, part 1: Types, effects, characterization methods, fouling mechanisms and interactions. *Membranes*, 11(10), 789.

De Paepe, J., Lindeboom, R. E., Vanoppen, M., Paepe, D., Demey, K., Coessens, D., & Vlaeminck, W. (2018). Refinery and concentration of nutrients from urine with electrodialysis enabled by upstream precipitation and nitrification. *Water Research*, 144, 76–86.

De Vrieze, J., Gildemyn, S., Arends, J. B., Vanwonterghem, I., Verbeken, K., Boon, N., ... & Rabaey, K. (2014). Biomass retention on electrodes rather than electrical current enhances stability in anaerobic digestion. *Water Research*, 54, 211–221.

ElMekawy, A., Srikanth, S., Bajracharya, S., Hegab, H. M., Nigam, P. S., Singh, A., Mohan, S. V., & Pant, D. (2015). Food and agricultural wastes as substrates for bioelectrochemical system (BES): The synchronized recovery of sustainable energy and waste treatment. *Food Research International*, 73, 213–225.

Escapa, A., Gómez, X., Tartakovsky, B., & Morán, A. (2012). Estimating microbial electrolysis cell (MEC) investment costs in wastewater treatment plants: Case study. *International Journal of Hydrogen Energy*, 37(24), 18641–18653.

Feng, Q., Song, Y. C., Kim, D. H., Kim, M. S., & Kim, D. H. (2019). Influence of the temperature and hydraulic retention time in bioelectrochemical anaerobic digestion of sewage sludge. *International Journal of Hydrogen Energy*, 44(4), 2170–2179.

Ferrer, J. J., Laborie, S., Durand, G., & Rakib, M. (2006). Formic acid regeneration by electromembrane processes. *Journal of Membrane Science*, 280(1–2), 509–516.

Fudge, T., Bulmer, I., Bowman, K., Pathmakanthan, S., Gambier, W., Dehouche, Z., ... & Constantinou, A. (2021). Microbial electrolysis cells for decentralised wastewater treatment: The next steps. *Water*, 13(4), 445.

Gajaraj, S., Sun, X., Zhang, C., & Hu, Z. (2019). Improved chromium reduction and removal from wastewater in continuous flow bioelectrochemical systems. *Environmental Science and Pollution Research*, 26(31), 31945–31955.

Garcia-Vasquez, W., Ghalloussi, R., Dammak, L., Larchet, C., Nikonenko, V., & Grande, D. (2014). Structure and properties of heterogeneous and homogeneous ion-exchange membranes subjected to ageing in sodium hypochlorite. *Journal of Membrane Science*, 452, 104–116.

Ge, Z., & He, Z. (2016). Long-term performance of a 200 liter modularized microbial fuel cell system treating municipal wastewater: Treatment, energy, and cost. *Environmental Science: Water Research & Technology*, 2(2), 274–281.

Goode, K. R., Asteriadou, K., Robbins, P. T., & Fryer, P. J. (2013). Fouling and cleaning studies in the food and beverage industry classified by cleaning type. *Comprehensive Reviews in Food Science and Food Safety*, 12(2), 121–143.

Government of Canada. (2012). Wastewater systems effluent Regulations (SOR/2019-139).
Gude, V. G. (2018). Integrating bioelectrochemical systems for sustainable wastewater treatment. *Clean Technologies and Environmental Policy*, 20(5), 911–924.
Guo, J., Dai, X., Xu, W., & Ma, M. (2008). Over expressing GSHI and AsPCSI simultaneously increases the tolerance and accumulation of cadmium and arsenic in Arabidopsis thaliana. *Chemosphere*, 72, 1020–1026.
Guo, H., & Kim, Y. (2018). Scalable multi-electrode microbial electrolysis cells for high electric current and rapid organic removal. *Journal of Power Sources*, 391, 67–72.
Guo, H., & Kim, Y. (2019). Stacked multi-electrode design of microbial electrolysis cells for rapid and low-sludge treatment of municipal wastewater. *Biotechnology for Biofuels*, 12(1), 1–10.
Guo, H., & Kim, Y. (2020). Mechanisms of heavy metal separation in bioelectrochemical systems and relative significance of precipitation. *Microbial Electrochemical Technologies*, 128–144.
Guo, H., & Kim, Y. (2021). Membrane scaling in electrodialysis fed with high-strength wastewater. *Environmental Engineering Science*, 38(9), 832–840.
Guo, X., Liu, J., & Xiao, B. (2013). Bioelectrochemical enhancement of hydrogen and methane production from the anaerobic digestion of sewage sludge in single-chamber membrane-free microbial electrolysis cells. *International Journal of Hydrogen Energy*, 38(3), 1342–1347.
Guo, Z., Liu, W., Yang, C., Gao, L., Thangavel, S., Wang, L., ... & Wang, A. (2017). Computational and experimental analysis of organic degradation positively regulated by bioelectrochemistry in an anaerobic bioreactor system. *Water Research*, 125, 170–179.
Guo, W., Ngo, H. H., & Li, J. (2012). A mini-review on membrane fouling. *Bioresource Technology*, 122, 27–34.
Guo, H., Yuan, P., Pavlovic, V., Barber, J., & Kim, Y. (2021). Ammonium sulfate production from wastewater and low-grade sulfuric acid using bipolar-and cation-exchange membranes. *Journal of Cleaner Production*, 285, 124888.
Gupta, S., Srivastava, P., Patil, S. A., & Yadav, A. K. (2021). A comprehensive review on emerging constructed wetland coupled microbial fuel cell technology: Potential applications and challenges. *Bioresource Technology*, 320, 124376.
Habibul, N., Hu, Y., Wang, Y. K., Chen, W., Yu, H. Q., & Sheng, G. P. (2016). Bioelectrochemical chromium (VI) removal in plant-microbial fuel cells. *Environmental Science & Technology*, 50(7), 3882–3889.
Hamelers, H. V., Ter Heijne, A., Sleutels, T. H., Jeremiasse, A. W., Strik, D. P., & Buisman, C. J. (2010). New applications and performance of bioelectrochemical systems. *Applied Microbiology and Biotechnology*, 85(6), 1673–1685.
Heidrich, E. S., Dolfing, J., Scott, K., Edwards, S. R., Jones, C., & Curtis, T. P. (2013). Production of hydrogen from domestic wastewater in a pilot-scale microbial electrolysis cell. *Applied Microbiology and Biotechnology*, 97(15), 6979–6989.
Hirano, S. I., & Matsumoto, N. (2018). Analysis of a bio-electrochemical reactor containing carbon fiber textiles for the anaerobic digestion of tomato plant residues. *Bioresource Technology*, 249, 809–817.
Ho, N. A. D., & Babel, S. (2019). Electrochemical reduction of different Ag (I)-containing solutions in bioelectrochemical systems for recovery of silver and simultaneous power generation. *RSC Advances*, 9(52), 30259–30268.
Ho, N. A. D., Babel, S., & Sombatmankhong, K. (2018). Bio-electrochemical system for recovery of silver coupled with power generation and wastewater treatment from silver (I) diammine complex. *Journal of Water Process Engineering*, 23, 186–194.
Huang, L., Li, T., Liu, C., Quan, X., Chen, L., Wang, A., & Chen, G. (2013). Synergetic interactions improve cobalt leaching from lithium cobalt oxide in microbial fuel cells. *Bioresource Technology*, 128, 539–546.
Huang, L., Liu, Y., Yu, L., Quan, X., & Chen, G. (2015). A new clean approach for production of Cobalt dihydroxide from aqueous Co (II) using oxygen-reducing biocathode microbial fuel cells. *Journal of Cleaner Production*, 86, 441–446.
Huang, T., Song, D., Liu, L., & Zhang, S. (2019). Cobalt recovery from the stripping solution of spent lithium-ion battery by a three-dimensional microbial fuel cell. *Separation and Purification Technology*, 215, 51–61.
Huang, C., & Xu, T. (2006). Electrodialysis with bipolar membranes for sustainable development. *Environmental Science & Technology*, 40(17), 5233–5243.
Huang, C., Xu, T., & Jacobs, M. L. (2006). Regenerating flue-gas desulfurizing agents by bipolar membrane electrodialysis. *AIChE Journal*, 52(1), 393–401.
Huang, L., Yao, B., Wu, D., & Quan, X. (2014). Complete cobalt recovery from lithium cobalt oxide in self-driven microbial fuel cell–microbial electrolysis cell systems. *Journal of Power Sources*, 259, 54–64.

Hu, H., Fan, Y., & Liu, H. (2009). Hydrogen production in single-chamber tubular microbial electrolysis cells using non-precious-metal catalysts. *International Journal of Hydrogen Energy*, 34(20), 8535–8542.

Ippersiel, D., Mondor, M., Lamarche, F., Tremblay, F., Dubreuil, J., & Masse, L. (2012). Nitrogen potential recovery and concentration of ammonia from swine manure using electrodialysis coupled with air stripping. *Journal of Environmental Management*, *95*, S165–S169.

Jiang, L., Huang, L., & Sun, Y. (2014). Recovery of flakey Cobalt from aqueous Co (II) with simultaneous hydrogen production in microbial electrolysis cells. *International Journal of Hydrogen Energy*, 39(2), 654–663.

Jingyu, H., Miwornunyuie, N., Ewusi-Mensah, D., & Koomson, D. A. (2020). Assessing the factors influencing the performance of constructed wetland–microbial fuel cell integration. *Water Science and Technology*, 81(4), 631–643.

Jwa, E., Yun, Y. M., Kim, H., Jeong, N., Park, S. C., & Nam, J. Y. (2019). Domestic wastewater treatment in a tubular microbial electrolysis cell with a membrane electrode assembly. *International Journal of Hydrogen Energy*, 44(2), 652–660. 23.

Kelleher, B. P., Leahy, J. J., Henihan, A. M., O'dwyer, T. F., Sutton, D., & Leahy, M. J. (2002). Advances in poultry litter disposal technology—a review. *Bioresource Technology*, 83(1), 27–36.

Kim, J., Kim, B., An, J., Lee, Y. S., & Chang, I. S. (2016). Development of anode zone using dual-anode system to reduce organic matter crossover in membraneless microbial fuel cells. *Bioresource Technology*, 213, 140–145.

Koter, S., & Warszawski, A. (2000). Electromembrane processes in environment protection. *Polish Journal of Environmental Studies*, 9(1), 45–56.

Kotsyurbenko, O. R., Chin, K. J., Glagolev, M. V., Stubner, S., Simankova, M. V., Nozhevnikova, A. N., & Conrad, R. (2004). Acetoclastic and hydrogenotrophic methane production and methanogenic populations in an acidic West-Siberian peat bog. *Environmental Microbiology*, 6(11), 1159–1173.

Lalaurette, E., Thammannagowda, S., Mohagheghi, A., Maness, P. C., & Logan, B. E. (2009). Hydrogen production from cellulose in a two-stage process combining fermentation and electrohydrogenesis. *International Journal of Hydrogen Energy*, 34(15), 6201–6210.

Lee, H. J., Hong, M. K., & Moon, S. H. (2012). A feasibility study on water softening by electrodeionization with the periodic polarity change. *Desalination*, 284, 221–227.

Lee, H. J., Park, J. S., & Moon, S. H. (2002). A study on fouling mitigation using pulsing electric fields in electrodialysis of lactate containing BSA. *Korean Journal of Chemical Engineering*, 19(5), 880–887.

Lei, C., Li, Z., Gao, Q., Fu, R., Wang, W., Li, Q., & Liu, Z. (2020). Comparative study on the production of gluconic acid by electrodialysis and bipolar membrane electrodialysis: Effects of cell configurations. *Journal of Membrane Science*, 608, 118192.

Li, Y., Chen, Y., & Wu, J. (2019). Enhancement of methane production in anaerobic di- gestion process: A review. *Appl. Energy*, 240, 120–137.

Li, Y., Shi, S., Cao, H., Wu, X., Zhao, Z., & Wang, L. (2016). Bipolar membrane electrodialysis for generation of hydrochloric acid and ammonia from simulated ammonium chloride wastewater. *Water Research*, 89, 201–209.

Li, Y., Shi, S., Cao, H., Zhao, Z., & Wen, H. (2018). Modification and properties characterization of heterogeneous anion-exchange membranes by electrodeposition of graphene oxide (GO). *Applied Surface Science*, 442, 700–710.

Li, J., Yuan, S., Wang, J., Zhu, J., Shen, J., & Van der Bruggen, B. (2018). Mussel-inspired modification of ion exchange membrane for monovalent separation. *Journal of Membrane Science*, 553, 139–150.

Li, Y., Zhang, Y., Liu, Y., Zhao, Z., Zhao, Z., Liu, S., & Quan, X. (2016). Enhancement of anaerobic methanogenesis at a short hydraulic retention time via bioelectrochemical enrichment of hydrogenotrophic methanogens. *Bioresource Technology*, 218, 505–511.

Lim, S. S., Fontmorin, J. M., Pham, H. T., Milner, E., Abdul, P. M., Scott, K., ... & Yu, E. H. (2021). Zinc removal and recovery from industrial wastewater with a microbial fuel cell: Experimental investigation and theoretical prediction. *Science of The Total Environment*, *776*, 145934.

Lin, C., Wu, P., Liu, Y., Wong, J. W., Yong, X., Wu, X., ... & Zhou, J. (2019). Enhanced biogas production and biodegradation of phenanthrene in wastewater sludge treated anaerobic digestion reactors fitted with a bioelectrode system. *Chemical Engineering Journal*, 365, 1–9.

Liu, H., Cheng, S., & Logan, B. E. (2005). Production of electricity from acetate or butyrate using a single chamber microbial fuel cell. *Environmental Science & Technology*, 39(2), 658–662.

Liu, S., Deng, Z., Li, H., & Feng, K. (2019). Contribution of electrodes and electric current to process stability and methane production during the electro-fermentation of food waste. *Bioresource Technology*, 288, 121536.

Liu, R., Wang, Y., Wu, G., Luo, J., & Wang, S. (2017). Development of a selective electrodialysis for nutrient recovery and desalination during secondary effluent treatment. *Chemical Engineering Journal*, 322, 224–233.

Liu, D., Zhang, L., Chen, S., Buisman, C., & Ter Heijne, A. (2016). Bioelectrochemical enhancement of methane production in low temperature anaerobic digestion at 10°C. *Water Research*, 99, 281–287.

Logan, B. E., Call, D., Cheng, S., Hamelers, H. V., Sleutels, T. H., Jeremiasse, A. W., & Rozendal, R. A. (2008). Microbial electrolysis cells for high yield hydrogen gas production from organic matter. *Environmental Science & Technology*, 42(23), 8630–8640.

Logan, B. E., Hamelers, B., Rozendal, R., Schröder, U., Keller, J., Freguia, S., ... & Rabaey, K. (2006). Microbial fuel cells: Methodology and technology. *Environmental Science & Technology*, 40(17), 5181–5192.

Lyberatos, G., & Skiadas, I. V. (1999). Modelling of anaerobic digestion—A review. *Global Nest International Journal*, 1(2), 63–76.

Manahan, S. (2017). *Environmental Chemistry*. CRC Press.

Metcalf, E., Abu-Orf, M., Bowden, G., Burton, F. L., Pfrang, W., Stensel, H. D., ... & AECOM (Firm). (2014). *Wastewater Engineering: Treatment and Resource Recovery*. McGraw Hill Education.

Mikhaylin, S., & Bazinet, L. (2016). Fouling on ion-exchange membranes: Classification, characterization and strategies of prevention and control. *Advances in Colloid and Interface Science*, 229, 34–56.

Modin, O., Fuad, N., & Rauch, S. (2017). Microbial electrochemical recovery of zinc. *Electrochimica Acta*, 248, 58–63.

Mohammadi, M., Guo, H., Yuan, P., Pavlovic, V., Barber, J., & Kim, Y. (2021a). Ammonia separation from wastewater using bipolar membrane electrodialysis. *Electrochemical Science Advances*, 1(4), e2000030.

Mohammadi, R., Ramasamy, D. L., & Sillanpää, M. (2021b). Enhancement of nitrate removal and recovery from municipal wastewater through single-and multi-batch electrodialysis: Process optimisation and energy consumption. *Desalination*, 498, 114726.

Mohanpuria, P., Rana, N. K., & Yadav, S. K. (2007). Cadmium induced oxidative stress influence on glutathione metabolic genes of *Camella sinensis* (L.). O Kuntze. *Environmental Toxicology*, 22, 368–374.

Mondor, M., Masse, L., Ippersiel, D., Lamarche, F., & Masse, D. I. (2008). Use of electrodialysis and reverse osmosis for the recovery and concentration of ammonia from swine manure. *Bioresource Technology*, 99(15), 7363–7368.

Mook, W. T., Aroua, M. K. T., Chakrabarti, M. H., Noor, I. M., Irfan, M. F., & Low, C. T. J. (2013). A review on the effect of bio-electrodes on denitrification and organic matter removal processes in bio-electrochemical systems. *Journal of Industrial and Engineering Chemistry*, 19(1), 1–13.

More, A. G., & Gupta, S. K. (2018). Evaluation of chromium removal efficiency at varying operating conditions of a novel bioelectrochemical system. *Bioprocess and Biosystems Engineering*, 41(10), 1547–1554.

More, A. G., & Gupta, S. K. (2021). Removal of chromium from electroplating industry wastewater using bioelectrochemical system: Kinetic study and statistical analysis. *Journal of Hazardous, Toxic, and Radioactive Waste*, 25(2), 04020069.

Nagajyoti, P. C., Lee, K. D., & Sreekanth, T. V. M. (2010). Heavy metals, occurrence and toxicity for plants: A review. *Environmental Chemistry Letters*, 8(3), 199–216.

Nancharaiah, Y. V., Mohan, S. V., & Lens, P. N. L. (2015). Metals removal and recovery in bioelectrochemical systems: A review. *Bioresource Technology*, 195, 102–114.

Nevin, K. P., Hensley, S. A., Franks, A. E., Summers, Z. M., Ou, J., Woodard, T. L., ... & Lovley, D. R. (2011). Electrosynthesis of organic compounds from carbon dioxide is catalyzed by a diversity of acetogenic microorganisms. *Applied and Environmental Microbiology*, 77(9), 2882–2886.

Newman, J., & Thomas-Alyea, K. E. (2012). *Electrochemical Systems*. John Wiley & Sons.

Park, J. G., Jiang, D., Lee, B., & Jun, H. B. (2020). Towards the practical application of bioelectrochemical anaerobic digestion (BEAD): Insights into electrode materials, reactor configurations, and process designs. *Water Research*, 184, 116214.

Park, J. G., Lee, B., Kwon, H. J., & Jun, H. B. (2019). Contribution analysis of methane production from food waste in bulk solution and on bio-electrode in a bio-electrochemical anaerobic digestion reactor. *Science of the Total Environment*, 670, 741–751.

Pham, T., Rabaey, K., Aelterman, P., Clauwaert, P., De Schamphelaire, L., Boon, N., et al. (2006). Microbial fuel cells in relation to conventional anaerobic digestion technology. *Engineering in Life Sciences*, 6(3), 285–292.

Pourcelly, G. (2002). Electrodialysis with bipolar membranes: Principles, optimization, and applications. *Russian Journal of Electrochemistry*, 38(8), 919–926.

Pronk, W., Biebow, M., & Boller, M. (2006). Electrodialysis for recovering salts from a urine solution containing micropollutants. *Environmental Science & Technology*, 40(7), 2414–2420.

Qin, B., Luo, H., Liu, G., Zhang, R., Chen, S., Hou, Y., & Luo, Y. (2012). Nickel ion removal from wastewater using the microbial electrolysis cell. *Bioresource Technology*, 121, 458–461.

Qiu, R., Zhang, B., Li, J., Lv, Q., Wang, S., & Gu, Q. (2017). Enhanced vanadium (V) reduction and bioelectricity generation in microbial fuel cells with biocathode. *Journal of Power Sources*, 359, 379–383.

Rabaey, K., Bützer, S., Brown, S., Keller, J., & Rozendal, R. A. (2010). High current generation coupled to caustic production using a lamellar bioelectrochemical system. *Environmental Science & Technology*, 44(11), 4315–4321.

Raikos, N. (2012). Heavy metals toxicity. *Aristotle University Medical Journal*, 39(1), 29–34.

Saad, D. R., Alismaeel, Z. T., & Abbar, A. H. (2020). Cobalt removal from simulated wastewaters using a novel flow-by fixed bed bio-electrochemical reactor. *Chemical Engineering and Processing-Process Intensification*, 156, 108097.

Saheb-Alam, S., Singh, A., Hermansson, M., Persson, F., Schnürer, A., Wilén, B. M., & Modin, O. (2018). Effect of start-up strategies and electrode materials on carbon dioxide reduction on biocathodes. *Applied and Environmental Microbiology*, 84(4), e02242–e17.

Selembo, P. A., Perez, J. M., Lloyd, W. A., & Logan, B. E. (2009). High hydrogen production from glycerol or glucose by electrohydrogenesis using microbial electrolysis cells. *International Journal of Hydrogen Energy*, 34(13), 5373–5381.

Shi, L., Hu, Y., Xie, S., Wu, G., Hu, Z., & Zhan, X. (2018). Recovery of nutrients and volatile fatty acids from pig manure hydrolysate using two-stage bipolar membrane electrodialysis. *Chemical Engineering Journal*, 334, 134–142.

Shi, L., Xiao, L., Hu, Z., & Zhan, X. (2020). Nutrient recovery from animal manure using bipolar membrane electrodialysis: Study on product purity and energy efficiency. *Water Cycle*, 1, 54–62.

Simons, R. (1993). Preparation of a high performance bipolar membrane. *Journal of Membrane Science*, 78(1–2), 13–23.

Song, Y. C., Feng, Q., & Ahn, Y. (2016). Performance of the bio-electrochemical anaerobic digestion of sewage sludge at different hydraulic retention times. *Energy & Fuels*, 30(1), 352–359.

Sravan, J. S., Nancharaiah, Y. V., Lens, P. N. L., & Mohan, S. V. (2020). Cathodic selenium recovery in bioelectrochemical system: Regulatory influence on anodic electrogenic activity. *Journal of Hazardous Materials*, *399*, 122843.

Srivastava, N. K., & Majumder, C. B. (2008). Novel biofiltration methods for the treatment of heavy metals from industrial wastewater. *Journal of Hazardous Materials*, 151(1), 1–8.

Strathmann, H. (2004). *Ion-Exchange Membrane Separation Processes*. Elsevier.

Strathmann, H. (2010). Electrodialysis, a mature technology with a multitude of new applications. *Desalination*, 264(3), 268–288.

Sui, P., Wen, X., & Huang, X. (2008). Feasibility of employing ultrasound for on-line membrane fouling control in an anaerobic membrane bioreactor. *Desalination*, 219(1–3), 203–213.

Tanaka, N., Nagase, M., & Higa, M. (2012). Organic fouling behavior of commercially available hydrocarbon-based anion-exchange membranes by various organic-fouling substances. *Desalination*, 296, 81–86.

Tao, H. C., Gao, Z. Y., Ding, H., Xu, N., & Wu, W. M. (2012). Recovery of silver from silver (I)-containing solutions in bioelectrochemical reactors. *Bioresource Technology*, 111, 92–7.

Tartakovsky, B., Mehta, P., Santoyo, G., & Guiot, S. R. (2011). Maximizing hydrogen production in a microbial electrolysis cell by real-time optimization of applied voltage. *International Journal of Hydrogen Energy*, 36(17), 10557–10564.

Tchobanoglus, G., Burton, F., & Stensel, H. D. (2003). Wastewater engineering: treatment and reuse. *American Water Works Association. Journal*, 95(5), 201.

Trivedi, G. S., Shah, B. G., Adhikary, S. K., Indusekhar, V. K., & Rangarajan, R. (1997). Studies on bipolar membranes. Part II—Conversion of sodium acetate to acetic acid and sodium hydroxide. *Reactive and Functional Polymers*, 32(2), 209–215.

USEPA, M. (2009). National primary drinking water regulations. EPA 816-F-09-004, 1–6.

van Linden, N., Bandinu, G. L., Vermaas, D. A., Spanjers, H., & van Lier, J. B. (2020). Bipolar membrane electrodialysis for energetically competitive ammonium removal and dissolved ammonia production. *Journal of Cleaner Production*, 259, 120788.

Vaselbehagh, M., Karkhanechi, H., Mulyati, S., Takagi, R., & Matsuyama, H. (2014). Improved antifouling of anion-exchange membrane by polydopamine coating in electrodialysis process. *Desalination*, 332(1), 126–133.

Villano, M., Aulenta, F., Ciucci, C., Ferri, T., Giuliano, A., & Majone, M. (2010). Bioelectrochemical reduction of CO_2 to CH_4 via direct and indirect extracellular electron transfer by a hydrogenophilic methanogenic culture. *Bioresource Technology*, 101(9), 3085–3090.
Wang, J. (2014). Decentralized biogas technology of anaerobic digestion and farm ecosystem: Opportunities and challenges. *Frontiers in Energy Research*, 2, 10.
Wang, Q., Huang, L., Pan, Y., Zhou, P., Quan, X., Logan, B. E., & Chen, H. (2016). Cooperative cathode electrode and in situ deposited copper for subsequent enhanced cd (II) removal and hydrogen evolution in bioelectrochemical systems. *Bioresource Technology*, 200, 565–571.
Wang, Q., Huang, L., Pan, Y., Quan, X., & Puma, G. L. (2017). Impact of Fe (III) as an effective electron-shuttle mediator for enhanced Cr (VI) reduction in microbial fuel cells: Reduction of diffusional resistances and cathode overpotentials. *Journal of Hazardous Materials*, 321, 896–906.
Wang, H., Luo, H., Fallgren, P. H., Jin, S., & Ren, Z. J. (2015). Bioelectrochemical system platform for sustainable environmental remediation and energy generation. *Biotechnology Advances*, 33(3), 317–334.
Wang, H., & Ren, Z. J. (2014). Bioelectrochemical metal recovery from wastewater: A review. *Water Research*, 66, 219–232.
Wang, H., Wang, H., Gao, C., & Liu, L. (2020). Enhanced removal of copper by electroflocculation and electroreduction in a novel bioelectrochemical system assisted microelectrolysis. *Bioresource Technology*, 297, 122507.
Wang, Y. H., Wang, B. S., Pan, B., Chen, Q. Y., & Yan, W. (2013a). Electricity production from a bio-electrochemical cell for silver recovery in alkaline media. *Applied Energy*, 112, 1337–1341.
Wang, X., Wang, Y., Zhang, X., Feng, H., Li, C., & Xu, T. (2013b). Phosphate recovery from excess sludge by conventional electrodialysis (CED) and electrodialysis with bipolar membranes (EDBM). *Industrial & Engineering Chemistry Research*, 52(45), 15896–15904.
Wang, Q., Yang, P., & Cong, W. (2011). Cation-exchange membrane fouling and cleaning in bipolar membrane electrodialysis of industrial glutamate production wastewater. *Separation and Purification Technology*, 79(1), 103–113.
Wang, L., Yang, C., Sangeetha, T., He, Z., Guo, Z., Gao, L., … & Liu, W. (2020). Methane production in a bioelectrochemistry integrated anaerobic reactor with layered nickel foam electrodes. *Bioresource Technology*, 313, 123657.
Wang, H., Zhang, S., Wang, J., Song, Q., Zhang, W., He, Q., … & Ma, F. (2018). Comparison of performance and microbial communities in a bioelectrochemical system for simultaneous denitrification and chromium removal: Effects of pH. *Process Biochemistry*, 73, 154–161.
Wang, X., Zhang, X., Wang, Y., Du, Y., Feng, H., & Xu, T. (2015). Simultaneous recovery of ammonium and phosphorus via the integration of electrodialysis with struvite reactor. *Journal of Membrane Science*, 490, 65–71.
Ward, A. J., Arola, K., Brewster, E. T., Mehta, C. M., & Batstone, D. J. (2018). Nutrient recovery from wastewater through pilot scale electrodialysis. *Water Research*, 135, 57–65.
WHO (2011). Guidelines for drinking-water quality. *World Health Organization*, 216, 303–304.
Xiao, B., Chen, X., Han, Y., Liu, J., & Guo, X. (2018). Bioelectrochemical enhancement of the anaerobic digestion of thermal-alkaline pretreated sludge in microbial electrolysis cells. *Renewable Energy*, 115, 1177–1183.
Xu, T., & Huang, C. (2008). Electrodialysis-based separation technologies: A critical review. *AIChE Journal*, 54(12), 3147–3159.
Yadav, A. K., Dash, P., Mohanty, A., Abbassi, R., & Mishra, B. K. (2012). Performance assessment of innovative constructed wetland-microbial fuel cell for electricity production and dye removal. *Ecological Engineering*, 47, 126–131.
Yazdi, H., Alzate-Gaviria, L., & Ren, Z. J. (2015). Pluggable microbial fuel cell stacks for septic wastewater treatment and electricity production. *Bioresource Technology*, 180, 258–263.
Yu, Z., Leng, X., Zhao, S., Ji, J., Zhou, T., Khan, A., … & Li, X. (2018). A review on the applications of microbial electrolysis cells in anaerobic digestion. *Bioresource Technology*, 255, 340–348.
Zakaria, B. S., & Dhar, B. R. (2019). Progress towards catalyzing electro-methanogenesis in anaerobic digestion process: Fundamentals, process optimization, design and scale-up considerations. *Bioresource Technology*, 289, 121738.
Zamalloa, C., Arends, J. B., Boon, N., & Verstraete, W. (2013). Performance of a lab-scale bio-electrochemical assisted septic tank for the anaerobic treatment of black water. *New Biotechnology*, 30(5), 573–580.
Zhang, X., Li, C., Wang, Y., Luo, J., & Xu, T. (2011). Recovery of acetic acid from simulated acetaldehyde wastewaters: Bipolar membrane electrodialysis processes and membrane selection. *Journal of Membrane Science*, 379(1–2), 184–190.

Zhang, Y., Van der Bruggen, B., Pinoy, L., & Meesschaert, B. (2009). Separation of nutrient ions and organic compounds from salts in RO concentrates by standard and monovalent selective ion-exchange membranes used in electrodialysis. *Journal of Membrane Science*, 332(1–2), 104–112.

Zhang, Baogang, et al. (2009). Simultaneous removal of sulfide and organics with vanadium (V) reduction in microbial fuel cells. *Journal of Chemical Technology & Biotechnology*, 84(12), 1780–1786.

Zhang, L. J., Tao, H. C., Wei, X. Y., Lei, T., Li, J. B., Wang, A. J., & Wu, W. M. (2012). Bioelectrochemical recovery of ammonia–copper (II) complexes from wastewater using a dual chamber microbial fuel cell. *Chemosphere*, 89(10), 1177–1182.

Zhang, L., Xu, Z., & He, Z. (2020). Selective recovery of lead and zinc through controlling cathodic potential in a bioelectrochemically-assisted electrodeposition system. *Journal of Hazardous Materials, 386*, 121941.

Zhao, Y., Shi, W., Van der Bruggen, B., Gao, C., & Shen, J. (2018). Tunable nanoscale interlayer of graphene with symmetrical polyelectrolyte multilayer architecture for lithium extraction. *Advanced Materials Interfaces*, 5(6), 1701449.

Zhao, Y., Tang, K., Liu, H., Van der Bruggen, B., Díaz, A. S., Shen, J., & Gao, C. (2016). An anion exchange membrane modified by alternate electro-deposition layers with enhanced monovalent selectivity. *Journal of Membrane Science*, 520, 262–271.

6 Microbial Electrochemical Technologies for Decentralized Wastewater Treatment and Resource Recovery

Basem S. Zakaria, Tae Hyun Chung, Zexuan Ju, and Bipro Ranjan Dhar
University of Alberta, Edmonton, Canada

6.1 INTRODUCTION

Implementing extensive wastewater collection and treatment systems to provide efficient and cost-effective treatment has been challenging in many developing countries as well as remote communities in developed countries. The centralized wastewater collection and treatment systems are expensive to build and operate, particularly in remote low-density communities with scattered households. Therefore, centralized wastewater treatment facilities mostly cover urban areas in most parts of the world. In contrast, small communities are often forced to discharge wastewater into the environment without adequate treatment, posing significant environmental and public health risks. As a result, developing sustainable decentralized wastewater infrastructure is being progressively considered (Capodaglio et al., 2021; Van Afferden et al., 2015). In this regard, microbial electrochemical technologies (METs) have shown great potential toward revamping existing wastewater management practices with advanced and innovative engineered approaches by promoting a circular economy approach in decentralized communities.

METs deploy unique microbial biocatalysts, known as electroactive bacteria (EAB), to capture electrons from wastewater and divert them toward various value-added products. EAB form conductive biofilms on anode electrodes in METs, and electrons released from microbial oxidation of organics are transferred to the anode surface through extracellular electron transport (Dhar et al., 2016; Malvankar et al., 2012). The electrons are transported to the terminal electron acceptor on the cathode due to the potential gradient between the electrodes, which enables the capture of chemical energy stored in organics. The commonly used microbial electrochemical systems for domestic wastewater include microbial fuel cells (MFCs) and microbial electrolysis cells (MECs). Briefly, MFCs consist of anode and cathode chambers, often separated by membranes. The cathode chamber is aerated using external O_2 by bubbling the water or using an air cathode to act as an electron acceptor. Then, the electrons and protons could be transported through an external wire and membrane, respectively, to the cathode chamber to be used in oxygen-reduction-to-water reaction, and this process generates bioelectricity. The mechanism of MECs is similar to MFCs, except they are operated with an externally applied voltage/potential to enable other value-added products, such as hydrogen, methane, nutrients (fertilizer), and other platform chemicals (Al-Mamun et al., 2017; Barua et al., 2019; Chung et al., 2020; Zakaria and Dhar, 2019). METs have also been combined with conventional wastewater technologies, such as constructed wetlands (CWs), biofilters, and septic tanks, used for decentralized wastewater treatment (Bolton and Randall, 2019; Dhadwal et al., 2021; Verma et al., 2022). Moreover, MECs have been frequently combined with the

DOI: 10.1201/9781003381419-8

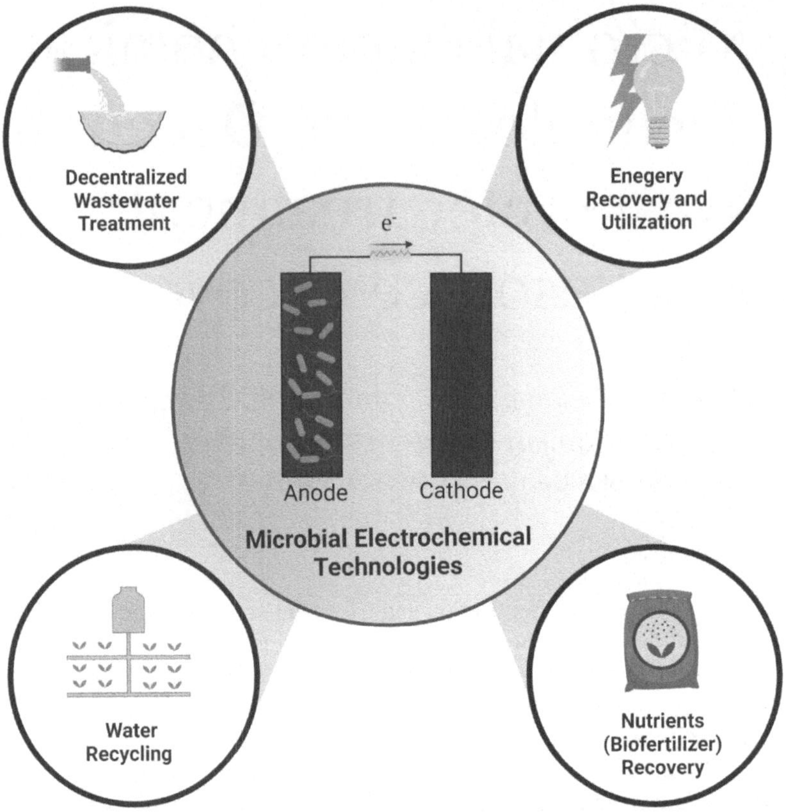

FIGURE 6.1 Different aspects of implementing microbial electrochemical technologies for wastewater treatment and resource recovery (created with BioRender).

conventional anaerobic digestion (AD) process for methane recovery from organic waste and high-strength wastewater (Huang et al., 2020; Zakaria and Dhar, 2019). Such approaches can potentially achieve energy-neutral or energy-positive wastewater treatment opportunities for remote communities. Besides, engineering METs with other physicochemical processes, such as membranes and chemical precipitation, have been explored for water reclamation and nutrient recovery (nitrogen, phosphorus, and potassium [NPK]) (Abourached et al., 2016; Monetti et al., 2021; Zinadini et al., 2017). Thus, METs can be low-cost wastewater solutions, which can offer exciting opportunities to secure water resources and to reduce the dependence on chemical fertilizer in remote communities (see Figure 6.1). More importantly, a few field-scale demonstrations of these approaches have already been reported in the literature (Jadhav et al., 2020; Sharma et al., 2021). This chapter provides an overview of various METs used for decentralized wastewater treatment with various resource recovery options. Moreover, current challenges and future research perspectives are highlighted at the end of this chapter.

6.2 WASTEWATER TREATMENT

To date, METs have been investigated for treating domestic and source-separated wastewater. Table 6.1 summarizes the key findings from relevant studies. The following subsections will provide detailed insights into the effectiveness and potential METs for decentralized wastewater treatment.

TABLE 6.1
Summary of Studies Using Microbial Electrochemical Technologies for Decentralized Wastewater Treatment and Resource Recovery

Wastewater	System Design	Key Findings	References
Graywater	MEC[a] followed by a biofilter	• 99.3% removal of TCOD[b] and 98.7% removal of anionic surfactants were achieved.	Dhadwal et al. (2021)
Graywater	MFC[c]	• A 77.6% removal of COD[d] was achieved. • A maximum power density reached 307.7 mW/m^2.	Sajithkumar and Ramasamy (2015)
Graywater	CW-MFC[e] followed by sand filtration	• A complete bacterial removal for a 4-log *Escherichia coli* influent load. • 99% COD[d] removal and a maximum power density of 4.33 mW/m^3 were achieved.	Bolton and Randall (2019)
Graywater	CW-MFC[e]	• A 98.5% removal of COD[d], 90.4% removal of NH_4^+–N, and 86.9% removal of NO_3–N were achieved. • A maximum power density reached 30.9 mW/m^3.	Tamta et al. (2020)
Graywater	CW-MFC[e]	• A 91.7% removal of soluble COD[d] was achieved. • A phosphate removal efficiency reached 56.3%. • A maximum power density reached 719.6 mW/m^3. • TSS[f] removal efficiency reached 78.4%.	Araneda et al. (2018)
Graywater	CW-MFC[e]	• COD[d] removal efficiency reached 93% at OLR[g] of 20.6 g COD/m^2-d. • NH_4^+-N removal efficiency reached 80%. • A maximum power density reached 11.6 mW/m^3.	Arvaniti and Fountoulakis (2021)
Blackwater	MEC[a]	• The methane generation increased gradually from 117.6 to 358.5 mL/L-d with the increase of OLR[g]. • A COD[d] removal efficiency reached 82.6% at low OLR[g] (0.77 g COD/L-d), while it decreased to 47.4% at higher OLR[g] (3.03 g COD/L-d).	Huang et al. (2021)
Blackwater	MFC[c]	• A removal of 65% of TOC[h] was achieved. • The maximum power density reached 384 mW/m^2.	Vogl et al. (2016)
Blackwater	MEC[a]	• >20% of COD[d] was removed. • The current density reached 9 A/m^3 at high TAN[i] concentration, while it decreased to 7 A/m^3 at low TAN[i]. However, the power density reached around 15 A/m^3 at all cases on day 33.	Deaver et al. (2022)
Blackwater	CW-MFC[e]	• The removal efficiencies of 89% of COD[d], 36% of total nitrogen, 70% of TAN[i], and 87% of ortho-phosphate were achieved. • 5-log unit reduction in coliform was achieved.	Talekar et al. (2018)
Blackwater	MFC[c]	• COD[d] removal reached 87% at HRT[j] of 36 h with final COD[d] effluent of 303 mg/L. • A maximum power of 61 mW was achieved.	Das et al. (2020)
Domestic wastewater	MFC[c]	• The system could efficiently remove >96% COD[d]. • The shock load of copper affected the MFC[c] performance.	Feng et al. (2013)
Domestic wastewater	MFC[c]	• A 94.5% COD[d] removal efficiency was achieved. • A maximum power density reached 142 mW/m^2.	Yazdi et al. (2015)
Mineral-rich wastewater	MFC[c]	• 70% of the COD[d], and 95% of nitrogen and phosphorus were removed. • A maximum power reached 2.08 mW.	Kim et al. (2021)

(Continued)

TABLE 6.1 (*Continued*)
Summary of Studies Using Microbial Electrochemical Technologies for Decentralized Wastewater Treatment and Resource Recovery

Wastewater	System Design	Key Findings	References
High-salinity mustard tuber wastewater	MFC[c]	• 86% of carbon and 97% of nitrogen were removed. • 8.3 W/m^3 of power density was achieved.	Zhang et al. (2020)
Domestic wastewater	CW-MFC[e]	• 81% of COD[d] and 98% of ammonia-nitrogen have been removed. • A maximum power density reached 8.6 mW/m^2.	González et al. (2021)
Domestic wastewater contaminated with azo dye	CW-MFC[e] amended with GAC[k]	• The highest removal of COD[d] reached 85.7%, and azo dye reached 91.2%. • A maximum power density of 0.302 W/m^3.	Fang et al. (2013a)
Urban wastewater	CW-MFC[e] coupled with anaerobic biofilter, primary settler, and reactive bed filter	• The integrated system efficiently removed the COD[d] (65%) and BOD[l] (82%) with nitrogen removal of 93%. • The system achieved a maximum power density of 0.33 mW/m^2.	Colares et al. (2022)
Domestic wastewater	MFC[c] coupled with septic tank primary treatment and disinfection unit	• The COD[d] and BOD[l] were removed by 86% and 87%, respectively. • 84% of total nitrogen and 64% of P were removed. • A maximum power reached 136.6 mW.	Linares et al. (2019)
Domestic wastewater	MEC[a]	• The system efficiently achieved 36% of TCOD[b]. • 0.015 L H_2/L-day was recovered.	Heidrich et al. (2013)
Domestic wastewater	MEC[a]	• At continuous mode, the COD[d] removal efficiency reached 33% at HRT[j] 4 h and 87% at HRT[j] 24 h. • A maximum of 0.31 kWh/kg-CODr at HRT[j] 4 h, and 0.69 kWh/kg-CODr at HRT[j] 24 h. • Methane generation reached ~0.06 L/L-d at HRT[j] 8 h.	Moreno et al. (2016)
Domestic wastewater	MEC[a]	• A 92% of COD[d] was removed. • 223 mL H_2 was generated.	Escapa et al. (2015)
Domestic wastewater	MEC[a] coupled with anaerobic fluidized bed membrane bioreactor	• 92.5% of TCOD[b] was removed at HRT[j] 9 h. • Complete removal of TSS[f] was achieved from 45 to <1 mg/L. • An electrical energy produced 0.0197 kWh/m^3.	Ren et al. (2014)
Fruit processing wastewater	MFC[c]	• An economic analysis was done for applying MFCs[c] as a pre-irrigation treatment method. • MFC's[c] total capital and operation costs (20 years) were estimated at 9% of the conventional aerobic treatment costs.	Abourached et al. (2016)
Dairy wastewater	MFC[c]-membrane process system (MPS)	• A low-energy water reuse system was developed combining an MFC with membrane separation system with antifouling nanofiltration and ultrafiltration membranes. • The process could remove 99% COD[d].	Zinadini et al. (2017)
Synthetic wastewater	Osmotic microbial fuel cell (OsMFC)	• OsMFC[m] was developed by using a forward osmosis (FO) membrane. • Produced a maximum power density of 2.39 W/m^3.	Zhang et al. (2011)

(*Continued*)

TABLE 6.1 (*Continued*)
Summary of Studies Using Microbial Electrochemical Technologies for Decentralized Wastewater Treatment and Resource Recovery

Wastewater	System Design	Key Findings	References
Source-separated urine	Bio-electro-concentration cells (BECs)	• The system achieved nitrogen recovery rate up to 4.9 kg N/m^3-d with a specific energy input of 4.1 kWh/kg-N. • The liquid fertilizer from the system contained 21.2–23.5 g nitrogen/L, 1.1 g P/L, and 5.4–8.9 g K/L.	Monetti et al. (2021)
Source-separated urine	Electrodialysis integrated within an electrochemical membrane bioreactor (EDMBR)	• The system could provide NH_4^+ and PO_4^{3-} removal efficiencies up to 97.4 and 76.7%, respectively. • The system achieved a maximum power density of 23.5 W/m^3.	Wang et al. (2017)
Source-separated urine + glucose	7-chamber urine-powered microbial nutrient recovery cell	• The average organics, nitrogen, and phosphorus removal efficiencies were 93.8, 73.1, and 86.2%, respectively. • Recovered nitrogen and phosphorus concentrations reached up to 1234 and 101 mg/L, respectively. • The maximum power density was 21.3 W/m^3.	Gao et al. (2018)
Source-separated urine	MEC connected to a membrane-aerated biofilm reactor (MABR)	• MEC provided 85% COD[d] and 100% urea removal from fresh urine. • The system achieved nitrogen recovery rate up to 700 mg N/L-d.	De Paepe et al. (2020)
Source-separated urine	Three-stage reactor with stacked MFCs[c] with a stirred tank reactor	• A three-stage-system incorporating a total of 10 MFC[c] units was tested for on-site treatment of source-separated urine. • 75.55% COD[d] removal was achieved. • The maximum power density was 14.5 mW/m^2.	Sharma et al. (2021)

[a] MEC: Microbial electrolysis cell.
[b] TCOD: Total chemical oxygen demand.
[c] MFC: Microbial fuel cell.
[d] COD: Chemical oxygen demand.
[e] CW-MFC: Constructed wetland-microbial fuel cell.
[f] TSS: Total suspended solids.
[g] OLR: Organic loading rate.
[h] TOC: Total organic carbon.
[i] TAN: Total ammonia-nitrogen.
[j] HRT: Hydraulic retention time.
[k] GAC: Granular activated carbon.
[l] BOD: Biological oxygen demand.
[m] OsMFC: Osmotic microbial fuel cell.

6.2.1 Graywater

Graywater refers to the wastewater generated from baths, washing machines, dishwashers, and kitchen sinks, contributing to 60–70% of domestic wastewater (Dhadwal et al., 2021; Friedler, 2010). Efforts have been made to treat graywater via several biological technologies, such as AD, microbial electrochemical system, and up-flow anaerobic sludge blanket (UASB), which achieved an efficient wastewater treatment and generate useful bioproducts such as H_2/CH_4 gases, chemicals, and electricity (Dhadwal et al., 2021; Leal et al., 2010). Typically, graywater contains various contaminants, such as heavy metals, fats, oil and grease, and surfactants (Dhadwal et al., 2021; Shaikh and Ahammed, 2020). The low-strength and biological contaminants of graywater make it attractive for water recycling, reuse, and reclaiming where the population faces acute water shortages (Shaikh and Ahammed, 2020). The reuse of graywater in flushing toilets could save up to 30% on daily water consumption in the household (Dhadwal et al., 2021). METs can provide attractive simultaneous organic removal and energy generation approaches with the low-strength characteristics of graywater (Dhadwal et al., 2021).

Several approaches have been reported in the literature on combining METs with traditional graywater treatment systems (e.g., CWs and biofilters) that could provide efficient removal of organic matters, pollutants, suspended solids, and surfactants (Bolton and Randall, 2019; Dhadwal et al., 2021; Sajithkumar and Ramasamy, 2015; Tamta et al., 2020). A recent study used an integrated system of MEC and granular activated carbon (GAC) biofilter for the treatment of synthetic graywater (Dhadwal et al., 2021). Their system achieved a removal efficiency of 99.3% chemical oxygen demand (COD) and 98.7% of anionic surfactants at a hydraulic residence time (HRT) of 25 h with a maximum volumetric current density of 0.65 A/m^3 at HRT of 1 day. Another study explored the efficiency of an MFC system for graywater treatment and achieved a COD removal efficiency of 77.6% and a maximum power density of 307.69 mW/m^2 (Sajithkumar and Ramasamy, 2015).

Moreover, integrating MFCs with CWs (CW-MFCs) could improve COD removal efficiency with electricity production (Bolton and Randall, 2019; Tamta et al., 2020). A combined system of sand biofilter with a CW-MFC system provided an efficient removal of COD of 99% from graywater at an HRT of 2.2 days (Bolton and Randall, 2019). Also, Tamta et al. (2020) designed a system of 12 vertically arranged CW-MFC systems, which provided a COD removal efficiency of 98.5% with a maximum power density of 30.85 mW/m^3. In contrast, a few studies showed no significant differences in COD removal efficiencies of CW-MFC compared to the control CW (Araneda et al., 2018; Arvaniti and Fountoulakis, 2021). However, these studies showed that integrating CW with MFC allows energy recovery while providing effective graywater treatment. Araneda et al. (2018) explored the decentralized graywater treatment using the CW-MFC system for 152 days. Despite the fact that similar COD removal efficiency compared to control CW (91.7 vs. 90%), the system achieved a maximum power density of 719.57 mW/m^3.

6.2.2 Blackwater

Blackwater from household toilets contains about 70% of the organic matter and less than 30% of the volume (Bose et al., 2021; Huang et al., 2021, 2020). AD is considered an attractive treatment technology for blackwater due to its high strength (>10 g COD/L), high levels of organic matter (50–70%), high nutrient proportion (80%), and small volume (<30%) (Bose et al., 2021; Huang et al., 2021, 2020). However, blackwater contains large amounts of complex organics, including fiber, protein, carbohydrates, fats, and biomass which results in lower biodegradability performance (Rose et al., 2015; Vogl et al., 2016). The effective AD of blackwater is mostly feasible under low organic loading rates (OLRs) of <1 g COD/L-d (de Graaff et al., 2010; Gao et al., 2019; Tervahauta et al., 2014). A recent study explored MEC-assisted anaerobic digester (MEC-AD) at a wide range of OLRs (0.77–3.03 g COD/L-d) for 420 days (Huang et al., 2021). A high COD removal efficiency (82.6%) was achieved at a low OLR of 0.77 g COD/L-d, while the removal efficiency dropped to 47.4% at a higher OLR of 3.03 g COD/L-d. However, the methane generation rate increased gradually from 117.6 to 358.5 mL/L-d with an increase of OLR from 0.77 to 3.03 g COD/L-d (Huang et al., 2021). The authors suggested that most of the remained effluent total COD (TCOD) was non-biodegradable and would require further post-treatment using aerobic or physicochemical processes. The successful treatment of complex compounds in blackwater using MEC-AD might be because of the syntrophic interactions between fermentative bacteria and hydrogenotrophic methanogens (Huang et al., 2021). Stand-alone MFC systems have also been explored for blackwater treatment. A single-chamber MFC system with a gas diffusion cathode showed total organic carbon (TOC) removal of 65% with an energy recovery of 384 mW/m^2 (Vogl et al., 2016). Das et al. (2020) demonstrated a field-scale stacked MFC system (consisting of six MFC modules with a total volume of 720 L) for the on-site treatment of waste generated from toilets. The COD removal reached 87% at HRT of 36 h with final COD effluent of 303 mg/L and a maximum power of 61 mW. This power was used to illuminate the toilet premises at night.

6.2.3 Urban and Domestic Wastewater

Domestic sewage is any type of wastewater generated from households or office buildings, including black and graywater (Paulo et al., 2013). However, urban wastewater refers to a mixture of domestic and industrial wastewater (if allowed to discharge to the sewers) and/or stormwater (Katuri et al., 2019). Due to urbanization, industrial organizations expanded to build their facilities in urban areas, which led to the discharge of industrial wastewater into the public sewer. Most notably, various industrial wastewater contains high COD concentrations with complex organic matter and large amounts of toxic compounds (Kanu and Achie, 2011; Mousazadeh et al., 2021). Commonly used technologies for decentralized wastewater sewage are septic tanks, which mainly lack efficient treatment for nutrients, micropollutants, and pathogenic organisms (Withers et al., 2014). However, several approaches have been explored to develop systems for harvesting bioenergy while treating wastewater (Feng et al., 2013; Katuri et al., 2019; Kim et al., 2021).

Based on previous studies, MFCs could simultaneously remove carbonaceous and nitrogenous pollutants (Feng et al., 2013; Yu et al., 2011). For instance, Feng et al. (2013) designed a bench-scale MFC for synthetic wastewater; acetate and ammonium as sole carbon and nitrogen sources, respectively. The system could efficiently remove >96% COD with a 60–80% reduction of total nitrogen. Also, this study explored the shock load of different concentrations of copper (Cu) (25–125 mg/L) as a heavy metal. Although the treatment efficiency was affected by the load of Cu at the beginning of the operation, the microbial community restored its ability to treat wastewater after around 30 days. Yazdi et al. (2015) examined the performance of an individual and stacked MFCs on synthetic wastewater (acetate) for decentralized treatment. Their results demonstrated that parallel connections of stacked MFC units showed higher performance in terms of wastewater treatment and electricity production. A recent study designed a decentralized system for synthetic mineral-rich wastewater treatment (ammonium dihydrogen phosphate) to simulate the wastewater obtained from agricultural waste for the decomposition of organic substances and resource recovery (struvite) (Kim et al., 2021). The MFC module (1 L) was coupled with a struvite precipitation tank (0.16 L). More than 70% of the organics and 95% of N and P have been removed. Thus, MFC systems can be considered efficient for organic removal and resource recovery to realize the circular economy in agriculture and livestock wastewater treatment. Another recent study demonstrated simultaneous power generation and carbon and nitrogen removal using a single-chambered MFC module (Zhang et al., 2020). This study treated a high-salinity mustard tuber wastewater with carbon and nitrogen removal efficiencies of 86% and 97%, respectively.

CWs are considered one of the efficient systems for low-cost wastewater treatment (Kivaisi, 2001). As discussed earlier, CW-MFCs are one of the efficient methods for the simultaneous treatment of wastewater and harvesting of bioenergy (Wang et al., 2016; Xu et al., 2018). A recent study explored the efficiency of the CW-MFC system planted with *Schoenoplectus californicus* for synthetic domestic wastewater treatment (glucose and acetate) (González et al., 2021). The COD removal efficiency reached 81% compared to the control CW of 72%. Also, the ammonia-nitrogen was removed >98% compared to 83% of the control CW. The maximum power density reached 8.6 mW/m^2. Also, a study showed that the integrated CW-MFCs system is limited not only to better COD removal efficiency of synthetic domestic wastewater (glucose) by around 13% but also to azo dye (150 mg/L) decolorization by 15% compared to the control CW (Fang et al., 2013a). In this study, the addition of GAC anode electrode adsorbed the azo dye and subsequently degraded by the anodic biofilm attached to the GAC while the cathode electrode degraded the azo dye by-products. Also, the electricity generation reached a maximum power density of 0.302 W/m^3. Another study demonstrated a pilot scale of CW-MFCs integrated with other treatment technologies (anaerobic biofilter, primary settler, and reactive bed filter) for efficient treatment of urban wastewater (Colares et al., 2022). The integrated system efficiently removed the COD (65%) and biological oxygen demand (BOD) (82%) with nitrogen removal of 93%. Also, the bioelectricity generation reached a maximum power density of 0.33 mW/m^2.

A few pilot scale MET studies (100–175 L) for domestic wastewater treatment have been reported. Heidrich et al. (2013) designed a 120 L of MEC system for the on-site treatment of domestic wastewater in Northern England with a volumetric OLR of 0.14 kg COD/m^3-d. The system efficiently achieved 36% of TCOD, reaching the UK discharge standard of 125 mg COD/L. Also, the system recovered around 0.015 L H_2/L-day. Another study designed a full-scale treatment of real domestic wastewater by adding a primary septic tank (primary treatment) and gravity-driven disinfection unit (tertiary treatment) before and after the MFC module (secondary treatment) (Linares et al., 2019). In this study, the secondary treatment consists of 18 MFC modules with an energy-harvesting circuit unit. The COD and BOD were removed by 86 and 87%, respectively. Also, this system showed a high performance in nutrient removal; total nitrogen (N) and phosphorus (P) were removed with 84 and 64%, respectively.

6.3 ENERGY RECOVERY AND ON-SITE UTILIZATION

To promote METs as low-cost decentralized treatment options, capturing and using renewable energies from wastewater would be vital. For instance, biomethane produced from METs can be used to generate heat and electricity and can be used on-site. In fact, many rural communities in developing countries (e.g., India, Nepal, Bangladesh, and Vietnam) use small-scale anaerobic digesters to produce biogas from household waste for cooking (Surendra et al., 2014). Furthermore, biohydrogen can be converted into electricity using fuel cell technology. A few recent studies suggested using off-grid hydrogen-based renewable energy systems for rural communities (Ayodele et al., 2021; Luta and Raji, 2018). As shown in Table 6.1, most studies on decentralized wastewater treatment with METs considered using MFCs, where energy will be recovered as electricity. A wide range of power densities (as mW/m^2 or mW/m^3) has been reported in previous studies (see Table 6.1). Although it is often believed that electricity produced from a single MFC unit would be inadequate for a practical application like directly powering electronic devices, recent studies demonstrated several approaches for the on-site use of electricity from stacked MFCs (Jadhav et al., 2020; Verma et al., 2022; Walter et al., 2020).

The electricity generated from MFCs has been successfully used to light electric bulbs, charge mobile phones, and operate small electronic gadgets and sensors (Jadhav et al., 2020; Mukherjee et al., 2022; Walter et al., 2020). For instance, Walter et al. (2020) developed two separate stacks of MFCs (one with 15 units and another with 18 units) operated by urine. These MFC stacks were connected with LED spotlights. After 20 h of inoculation, the stacks could generate adequate energy to directly power the lighting system. Furthermore, Walter et al. (2017) developed a stack of six MFCs connected in series, which could produce adequate electricity to provide 105 min of call time for every 3 h of charging cell phones while consuming only 600 mL of urine every 6 h. These results indicated that the MFCs could be implemented as an off-grid power source in decentralized areas. However, an advanced power management system (PMS) would be necessary for the effective management and on-site utilization of electricity produced by MFCs (Jadhav et al., 2020; Kim et al., 2019; Verma et al., 2022). Briefly, PMS stores the charge produced by individual MFC units in a stack, amplifies, and releases it on an on-demand basis (Figure 6.2). Various electrical components of PMS include charge pumps, capacitors, boost converters, charge control circuits, rectifiers, and regulators (Jadhav et al., 2020; Kim et al., 2019; Mukherjee et al., 2022; Verma et al., 2022). Each component has its function. For instance, capacitors and boosters can take care of low current/charge

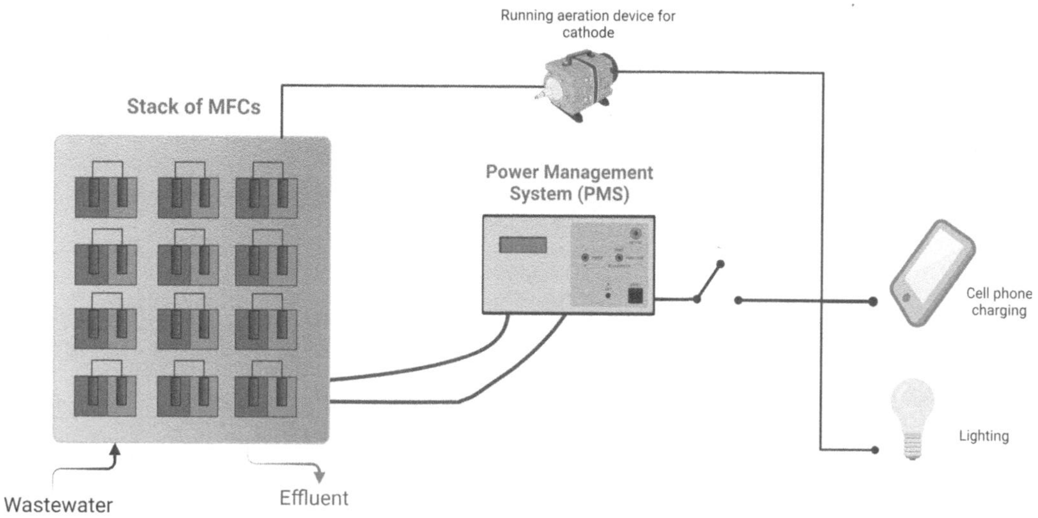

FIGURE 6.2 A schematic representation of applying a power management system (PMS) for on-site utilization of electricity produced by a stack of microbial fuel cells (created with BioRender).

generation and its fluctuations in MFCs over time (Jadhav et al., 2020). Recently, Srivastava et al. (2021) demonstrated a self-automated CW-MFC system, where a PMS was applied for harvesting and storing low power generated by the CW-MFC and used to intermittently operate an air pump in the cathode as a part of the treatment process.

6.4 WATER RECYCLING

Reclamation and reusing wastewater are increasingly viewed as a practical approach for better managing and conserving water resources in remote communities and/or communities with limited freshwater resources. For instance, relying on groundwater sources for domestic and agricultural water-related activities, large withdrawals, and decreasing recharge rates threaten the sustainability of groundwater sources. In this context, a few studies demonstrated the effectiveness of METs for water reclamation (Abourached et al., 2016; Bolton and Randall, 2019; Jadhav et al., 2020; Zinadini et al., 2017), which can potentially decrease water demand in urban areas, as well as remote communities with limited freshwater resources. Abourached et al. (2016) performed an economic analysis of using MFC as a method for fruit processing wastewater recycling to non-potable water for irrigation purposes. Based on their research, over 20 years, MFC's capital and operating costs would be \$6.4 million, which would be as high as \$68.2 million for conventional aerobic treatment. Their assessments also suggested that water recycling with MFCs can provide an attractive cost-saving associated with irrigation for farmers in semi-arid regions.

However, wastewater reuse has challenges associated with pathogens and other emerging contaminants (e.g., micropollutants) removal, which is critical for mitigating environmental and public health risks. Previous studies suggested that integrating METs with other physicochemical processes might be necessary to overcome these challenges (see Figure 6.3). For instance, Zinadini et al. (2017) reported that integrating advanced membrane technologies (e.g., nanofiltration) would be needed to produce hygienic water from wastewater using MFCs. In their study, adding nanofiltration could completely remove total coliform and fecal coliform. In contrast, the effluent from MFCs contained high total coliform and fecal coliform (≥1600 most probable number [MPN]). Furthermore, the combined treatment could provide 99.7% of COD removal as compared to a stand-alone MFC system. Moreover, Bolton and Randall (Bolton and Randall, 2019) proposed a CW-MFC followed by the biological sand filter (BSF) for recycling graywater to potable water,

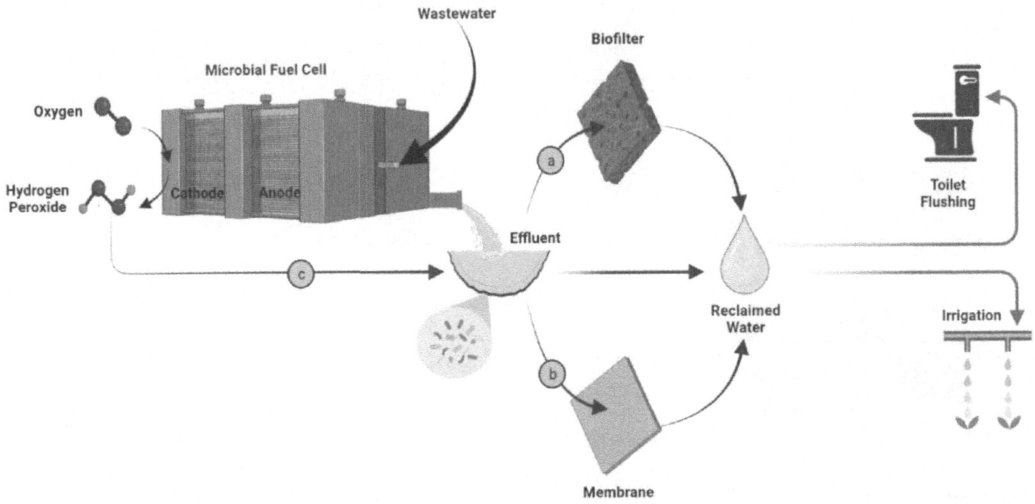

FIGURE 6.3 Recycling wastewater for toilet flushing and irrigation using a microbial fuel cell combined with (a) biofilter, (b) membrane, and (c) disinfection using hydrogen peroxide produced on the cathode (created with BioRender).

complying with the South African standards for drinking water quality. Notably, their proposed system could achieve 4-log removal of *Escherichia coli* with 99% COD removal. However, stand-alone METs have been successfully demonstrated for the on-site production of chemical oxidants, such as hydrogen peroxide, for the potential wastewater disinfection application (Chung et al., 2020). For instance, Arends et al. (2014) proposed H_2O_2 production in MFCs to disinfect wetland effluent. The produced H_2O_2 (0.1%) could successfully reduce the total bacterial counts and total coliforms to less than <75 colony forming unit (CFU) mL. Also, a recent review article suggested that MFCs, as a stand-alone process, can effectively degrade a wide range of antibiotic compounds found in wastewater (Zakaria and Dhar, 2022). Thus, these findings suggest that stand-alone microbial electrochemical systems can still be engineered and deployed for recycling water for safe reuse.

A few studies also demonstrated osmotic MFCs (OsMFCs) for simultaneous wastewater treatment, removal of salts from saline water to recycle water, and energy recovery (Al-Mamun et al., 2017; Zhang et al., 2011). The OsMFCs use the synergy between METs and the forward osmosis (FO) process. FO is an osmotically driven, membrane-based separation process in which water molecules pass from a diluted feed solution to a more concentrated draw solution through a synthetic polymeric membrane (Das et al., 2019). The driving force is the potential chemical difference between the concentrated draw solution and the less concentrated feed solution with lower osmotic pressure. Compared to the pressure-driven reverse osmosis (RO) process, FO requires no hydraulic pressure to recover water. Eliminating hydraulic pressure saves substantial energy, giving FO an edge over the RO process. A few recent studies have demonstrated that FO can effectively recycle graywater (Wang et al., 2018; Xiao et al., 2018). However, the remaining concentrates from the process need to be post-treated to remove organics, surfactants, and pathogens. Additionally, separation or regeneration of draw solute would be required, which is an energy-demanding process. Thus, further developments of MET-based systems are needed to provide an effective decentralized water system in coastal regions.

6.5 NUTRIENT RECOVERY

To date, a few studies investigated METs to recover nutrients, such as nitrogen (N), phosphorous (P), and potassium (K) from wastewater (e.g., source-separated urine), which can potentially be used as biofertilizer and reduce the dependence of chemical fertilizer (De Paepe et al., 2020;

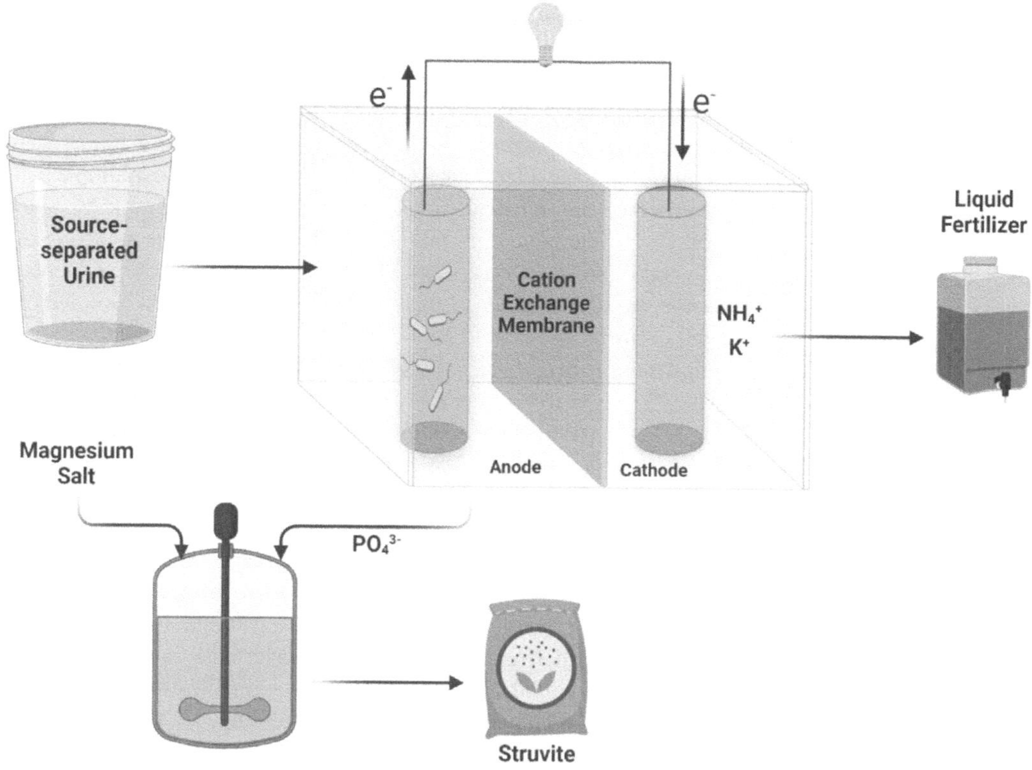

FIGURE 6.4 A schematic representation of NPK recovery using a microbial fuel cell (created with BioRender).

Monetti et al., 2021; Sharma et al., 2021). Figure 6.4 shows a typical approach to integrating METs with struvite precipitation for N, P, and K recovery. However, researchers have proposed various advanced systems over the years. For instance, Monetti et al. (2021) fabricated three bio-electro-concentration cells (BECs) by combining MECs with an electrodialysis process (three compartments separated by cation-exchange and anion-exchange membranes) to investigate their performance on nutrient recovery. Specifically, their BEC systems designed with catholyte recirculation provided additional buffering capacity for anode biofilm and augmented the current output and proton production, which led to local pH drops for promoting ammonium production. Overall, their study showed a very high N recovery efficiency, achieving a volumetric N recovery rate of 4.9 ± 0.1 kg-N/m^3/d and a specific energy consumption of 4.1 ± 0.0 kWh/kg-N. The energy input of BECs was superior compared to other N-removal methods, such as the two-stage nitrification-denitrification process (12.5 kWh/kg-N), anaerobic ammonia oxidation process (Anammox®; 4.2 kWh/kg-N) (Kuntke et al., 2018; Lackner et al., 2014), and the Haber-Bosch process for industrial-scale ammonia synthesis (10–12.5 kWh/kg-N) (Christiaens et al., 2017; Xie et al., 2014).

An *in situ* nutrient recovery system has been developed by Wang et al. (2017) for urine treatment where they integrated the electrodialysis process with an electrochemical membrane bioreactor (EDMBR). With such a unique system, phosphate and sulfate were selectively recovered while simultaneously generating electricity. Additionally, enhanced biological nutrient removal was achieved by their EDMBR system, which was capable of removing 75.5–85.4% of total N and P from real wastewater and source-separated urine amended with acetate. The authors further highlighted that the acetate residual in the anode and the amount of electrons and dissolved oxygen gradient in cathode biofilm played a role in ensuring real-time nitrification and denitrification, which contributed to the more efficient removal of total N and ammonia. Moreover, their EDMBR system was capable of performing desalination to target various mineral ions (Na^+, K^+, Ca^{2+}, Mg^{2+}, and Cl^-) in urine.

Gao et al. (2018) developed a seven-chamber urine-powered microbial nutrient recovery cell (called U-power) for *in situ* nutrient recovery. The system consisted of an anode chamber filled with GAC, a cathode chamber, two dilution chambers, and three recovery chambers, which were separated by cation-exchange/anion-exchange membranes. The implementation of multiple ion-exchange resins in their system promoted the transfer of ions and decreased the internal resistance (e.g., increase in current generation boosting the ion migration), which ultimately contributed to the higher removal of total N (73.1%) and P (86.2%) along with COD (93.8%) in urine without the need for external energy input. Moreover, the performance of their system was quite robust as it demonstrated high recovery and removal performances for 12 h.

De Paepe et al. (2020) demonstrated a two-stage process combining an MEC connected to a membrane-aerated biofilm reactor (MABR) to convert real urine into a reliable nutrient-rich and COD-low nutrient solution. First, MEC in the first stage removed the incoming biodegradable COD, and then the membrane-aerated nitrification took place to remove nitrogen. After determining the optimal configuration of their MEC-MABR (e.g., high pH environment), it removed a substantial amount of total N (96–98% removal; all total ammonia-nitrogen converted to nitrate), and COD (>90%) in four to seven days of hydraulic retention time. This integrated biosystem was unique as it suppressed the denitrification by removing the biodegradable COD using anodic bio-oxidation in MEC in the first stage of the treatment process (e.g., less COD was available to heterotrophic bacteria responsible for denitrification). Furthermore, the formation of hydroxide ions (OH^-) in the cathode chamber led to a further increase in alkalinity for optimizing the nitrification process.

Recently, Sharma et al. (2021) reported a field demonstration of a three-stage MFC system with a total capacity of 10 L. The system was integrated with a urinal for energy and nutrient recovery from real urine. The two MFC units in the first stage produced effluent with high pH suitable for nutrient recovery as struvite in the second stage consisted of a stirred tank reactor. The third stage consisted of an MFC stack (eight units) to treat supernatant urine from the second stage with energy recovery. The system achieved a COD removal efficiency of 75.55% and a power density of up to 14.5 mW/m^2. The recovery efficiencies of ortho-phosphates and struvite were 90 and 46%, respectively.

In summary, MET systems can be used for decentralized treatment systems to achieve sustainable development via energy and resource recovery. P is generally recovered via struvite formation, and N is recovered from a membrane-assisted system. Innovations such as on-site production of base solution or recirculation of anodic effluent to cathodic tend to be the feasible way to adjust the pH of the solution to boost the alkalinity and current density. This often leads to full nitrification in the second stage and the enhancement of N recovery, making the system performance more efficient, in terms of chemical process and energy. The ability to handle the fluctuation of source intake urine is vital to the system's operation stability. If the hybrid system is incompetent to serve a large loading or perform adequate nutrient recovery or COD removal, the energy contained in the waste/wastewater, such as urine, would not be used, and an excessive amount of COD would flow downstream, resulting in downstream fouling and posing challenges to the subsequential treatment process. In most cases, synthetic urine was used during testing. Further studies should be conducted with real urine to justify the system's strength in processing complex intake sources. It's also necessary to further investigate the optimum operation conditions to escalate the recovery efficiencies and try to couple with other devices to reclaim more nutrients.

6.6 CHALLENGES IN THE PROCESS

Despite the potential benefits of using METs for decentralized wastewater treatment, their field application remains challenging. Previous studies (Dhar et al., 2016; Hiegemann et al., 2016; Liang et al., 2018; Sim et al., 2018) have made significant efforts to design scalable MET systems, but there is still a substantial gap between small-scale and larger scale systems, in terms of current/power density and recovery rate of value-added products. Specifically, the challenges of determining optimal electrode spacing, high internal resistance, and high energy input are the main arguments against the commercialization and scaling up of METs (Chung et al., 2020; Chung and Dhar, 2021a; Zakaria and Dhar, 2019). Moreover, microbiological restrictions (contaminant degradation, biochemical and electron transfer pathways), material cost and efficiency, and fluid dynamics (mass transfer, homogeneity of fluid flow, mixing intensity, etc.) also need to be considered when scaling up METs (Jadhav et al., 2020; Katuri et al., 2018; Liu and Cheng, 2014). Overall, the commercialization of METs is also very difficult considering their robustness, reproducibility, and system long-term stability (Katuri et al., 2018; Liu and Cheng, 2014; Pant et al., 2011). Therefore, developing a scalable process design and configuration is crucial for advancing decentralized wastewater-fed METs (Jadhav et al., 2020; Zakaria and Dhar, 2019).

Many research studies have combined METs with other technologies to improve wastewater treatment efficiency and recover valuable resources. Still, it is difficult to determine the best integrated MET platform for decentralized wastewater treatment, as this depends on the target contaminants or resources. Modifying or adding to the integrated MET system may result in more complicated treatment processes, requiring skilled labor, occupying more land, and increasing capital and operational costs. Therefore, a more adaptable and versatile modular MET design is needed for efficient resource recovery and decentralized wastewater treatment. For scaling up, stacking multiple MET units (connected in series or parallel) can be a feasible option over the simple enlargement of reactor size (Aiken et al., 2019; Jadhav et al., 2020). Increasing the reactor size of MET leads to many challenges, such as increased internal resistance and mass transfer limitation, leading to decreased system performance (Chung et al., 2021; Zakaria and Dhar, 2019). However, Aiken et al. (2019) indicated that stacked MFCs often experience voltage reversal, which could damage the bioanode, but this operational issue is prevented in MEC mode due to the applied voltage (Li et al., 2017).

As discussed previously, MEC mode generates various higher value products, such as hydrogen, methane, and other chemicals, whereas MFC is limited to bioelectricity production only (Gil-Carrera et al., 2013). Thus, MEC mode could be a more attractive option over MFC mode. Aiken et al. (Aiken et al., 2019) found that the cost of MET materials (electrodes contributing to 94% of overall fabrication cost) could be reduced by 90% by conducting a series of case studies and comparing different working scenarios (membrane and electrode replacement vs. applied voltage reduction vs. anode/current collector value return, etc.). Such a finding could be adopted for METs treating decentralized wastewater on a field scale. Verma and their team recently reviewed past published reports related to MFC practicability and summarized how MFC could practically contribute to typical sanitation practices, such as bioelectric toilets, green latrines, and pee-power urinals (Verma et al., 2022). These practical applications of METs could greatly assist developing countries and remote areas with high dependency on decentralized wastewater treatment plants.

Emerging manufacturing and computational tools can provide novel solutions for overcoming the abovementioned challenges (Chung and Dhar, 2021a–c; Chung et al., 2021). For instance, the application of the 3D-printing (3DP) technique enables more precise manufacturing of METs and their parts (reactor body, electrodes, bio-electrodes, chambers, and ion-exchange membranes), which could assist their fabrication on a larger scale (Chung and Dhar, 2021a; Theodosiou et al., 2020; You et al., 2017). To date, an increase in the system performance of METs was observed (biosensing capability, current/power generation) when they were fabricated via 3DP compared to other conventional subtractive manufacturing methods (Bian et al., 2018a,b; Quaglio et al., 2019). Considering the practical applications of decentralized wastewater-treating METs, 3DP could also offer time-efficient and labor-free manufacturing with minimal waste generation (Chung and Dhar, 2021a; Theodosiou et al., 2020).

Decentralized wastewater treatment and resource recovery using METs pose significant challenges due to their highly complex, time-variant, and non-linear nature (Chung and Dhar, 2021b; Fang et al., 2013b; Jadhav et al., 2020; Yewale et al., 2020). To date, we heavily rely on experimental research for optimizing METs, which is very time-consuming and laborious (Chung et al., 2021; Yewale et al., 2020). To overcome this challenge, machine learning (ML) algorithms such as artificial neural networks and deep learning can be implemented to model the relationships between various MET working conditions and their system performance. Recent successful applications of ML on METs have optimized their performance by identifying relationships among wastewater characteristics, biofilm communities, functional stability of the microbial community, feedstock types, and current output, among others (Cai et al., 2019; Lesnik et al., 2019; Lesnik and Liu, 2017; Tsompanas et al., 2019). Although few published reports on the use of METs on decentralized wastewater exist, ML models can still be applied with a small dataset to support the field application of METs (Tsompanas et al., 2019). Therefore, implementing ML algorithms can help optimize the design and operating parameters in METs for decentralized wastewater treatment and resource recovery.

6.7 CONCLUSIONS AND PERSPECTIVES

As the world recognizes the need for wastewater treatment solutions that are both environmentally sustainable and economically viable, decentralized wastewater systems have become more popular. However, we still need to find ways to make these systems more efficient for practical applications. In this context, METs offer a cost-effective and proficient alternative to conventional wastewater treatment methods and could be the key to solving our wastewater treatment challenges and protecting water resources and the environment. Various applications of METs on decentralized wastewater treatment, such as graywater, blackwater, and urban and domestic wastewater, were comprehensively reviewed in this chapter. Moreover, METs for recovering value-added products and recycling water from decentralized wastewater were discussed. However, several challenges are limiting the application of METs on decentralized wastewater treatment on a field scale, and currently, METs greatly need more practicality. Past research endeavors have aimed at designing scalable MET systems. However, a considerable gap persists between small-scale and larger scale systems concerning current/power density and value-added product recovery rates. Factors such as optimal electrode spacing, high internal resistance, and energy input present substantial hurdles against MET commercialization and scaling up. Microbiological constraints, material costs, efficiency, and fluid dynamics also pose obstacles to MET scaling. Robustness, reproducibility, and long-term stability further complicate the commercial viability of METs.

Overcoming these aforementioned challenges necessitates innovative approaches, such as innovative system designs, cost reduction strategies, and integration of advanced manufacturing and computational tools. For instance, developing a modular MET design that is adaptable and versatile is advocated, potentially involving stacking multiple MET units in series or parallel configurations. 3D printing can enhance manufacturing precision and efficiency, making MET fabrication on a larger scale more feasible. Meanwhile, incorporating ML algorithms can offer the potential to model the complex relationships between MET performance and various design parameters, optimize their performance, and aid in their practical implementation. By addressing obstacles toward field application, METs could emerge as valuable tools in addressing the challenges of decentralized wastewater treatment, particularly in regions with limited resources and infrastructure.

REFERENCES

Abourached, C., English, M.J., Liu, H., 2016. Wastewater treatment by microbial fuel cell (MFC) prior irrigation water reuse. J. Clean. Prod. 137, 144–149.

Aiken, D.C., Curtis, T.P., Heidrich, E.S., 2019. Avenues to the financial viability of microbial electrolysis cells [MEC] for domestic wastewater treatment and hydrogen production. Int. J. Hydrogen Energy 44, 2426–2434.

Al-Mamun, A., Baawain, M.S., Dhar, B.R., Kim, I.S., 2017. Improved recovery of bioenergy and osmotic water in an osmotic microbial fuel cell using micro-diffuser assisted marine aerobic biofilm on cathode. Biochem. Eng. J. 128, 235–242.

Araneda, I., Tapia, N.F., Allende, K.L., Vargas, I.T., 2018. Constructed wetland-microbial fuel cells for sustainable graywater treatment. Water (Switzerland) 10, 940.

Arends, J.B.A., Van Denhouwe, S., Verstraete, W., Boon, N., Rabaey, K., 2014. Enhanced disinfection of wastewater by combining wetland treatment with bioelectrochemical H_2O_2 production. Bioresour. Technol. 155, 352–358.

Arvaniti, I., Fountoulakis, M.S., 2021. Use of a graphite-cement composite as electrode material in upflow constructed wetland-microbial fuel cell for graywater treatment and bioelectricity generation. J. Environ. Chem. Eng. 9, 105158.

Ayodele, T.R., Mosetlhe, T.C., Yusuff, A.A., Ogunjuyigbe, A.S.O., 2021. Off-grid hybrid renewable energy system with hydrogen storage for South African rural community health clinic. Int. J. Hydrogen Energy 46, 19871–19885.

Barua, S., Zakaria, B.S., Chung, T., Hai, F.I., Haile, T., Al-Mamun, A., Dhar, B.R., 2019. Microbial electrolysis followed by chemical precipitation for effective nutrients recovery from digested sludge centrate in WWTPs. Chem. Eng. J. 361, 256–265.

Bian, B., Shi, D., Cai, X., Hu, M., Guo, Q., Zhang, C., Wang, Q., Sun, A.X., Yang, J., 2018a. 3D printed porous carbon anode for enhanced power generation in microbial fuel cell. Nano Energy 44, 174–180.

Bian, B., Wang, C., Hu, M., Yang, Z., Cai, X., Shi, D., Yang, J., 2018b. Application of 3D printed porous copper anode in microbial fuel cells. Front. Energy Res. 6, 50.

Bolton, C.R., Randall, D.G., 2019. Development of an integrated wetland microbial fuel cell and sand filtration system for graywater treatment. J. Environ. Chem. Eng. 7, 103249.

Bose, R.S., Zakaria, B.S., Kumar Tiwari, M., Ranjan Dhar, B., 2021. High-rate blackwater anaerobic digestion under septic tank conditions with the amendment of biosolids-derived biochar synthesized at different temperatures. Bioresour. Technol. 331, 125052.

Cai, W., Lesnik, K.L., Wade, M.J., Heidrich, E.S., Wang, Y., Liu, H., 2019. Incorporating microbial community data with machine learning techniques to predict feed substrates in microbial fuel cells. Biosens. Bioelectron. 133, 64–71.

Capodaglio, A.G., Bolognesi, S., Cecconet, D., 2021. Sustainable, decentralized sanitation and reuse with hybrid nature-based systems. Water (Switzerland) 13, 1583.

Christiaens, M.E.R., Gildemyn, S., Matassa, S., Ysebaert, T., De Vrieze, J., Rabaey, K., 2017. Electrochemical ammonia recovery from source-separated urine for microbial protein production. Environ. Sci. Technol. 51, 13143–13150.

Chung, T.H., Dhar, B.R., 2021a. A multi-perspective review on microbial electrochemical technologies for food waste valorization. Bioresour. Technol. 342, 125950.

Chung, T.H., Dhar, B.R., 2021b. Biosensors and bioelectronics paper-based platforms for microbial electrochemical cell-based biosensors: a review. Biosens. Bioelectron. 192, 113485.

Chung, T.H., Dhar, B.R., 2021c. A mini-review on applications of 3D printing for microbial electrochemical technologies. Front. Energy Res. 9, 222.

Chung, T.H., Meshref, M.N.A., Dhar, B.R., 2021. A review and roadmap for developing microbial electrochemical cell-based biosensors for recalcitrant environmental contaminants, emphasis on aromatic compounds. Chem. Eng. J. 424, 130245.

Chung, T.H., Meshref, M.N.A., Hai, F.I., Al-Mamun, A., Dhar, B.R., 2020. Microbial electrochemical systems for hydrogen peroxide synthesis: critical review of process optimization, prospective environmental applications, and challenges. Bioresour. Technol. 313, 123727.

Colares, G.S., Dell'Osbel, N., Paranhos, G., Cerentini, P., Oliveira, G.A., Silveira, E., Rodrigues, L.R., Soares, J., Lutterbeck, C.A., Rodriguez, A.L., Vymazal, J., Machado, ÊL., 2022. Hybrid constructed wetlands integrated with microbial fuel cells and reactive bed filter for wastewater treatment and bioelectricity generation. Environ. Sci. Pollut. Res. 29, 22223–22236.

Das, I., Ghangrekar, M.M., Satyakam, R., Srivastava, P., Khan, S., Pandey, H.N., 2020. On-site sanitary wastewater treatment system using 720-l stacked microbial fuel cell: case study. J. Hazardous, Toxic, Radioact. Waste. 24, 04020025.

Das, P., Singh, K.K.K., Dutta, S., 2019. Insight into emerging applications of forward osmosis systems. J. Ind. Eng. Chem. 72, 1–17.

de Graaff, M.S., Temmink, H., Zeeman, G., Buisman, C.J.N., 2010. Anaerobic treatment of concentrated black water in a UASB reactor at a short HRT. Water (Switzerland) 2, 101–119.

De Paepe, J., De Paepe, K., Gòdia, F., Rabaey, K., Vlaeminck, S.E., Clauwaert, P., 2020. Bio-electrochemical COD removal for energy-efficient, maximum and robust nitrogen recovery from urine through membrane aerated nitrification. Water Res. 185, 116223.

Deaver, J.A., Kerr, C.A., Popat, S.C., 2022. Primary sludge-based blackwater favors electrical current over methane production in microbial electrochemical cells. J. Water Process Eng. 47, 102848.

Dhadwal, M., Liu, Y., Dhar, B.R.B.R., 2021. Coupling microbial electrolysis cell and activated carbon biofilter for source-separated graywater treatment. Processes 9, 281.

Dhar, B.R., Ryu, H., Ren, H., Domingo, J.W.S., Chae, J., Lee, H.-S., 2016. High biofilm conductivity maintained despite anode potential changes in a Geobacter-enriched biofilm. ChemSusChem 9, 3485–3491.

Dhar, B.R., Ryu, H., Santo Domingo, J.W., Lee, H.-S., 2016. Ohmic resistance affects microbial community and electrochemical kinetics in a multi-anode microbial electrochemical cell. J. Power Sources 331, 315–321.

Escapa, A., San-Martín, M.I., Mateos, R., Morán, A., 2015. Scaling-up of membraneless microbial electrolysis cells (MECs) for domestic wastewater treatment: bottlenecks and limitations. Bioresour. Technol. 180, 72–78.

Fang, Z., Song, H.L., Cang, N., Li, X.N., 2013a. Performance of microbial fuel cell coupled constructed wetland system for decolorization of azo dye and bioelectricity generation. Bioresour. Technol. 144, 165–171.

Fang, F., Zang, G.L., Sun, M., Yu, H.Q., 2013b. Optimizing multi-variables of microbial fuel cell for electricity generation with an integrated modeling and experimental approach. Appl. Energy 110, 98–103.

Feng, C., Hu, A., Chen, S., Yu, C.P., 2013. A decentralized wastewater treatment system using microbial fuel cell techniques and its response to a copper shock load. Bioresour. Technol. 143, 76–82.

Friedler, E., 2010. Quality of individual domestic graywater streams and its implication for on-site treatment and reuse possibilities. Environ. Technol. 25, 997–1008.

Gao, M., Guo, B., Zhang, L., Zhang, Y., Liu, Y., 2019. Microbial community dynamics in anaerobic digesters treating conventional and vacuum toilet flushed blackwater. Water Res. 160, 249–258.

Gao, Y., Sun, D., Wang, H., Lu, L., Ma, H., Wang, L., Ren, Z.J., Liang, P., Zhang, X., Chen, X., Huang, X., 2018. Urine-powered synergy of nutrient recovery and urine purification in a microbial electrochemical system. Environ. Sci. Water Res. Technol. 4, 1427–1438.

Gil-Carrera, L., Escapa, A., Carracedo, B., Morán, A., Gómez, X., 2013. Performance of a semi-pilot tubular microbial electrolysis cell (MEC) under several hydraulic retention times and applied voltages. Bioresour. Technol. 146, 63–69.

González, T., Puigagut, J., Vidal, G., 2021. Organic matter removal and nitrogen transformation by a constructed wetland-microbial fuel cell system with simultaneous bioelectricity generation. Sci. Total Environ. 753, 142075.

Heidrich, E.S., Dolfing, J., Scott, K., Edwards, S.R., Jones, C., Curtis, T.P., 2013. Production of hydrogen from domestic wastewater in a pilot-scale microbial electrolysis cell. Appl. Microbiol. Biotechnol. 97, 6979–6989.

Hiegemann, H., Herzer, D., Nettmann, E., Lübken, M., Schulte, P., Schmelz, K.G., Gredigk-Hoffmann, S., Wichern, M., 2016. An integrated 45 L pilot microbial fuel cell system at a full-scale wastewater treatment plant. Bioresour. Technol. 218, 115–122.

Huang, Q., Liu, Y., Dhar, B.R., 2020. A critical review of microbial electrolysis cells coupled with anaerobic digester for enhanced biomethane recovery from high-strength feedstocks. Crit. Rev. Environ. Sci. Technol. 52, 50–89.

Huang, Q., Liu, Y., Dhar, B.R., 2021. Pushing the organic loading rate in electrochemically assisted anaerobic digestion of blackwater at ambient temperature: insights into microbial community dynamics. Sci. Total Environ. 781, 146694.

Jadhav, D.A., Das, I., Ghangrekar, M.M., Pant, D., 2020. Moving towards practical applications of microbial fuel cells for sanitation and resource recovery. J. Water Process Eng. 38, 101566.

Kanu, I., Achie, O.K., 2011. Industrial effluents and their impact on water quality of receiving rivers in Nigeria. J. Appl. Technol. Environ. Sanit. 1, 75–86.

Katuri, K.P., Ali, M., Saikaly, P.E., 2019. The role of microbial electrolysis cell in urban wastewater treatment: Integration options, challenges, and prospects. Curr. Opin. Biotechnol. 57, 101–110.

Katuri, K.P., Kalathil, S., Ragab, A., Bian, B., Alqahtani, M.F., Pant, D., Saikaly, P.E., 2018. Dual-function electrocatalytic and Macroporous hollow-fiber cathode for converting waste streams to valuable resources using microbial electrochemical systems. Adv. Mater. 30, 1707072.

Kim, B., Jang, N., Lee, M., Jang, J.K., Chang, I.S., 2021. Microbial fuel cell driven mineral rich wastewater treatment process for circular economy by creating virtuous cycles. Bioresour. Technol. 320, 124254.

Kim, T., Yeo, J., Yang, Y., Kang, S., Paek, Y., Kwon, J.K., Jang, J.K., 2019. Boosting voltage without electrochemical degradation using energy-harvesting circuits and power management system-coupled multiple microbial fuel cells. J. Power Sources 410–411, 171–178.

Kivaisi, A.K., 2001. The potential for constructed wetlands for wastewater treatment and reuse in developing countries: A review. Ecol. Eng. 16, 545–560.

Kuntke, P., Sleutels, T.H.J.A., Rodríguez Arredondo, M., Georg, S., Barbosa, S.G., ter Heijne, A., Hamelers, H.V.M., Buisman, C.J.N., 2018. (Bio)electrochemical ammonia recovery: Progress and perspectives. Appl. Microbiol. Biotechnol. 102, 3865–3878.

Lackner, S., Gilbert, E.M., Vlaeminck, S.E., Joss, A., Horn, H., van Loosdrecht, M.C.M., 2014. Full-scale partial nitritation/anammox experiences – an application survey. Water Res. 55, 292–303.

Leal, L.H., Temmink, H., Zeeman, G., Cees, C.J., 2010. Comparison of three systems for biological graywater treatment. Water (Switzerland). 2, 155–169.

Lesnik, K.L., Cai, W., Liu, H., 2019. Microbial community predicts functional stability of microbial fuel cells. Environ. Sci. Technol. 54, 427–436.

Lesnik, K.L., Liu, H., 2017. Predicting microbial fuel cell biofilm communities and bioreactor performance using artificial neural networks. Environ. Sci. Technol. 51, 10881–10892.

Liang, P., Duan, R., Jiang, Y., Zhang, X., Qiu, Y., Huang, X., 2018. One-year operation of 1000-L modularized microbial fuel cell for municipal wastewater treatment. Water Res. 141, 1–8.

Li, J., Li, H., Fu, Q., Liao, Q., Zhu, X., Kobayashi, H., Ye, D., 2017. Voltage reversal causes bioanode corrosion in microbial fuel cell stacks. Int. J. Hydrogen Energy. 42, 27649–27656.

Linares, R.V., Domínguez-Maldonado, J., Rodríguez-Leal, E., Patrón, G., Castillo-Hernández, A., Miranda, A., Romero, D.D., Moreno-Cervera, R., Camara-Chale, G., Borroto, C.G., Alzate-Gaviria, L., 2019. Scale up of microbial fuel cell stack system for residential wastewater treatment in continuous mode operation. Water (Switzerland) 11, 217.

Liu, W., Cheng, S., 2014. Microbial fuel cells for energy production from wastewaters: the way toward practical application. J. Zhejiang Univ. Sci. A 15, 841–861.

Luta, D.N., Raji, A.K., 2018. Decision-making between a grid extension and a rural renewable off-grid system with hydrogen generation. Int. J. Hydrogen Energy 43, 9535–9548.

Malvankar, N.S., Tuominen, M.T., Lovley, D.R., 2012. Biofilm conductivity is a decisive variable for high-current-density *Geobacter sulfurreducens* microbial fuel cells. Energy Environ. Sci. 5, 5790–5797.

Monetti, J., Ledezma, P., Freguia, S., 2021. Optimised operational parameters for improved nutrient recovery from hydrolysed urine by bio-electroconcentration. Sep. Purif. Technol. 279, 119793.

Moreno, R., San-Martín, M.I., Escapa, A., Morán, A., 2016. Domestic wastewater treatment in parallel with methane production in a microbial electrolysis cell. Renew. Energy 93, 442–448.

Mousazadeh, M., Niaragh, E.K., Usman, M., Khan, S.U., Sandoval, M.A., Al-Qodah, Z., Khalid, Z., Bin Gilhotra, V., Emamjomeh, M.M., 2021. A critical review of state-of-the-art electrocoagulation technique applied to COD-rich industrial wastewaters. Environ. Sci. Pollut. Res. 28, 43143–43172.

Mukherjee, A., Patel, V., Shah, M.T., Jadhav, D.A., Munshi, N.S., Chendake, A.D., Pant, D., 2022. Effective power management system in stacked microbial fuel cells for onsite applications. J. Power Sources 517, 230684.

Pant, D., Singh, A., Van Bogaert, G., Gallego, Y.A., Diels, L., Vanbroekhoven, K., 2011. An introduction to the life cycle assessment (LCA) of bioelectrochemical systems (BES) for sustainable energy and product generation: relevance and key aspects. Renew. Sustain. Energy Rev. 15, 1305–1313.

Paulo, P.L., Azevedo, C., Begosso, L., Galbiati, A.F., Boncz, M.A., 2013. Natural systems treating graywater and blackwater on-site: integrating treatment, reuse and landscaping. Ecol. Eng. 50, 95–100.

Quaglio, M., Massaglia, G., Vasile, N., Margaria, V., Chiodoni, A., Salvador, G.P., Marasso, S.L., Cocuzza, M., Saracco, G., Pirri, F.C., 2019. A fluid dynamics perspective on material selection in microbial fuel cell-based biosensors. Int. J. Hydrogen Energy 44, 4533–4542.

Ren, L., Ahn, Y., Logan, B.E., 2014. A two-stage microbial fuel cell and anaerobic fluidized bed membrane bioreactor (MFC-AFMBR) system for effective domestic wastewater treatment. Environ. Sci. Technol. 48, 4199–4206.

Rose, C., Parker, A., Jefferson, B., Cartmell, E., 2015. The characterization of feces and urine: a review of the literature to inform advanced treatment technology. Crit. Rev. Environ. Sci. Technol. 45, 1827–1879.

Sajithkumar, K.J., Ramasamy, E.V., 2015. Graywater treatment with simultaneous generation of energy using low-cost microbial fuel cells. Environ. Res. Eng. Manage. 71, 3, 11482.

Shaikh, I.N., Ahammed, M.M., 2020. Quantity and quality characteristics of graywater: a review. J. Environ. Manage. 261, 110266.

Sharma, P., Talekar, G.V., Mutnuri, S., 2021. Demonstration of energy and nutrient recovery from urine by field-scale microbial fuel cell system. Process Biochem. 101, 89–98.

Sim, J., Reid, R., Hussain, A., An, J., Lee, H.S., 2018. Hydrogen peroxide production in a pilot-scale microbial electrolysis cell. Biotechnol. Reports. 19, e00276.

Srivastava, P., Belford, A., Abbassi, R., Asadnia, M., Garaniya, V., Yadav, A.K., 2021. Low-power energy harvester from constructed wetland-microbial fuel cells for initiating a self-sustainable treatment process. Sustain. Energy Technol. Assessments 46, 101282.

Surendra, K.C., Takara, D., Hashimoto, A.G., Khanal, S.K., 2014. Biogas as a sustainable energy source for developing countries: opportunities and challenges. Renew. Sustain. Energy Rev. 31, 846–859.

Talekar, G.V., Sharma, P., Yadav, A., Clauwaert, P., Rabaey, K., Mutnuri, S., 2018. Sanitation of blackwater via sequential wetland and electrochemical treatment. NPJ Clean Water. 1, 14.

Tamta, P., Rani, N., Yadav, A.K., 2020. Enhanced wastewater treatment and electricity generation using stacked constructed wetland–microbial fuel cells. Environ. Chem. Lett. 18, 871–879.

Tervahauta, T., Bryant, I.M., Leal, L.H., Buisman, C.J.N., Zeeman, G., 2014. Improved energy recovery by anaerobic greywater sludge treatment with black water. Water (Switzerland) 6, 2436–2448.

Theodosiou, P., Greenman, J., Ieropoulos, I.A., 2020. Developing 3D-printable cathode electrode for monolithically printed microbial fuel cells (MFCs). Molecules 25, 3635.

Tsompanas, M.A., You, J., Wallis, L., Greenman, J., Ieropoulos, I., 2019. Artificial neural network simulating microbial fuel cells with different membrane materials and electrode configurations. J. Power Sources. 436, 226832.

Van Afferden, M., Cardona, J.A., Müller, R.A., Lee, M.Y., Subah, A., 2015. A new approach to implementing decentralized wastewater treatment concepts. Water Sci. Technol. 72, 1923–1930.

Verma, M., Verma, M.K., Singh, V., Singh, J., Singh, V., Mishra, V., 2022. Advancements in applicability of microbial fuel cell for energy recovery from human waste. Bioresour. Technol. Rep. 17, 100978.

Vogl, A., Bischof, F., Wichern, M., 2016. Single chamber microbial fuel cells for high strength wastewater and blackwater treatment—a comparison of idealized wastewater, synthetic human blackwater, and diluted pig manure. Biochem. Eng. J. 115, 64–71.

Walter, X.A., Stinchcombe, A., Greenman, J., Ieropoulos, I., 2017. Urine transduction to usable energy: a modular MFC approach for smartphone and remote system charging. Appl. Energy. 192, 575–581.

Walter, X.A., You, J., Winfield, J., Bajarunas, U., Greenman, J., Ieropoulos, I.A., 2020. From the lab to the field: self-stratifying microbial fuel cells stacks directly powering lights. Appl. Energy 277, 115514.

Wang, Y.K., Geng, Y.K., Pan, X.R., Sheng, G.P., 2017. In situ utilization of generated electricity for nutrient recovery in urine treatment using a selective electrodialysis membrane bioreactor. Chem. Eng. Sci. 171, 451–458.

Wang, J., Song, X., Wang, Y., Abayneh, B., Li, Y., Yan, D., Bai, J., 2016. Nitrate removal and bioenergy production in constructed wetland coupled with microbial fuel cell: establishment of electrochemically active bacteria community on anode. Bioresour. Technol. 221, 358–365.

Wang, J., Xiao, T., Bao, R., Li, T., Wang, Y., Li, D., Li, X., He, T., 2018. Zwitterionic surface modification of forward osmosis membranes using N-aminoethyl piperazine propane sulfonate for greywater treatment. Process Saf. Environ. Prot. 116, 632–639.

Withers, P.J.A., Jordan, P., May, L., Jarvie, H.P., Deal, N.E., 2014. Do septic tank systems pose a hidden threat to water quality? Front. Ecol. Environ. 12, 123–130.

Xiao, T., Dou, P., Wang, J., Song, J., Wang, Y., Li, X.M., He, T., 2018. Concentrating graywater using hollow fiber thin film composite forward osmosis membranes: fouling and process optimization. Chem. Eng. Sci. 190, 140–148.

Xie, M., Nghiem, L.D., Price, W.E., Elimelech, M., 2014. Toward resource recovery from wastewater: extraction of phosphorus from digested sludge using a hybrid forward osmosis–membrane distillation process. Environ. Sci. Technol. Lett. 1, 191–195.

Xu, F., Cao, F.-q., Kong, Q., Zhou, L.-l., Yuan, Q., Zhu, Y.-j., Wang, Q., Du, Y.-d., Wang, Z.-d., 2018. Electricity production and evolution of microbial community in the constructed wetland-microbial fuel cell. Chem. Eng. J. 339, 479–486.

Yazdi, H., Alzate-Gaviria, L., Ren, Z.J., 2015. Pluggable microbial fuel cell stacks for septic wastewater treatment and electricity production. Bioresour. Technol. 180, 258–263.

Yewale, A., Methekar, R., Agrawal, S., 2020. Multiple model-based control of multi variable continuous microbial fuel cell (CMFC) using machine learning approaches. Comput. Chem. Eng. 140, 106884.

You, J., Preen, R.J., Bull, L., Greenman, J., Ieropoulos, I., 2017. 3D printed components of microbial fuel cells: towards monolithic microbial fuel cell fabrication using additive layer manufacturing. Sustain. Energy Technol. Assessments. 19, 94–101.

Yu, C.P., Liang, Z., Das, A., Hu, Z., 2011. Nitrogen removal from wastewater using membrane aerated microbial fuel cell techniques. Water Res. 45, 1157–1164.

Zakaria, B.S., Dhar, B.R., 2019. Progress towards catalyzing electro-methanogenesis in anaerobic digestion process: fundamentals, process optimization, design and scale-up considerations. Bioresour. Technol. 289, 121738.

Zakaria, B.S., Dhar, B.R., 2022. A review of stand-alone and hybrid microbial electrochemical systems for antibiotics removal from wastewater. Processes 10, 714.

Zhang, F., Brastad, K.S., He, Z., 2011. Integrating forward osmosis into microbial fuel cells for wastewater treatment, water extraction and bioelectricity generation. Environ. Sci. Technol. 45, 6690–6696.

Zhang, L., Wang, J., Fu, G., Zhang, Z., 2020. Simultaneous electricity generation and nitrogen and carbon removal in single-chamber microbial fuel cell for high-salinity wastewater treatment. J. Clean. Prod. 276, 123203.

Zinadini, S., Zinatizadeh, A.A., Rahimi, M., Vatanpour, V., Bahrami, K., 2017. Energy recovery and hygienic water production from wastewater using an innovative integrated microbial fuel cell–membrane separation process. Energy 141, 1350–1362.

7 Sustainable, Decentralized Sanitation and Reuse Using Hybrid Systems

Bhoomika Yadav[1], *Anusha Atmakuri*[1], *Rajeshwar D. Tyagi*[2,3], *Patrick Drogui*[1], *and Jonathan W.C. Wong*[2]
[1] Université du Québec, Québec, Canada
[2] Dongguan University of Technology, Dongguan, China
[3] BOSK Bioproducts, Québec, Canada

7.1 INTRODUCTION

With the increase in global population, the demand for clean water is also increasing in addition to the pollution of water resources from domestic, agricultural and industrial sectors. To solve this, wastewater treatment and recycling reaching even the segregated and rural, low-income areas of the cities are required. It is estimated that globally only 20% of the wastewater is being treated with the remaining left in the environment (Kumar et al., 2023). Moreover, the centralized treatment systems are not feasible in many areas with small communities. In this context, there is an increase attention toward decentralized ecological treatment systems because they are cost-effective, environmentally friendly, feasible and easy to use along with the advantages of recovering energy and nutrients.

In 2012, the United States National Research Council implied wastewater reuse as an unused available water source. To decentralize the treatment systems, new approaches of planning, decision-making, infrastructure design and new arrangements in system operations are required with possible adoption of more efficient technologies for resource recycle and recovery and enhance the resilience and sustainability of ecosystems (Capodaglio et al., 2021).

Nature-based processes for wastewater treatment include constructed wetlands (CWs), lagoons, algae ponds, waste stabilization ponds, buffer zones and vegetated drainage ditches. They are based on the natural capacity of the microbes and plants to behave in synergy for wastewater treatment by using complex biochemical, physical and physiological processes. Nature-based systems like CWs have proven to be incredibly efficient in treating wastewater with high chemical oxygen demand (COD). However, there can be certain limitations such as low treatment performances due to predominance of anaerobic process because of lack of efficient electron acceptors and seasonal variability of pollutant removal efficiency (Kumar et al., 2023).

In the last few years, other eco-engineered technologies have been developed that focus on treating the wastewater and recovering the energy and resources from the wastewater simultaneously. Among them, bio electrochemical systems (BES) such as microbial fuel cells (MFCs) have emerged as potential candidates in decentralized renewable energy production for small-scale wastewater treatment and energy recovery (Kumar et al., 2023). MFCs have several benefits over the conventional wastewater treatment methods, including reduced aeration requirements, efficient sludge removal and electricity production, but they are mostly used at lab-scale and are difficult to scale-up.

In this line of research, this chapter focusses on the hybrid-based systems involving CW-MFC, algae-MFC and algae-CW systems for enhancing wastewater treatment and bioenergy production.

 DOI: 10.1201/9781003381419-9

The aim is to give insights on the significance of the existing nature-based technologies integrated with BES and highlight the challenges and future outlook.

7.2 CONSTRUCTED WETLANDS AND MICROBIAL FUEL CELLS

Recently, CW and MFCs have emerged as attractive technologies for sustainable treatment of wastewater. CW is a natural wastewater treatment technology in which pollutants are removed using techniques like filtration and adsorption by plants and aerobic/anaerobic degradation by microbes (Araneda et al., 2018). They could serve as an alternative technology to conventional wastewater systems such as activated sludge systems in small communities due to their low energy requirements and easy operation and maintenance. However, they are characterized by higher specific area, and therefore, several intensifying techniques such as forced aeration have been recently investigated (Corbella et al., 2017). But these techniques can result in increasing the energy consumption of CW systems.

MFCs are bioelectrical systems that transform chemical energy in organic compounds to electricity using exoelectrogenic bacteria. The organic compounds are oxidized releasing electrons and protons where the electrons are transferred to anode and transported through external circuit to cathode where along with protons, they used to reduce an electron acceptor (for example, oxygen). The MFC performance depends on the redox gradient between the cathodes. Both these technologies (CW and MFCs) have been used for wastewater treatment; however, their hybrid system is an emerging technology for sustainable wastewater treatment. The presence of organic matter in wastewater and the naturally generated redox between upper layers (aerobic conditions) and deeper layers (anaerobic conditions) of horizontal subsurface CWs can provide favorable conditions to apply MFCs in CW systems. Incorporation of MFC in CW can improve the COD removal efficiency as compared to the individual technologies (Doherty et al., 2015). This hybrid system provides surplus energy that can cover (partially) energy input required for wastewater treatment in the system. Implementation of MFCs in CW can improve the CW treatment efficiency and reduce their surface requirements. However, materials used for conventional MFC electrodes are expensive with poor environmental performance. So, even the energy inputs and area requirement are lower; the costs and environmental impacts of CW-MFC hybrid systems should be analyzed (Corbella et al., 2017). The setup of a general CW-MFC system, as demonstrated in Figure 7.1, consists of an anode, located in the anaerobic zone, and a cathode located in the anoxic/aerobic zone.

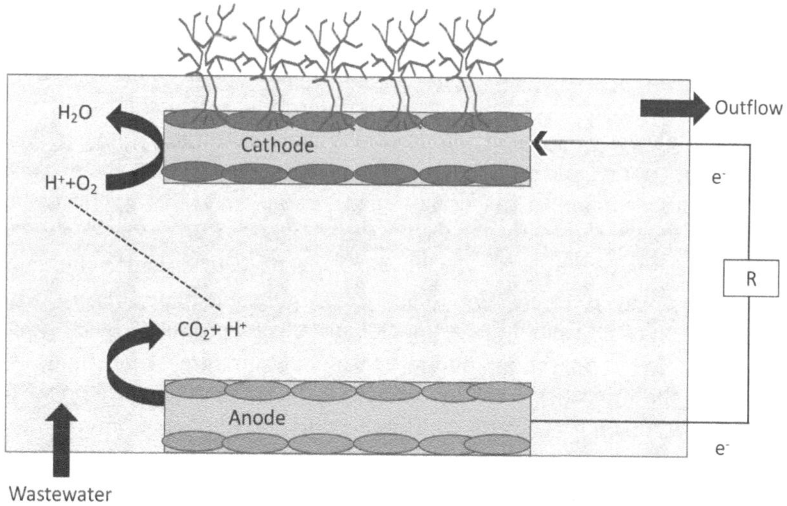

FIGURE 7.1 Schematic of constructed wetland-MFC system setup (R: Resistance).

Researchers have assessed the environmental impacts of CW-MFCs where conventional CW, CW with gravel-based anode MFC and CW with graphite-based anode MFC were compared (Corbella et al., 2017). The CW-MFCs had lower specific area requirement than conventional CW. Out of all three systems, CW with graphite-based anode MFC appeared to be most environmentally friendly solution that can reduce the CW system footprint by 20%. However, the CW-MFC system was found to be 1.5 times more expensive than conventional CW.

Researchers have focused to develop sustainable decentralized technologies at community level by treating the domestic wastewater sewage and reverse osmosis (RO) concentrate by CW-MFC. Two CW-MFCs systems were run, one with plants and one without plants to observe the wastewater treatability (Thakur et al., 2022). The highest COD removal was 74.29% for the CW-MFC with plants higher than CW-MFC without plants (65%) because of the presence of *Canna indica*. Also, the enrichment of nutrients by using RO concentrate enhanced the electrochemical performance of both the systems with power densities of 0.38 W/m^3 for system with plants and 0.33 W/m^3 for system without plants. This study shows the successful treatment of RO concentrate with sewage using up-flow CW-MFC that can be used in households as decentralized treatment system.

In another study, Kumar et al. (2023) evaluated pilot scale CW-MFC system for COD removal from synthetic wastewater and bioelectricity generation. It consisted of nine individual CW-MFC unites with two aerobic chambers where *Typha angustifolia* was used as wetland plant. Two modes fed-batch and continuous up-flow modes were run and to enhance the electrical performance in two auxiliary MFCs equipped with terracotta-based separator-electrode-assembly (SEA). In both modes of operation, the internal resistance of auxiliary MFCs was lower (two times) than conventional sediment-configures MFCs, which overall shows improved energetics of the system.

Araneda et al. (2018) incorporated MFCs into CWs for the treatment of graywater and energy recovery. Removal efficiencies for soluble COD, nitrate and phosphate were found to be 91.7, 86.5 and 56%, respectively. Energy generation reached the maximum power density of 33.5 and 719.57 mW/m^3 when anode potential was poised at −150 mV vs Ag/AgCl. Thus, hybrid systems such as CW-MFCs could serve as a sustainable method for graywater treatment with electricity production. However, further studies are required to implement it as decentralized technology and evaluate the long-term performance of this technology.

7.3 ALGAL AND MICROBIAL FUEL CELLS HYBRIDS

Algal-MFC is a system that can convert solar energy into electrical energy via biological pathways. Microalgae biomass contains around 32% proteins and 51% carbohydrates that can be degraded by electrogens to produce electricity (Rashid et al., 2013). Within the algal-MFC, the algae can be present either in the cathode chamber or both the cathode and anode chamber as anodic half-cell of the fuel cell. When algae are used with the wastewater or activated sludge, algae become part of the microbiome of the wastewater and help in oxidizing the substrate. However, the use of algae in anodic chamber does not significantly improve the performance. Hence, the focus shifted to the use of algae in the cathode chamber, for carbon sequestration and other applications (Kannan & Donnellan, 2021).

Conventionally, expensive catalysts like Pt, CuO are used to enhance the cathode performance. An alternative way is to circulate water containing adequate dissolved oxygen (DO) where the concentration of DO should be equivalent to oxygen produced by catalysts. However, this adds costs to the MFC operations due to continuous pumping. Biocathode is an innovative approach where the microbes work as catalyst to facilitate electrochemical reduction on the cathode surface. Replacing catalysts with photosynthetic algae is a promising approach to increase the performance of cathode. During their growth, microalgae release oxygen which serves as terminal electron acceptor. Another benefit from using microalgae as MFC's cathode is carbon fixation. In MFC system, the electrogens produce CO_2; microalgae at cathode use this CO_2 as carbon source and increase the biomass that can be further used to extract fuels like biohydrogen, biodiesel, bioethanol from it (Cui et al., 2014). Factors such as pH, temperature, substrate type and organic loading rate, photosynthetic efficiency and MFC design (electrodes, membrane, chambers etc.) can influence the bioelectricity production from algae-MFC (Shukla & Kumar, 2018).

Cui et al. (2014) developed an MFC to treat waste, produce electricity and cultivate algae simultaneously. They used dead microalgae biomass at anode as nutrient source and activated sludge as inoculum. Microbes from sludge used dead microalgae as nutrients to generate electrons. Live microalgae were grown at cathode. The CO_2 released at anode was directed toward cathode to be used by microalgae for the growth. The maximum power density of 1.93 W/m^2 and coulombic efficiency of 6.3% were obtained using 2500 mg COD/L of microalgae powder (0.5 g/L). A biomass concentration of 1.24 g/L was obtained by microalgae with 5–14% v/v CO_2.

Microalgae such as *Tetraselmis gracilis*, *Mougeotia*, *Chlorella* sp., *Scenedesmus* and blue-green algae have been successfully used for bioelectricity generation (Shukla & Kumar, 2018). Velasquez-Orta et al. (2009) compared bioelectricity production when *Chlorella vulgaris* (phytoplankton) and *Ulva lactuca* (macrophyte) were used as substrates in single chamber MFCs. *C. vulgaris* produced more energy per substrate mass, i.e., 2.5 kWh/kg. *U. lactuca* was degraded more over a batch cycle (73% COD). Maximum power densities of 0.98 and 0.76 W/m^2 were obtained for *C. vulgaris* and *U. lactuca*, respectively.

Introducing photosynthetic algae in single chamber MFC can lead to inefficient separation of bulk solution and high levels of oxygen production causing inhibition of the anodic performance. To solve this problem, researchers have also purged the nitrogen gas or added highly concentrated salt solution (Shukla & Kumar, 2018). In such cases, algae should have osmotic tolerance for normal function. Also, these options may not be commercially feasible and produce low current. Elmaadawy et al. (2020) introduced an innovative oxygen consuming unit (OCU) to the algal biofilm single chamber MFC to constrain the oxygen produced by algae to reduce the oxygen diffusion to anode. This system performed better in terms of COD, NH_4^+-N, TN removals as compared to algal-MFC. The system with OCU thickness of 2 cm removed more than 86% COD, 89.4% NH_4^+-N and 76.7% TN while producing maximum voltage of 0.39 V and biomass productivity of 1.23 g/L/day.

Hence, with the extraction of energy from wastewater and capture of energy from solar source, electrical energy and biomass can be obtain simultaneously from algal-MFC systems.

7.4 ALGAL AND CONSTRUCTED WETLANDS HYBRID SYSTEMS

Bioremediation provides an array of green technologies for the economic and efficient treatment of wastewater. For example, algae-based systems result in the noteworthy removal of nutrients. Algae have been utilized for wastewater treatment since 1957. In contrast, CWs have been proven to be excellent remediation tools for nutrients, organic and inorganic compounds from wastewater. As all treatment options, both systems have distinctive limitations like algae are efficient to a lower extent in removal of COD although good in phosphorus removal. CWs are prominent in wastewater treatment as they are economic, highly efficient and eco-friendly. CWs are stated to have high nitrogen and COD removal rate, however, poor efficiency in phosphorous removal and denitrification efficiency when used for tertiary wastewater treatment owing to the lack of carbon sources in the secondary effluent. Usually nitrification and denitrification are the key routes for nitrogen removal in CWs as illustrated in previous studies (Han et al., 2019). Denitrification is the restricting factor when wastewater with a low chemical oxygen demand/nitrogen (COD/N) ratio is treated in CWs, necessitating addition of external carbon source. The use of organic matter with a low molecular weight can improve the denitrification in CWs; however, it is expensive and raises the risk of high COD concentrations in the effluent.

Hence, an integration of microalgae with CWs in series may help to improve wastewater treatment rates. However, only a few attempts have been made for such integrated systems having algae and CW together (Chauhan & Prajapati, 2022). Figure 7.2 shows the integration of algae and CWs

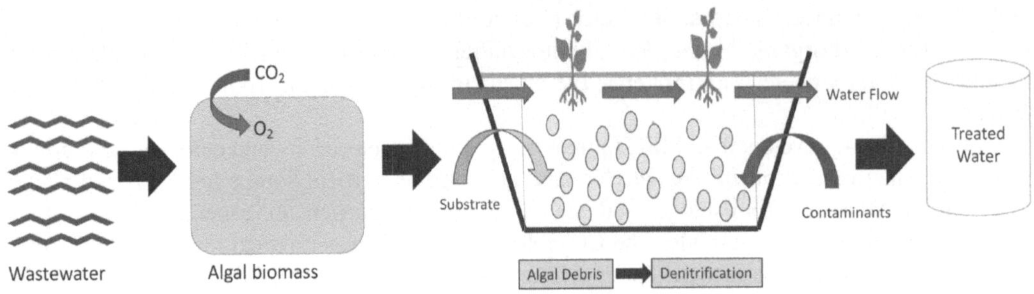

FIGURE 7.2 Algae coupled constructed wetland for wastewater treatment.

for wastewater treatment. As seen in the figure, algae-based systems lead to addition of external carbon, known to increase the process of denitrification.

One of the studies by Cheng et al. (2022) reported the effect of algae addition as an external carbon source in CWs, illustrated the nitrogen removal efficiency of CWs to be highly dependent on the dosage of algae. The study reported the potential of algae addition to provide enough carbon to boost denitrification during treatment of wastewater with a low COD/N ratio. Single-celled algae, such as *C. vulgaris* and *Scenedesmus* sp., include several proteins, fatty acids and polysaccharides, hence can be used as carbon sources for denitrification (Hume et al., 2002).

In the recent past, various researchers have utilized a combined system integrating a high-rate algal pond (HRAP) with a CW for high nitrogen synthetic wastewater treatment (Ding et al., 2016).

The results from a study illustrated that in addition to algae assimilating nitrogen in the algal pond and escalating the DO concentrations of the influent to the CW, it also acts as a carbon source in the CW (Milledge et al., 2019).

A previous study by Cheng and his group on algae carbon releasing demonstrated that the soluble COD attained the stationary phase after 10 days, with the maximum carbon release at 370 mg COD_{Cr}/g (dry weight, DW) (Cheng et al., 2022). Yet, no detailed studies on the role of microalgae as a carbon source are present so far. The main objective of this study was evaluation of nitrogen removal efficiency in CW with varied algal dosages and to determine the nitrogen transformations in response to the algae.

Impact of algae addition on oxidation-reduction potential (ORP) and volatile fatty acids (VFAs) in addition to nitrogen removal efficiency was studied and reported. Biodegradation of organics was achieved by both aerobic and anaerobic microorganisms, subject to availability of oxygen. According to the ORP results, with an increasing COD_{Cr}/N and an increasing HLR, the CWs became less favorable for aerobic bacteria. ORP plays an important role in VFA production, and the highest VFA production from wood waste was reported at ORP = −200 to −100 mv. VFA concentrations and types were reported to have significant influences on the denitrification rate. This study along with other studies has reported that the ratio of VFAs/available carbon sources (including reducing sugar, proteins and VFAs) has a strong positive correlation with the nitrate removal rates.

As in the case of nitrogen removal, oxygen concentrations above 1.5 mg/L have been reported as required for nitrification (Vymazal, 2007), and oxidation of organic matter controlled the activity of autotrophic ammonia oxidation bacteria by DO competition (Li et al., 2017). Furthermore, the addition of algae brought more NH_4^+-N loading into the CW. It is probable that the higher NH_4^+-N loading and lower ORP level might lead to the reduction in NH_4^+-N removal percentage of the CW (Cheng et al., 2022).

From the various studies, it can be summarized that the addition of algae into the CWs making them hybrid systems resulted in the anaerobic decomposition in them, thereby causing the accumulation of soluble organic compounds (typically VFAs), ultimately promoting denitrification throughout the CW (Chauhan & Prajapati, 2022).

7.5 CASE STUDIES OF THE HYBRID SYSTEMS

While these hybrid water supply systems seem to be promising, there are some inherent challenges such as public acceptance, particularly for new technologies and unfamiliar practices (e.g., graywater recycling), unknown impacts, understanding the interaction between centralized and decentralized services and diverse infrastructure management. Hybrid CW, even though natural (extensive) treatments such as free surface flow, horizontal and vertical flow CW (VFCW) require pumps and some skilled operation and maintenance to adjust the filter loads correctly. Hence, they are most suited to small communities organized enough to thoroughly plan and maintain the wetland for the duration of their life.

Hybrid water systems combine the merits of both the centralized and decentralized systems. They are novel and innovative approaches to respond to water scarcity because of population growth and uncertainties in climate change. The water security could be enhanced through local source diversification (Capodaglio et al., 2021). Such diversification of water sources using alternative local water sources minimizes the amount of water abstracted from waterways, groundwater or dams, thereby improving water security (Buller et al., 2022). They can build adaptive capacity by increasing the diversity and flexibility of water supply systems. According to (Weber et al., 2007), planning and construction time of such systems are lower compared to centralized systems. At the same time, they are more flexible as it is easier to adapt the design to local conditions. Hybrid systems could act as a bridge between centralized and decentralized systems, with their barriers, and provide a balance among them by acting as a transition toward a sustainable water system.

Hybrid systems might help defer augmentation of existing infrastructure. For instance, traditional water and wastewater pipe network design is driven to a large extent by the urge to cater for peak demands. Mitigation of these peaks might be achieved using decentralized solutions.

i. A case study carried out in China shows that the hybrid water system offers flexible answers to cope with the new demands wherever certain thresholds of population density are exceeded. Similar results were observed not only in high population cities like Qingdao, China, but also in various cities of East Germany with shrinking population (Bieker et al., 2010). Semi-centralized supply and treatment systems can unlock a wide range of possibilities in resource management, especially by minimizing freshwater demand of new urban areas. Further, it can reduce social, economic and environmental vulnerability to climate change by creating distributed water systems through infrastructure design choices at a range of scales from household to regional level, ensuring greater flexibility. In addition, the planning and operation of decentralized systems being of a smaller scale entail a lower level of implementation and operational risks. In addition, water infrastructure systems consisting of semi-centralized systems would be more resilient because a failure in these systems would only affect a small part of the urban area.
ii. Studies by researchers stated the capacity of hybrid water systems to minimize potable water use and energy consumption, along with increasing the potential for nutrient recovery on large urban areas (Daigger, 2009; Daigger & Crawford, 2007).
iii. A study from Mogi Mirim municipality, Brazil, demonstrated a hybrid system design based on photo-voltaic energy, biomass gasifier and electricity grid aimed to optimize the energy supply and the economics of a wastewater treatment plant (based on activated sludge system with extended aeration, high-energy demanding process). The optimal combination of energy technologies, with and without electric energy generation from biogas combustion obtained from anaerobic digestion of sewage sludge, was evaluated to define the better economic arrangement. The cost of energy for an off-grid biomass gasifier standing alone is USD 0.0426/kWh. Yet, in a grid-connected system involving photovoltaic energy and biomass gasifier, the cost diminished to USD 0.0298/kWh with an energy surplus available

to supply the whole facility requirement and enough to supply 788,000 residential power consumers (Buller et al., 2022).

iv. A case study from India focused on treatment of domestic wastewater including VFCW and microalgal treatment system (MTS), harvesting of the high value algal biomass and extraction of value-added bioproducts from the biomass. They used MTS instead of VFCW in the second stage of CW, as compared to previous studies. In brief, the gravels of different sizes were used as media (from bottom level; 20 cm [10–20 mm], 20 cm [10–15 mm] and 50 cm [2–8 mm]). The wetland was being used for 2.5 years continuously. The experimental trial was carried out for 30 days to understand the initial performance, characterization, feasibility and potential of VFCW and MTS system in integrative manner. The surface area of VFCW was 4.2 m^2 (dimension—length [2 m], width [2.1 m] and depth [1 m]). The study was started in March 2017. The VFCW was fed with 300 L/day of domestic wastewater in the morning (9 AM) using 1 HP water pump (flow rate—150 L/min), and the outlet was closed. In the evening (6 PM), the outlet was opened; the effluent from VFCW was collected in a collection tank. The optimum loading rate was found to be 60–70 g COD/m^2 per day. The study illustrated that at hydraulic loading rate of 150 L/person, the footprint required by integration of treatment vertical flow wetland and MTS can be reduced to 1.35 m^2/person (Chavan & Mutnuri, 2021).

v. A full scale application of the CW-MFC scheme is the microbial electrochemical-based CW (iMETland), which is the sole full-scale application of the CW-BES concept (Capodaglio et al., 2021). Units were placed in Spain, Denmark and Argentina. In an METland system, electrons are transferred to an electro-conductive biofilter (e.g., coke) that connects anoxic and oxic zones, acting as an unlimited acceptor to maximize substrate consumption.

7.6 CHALLENGES IN THE PROCESS

Hybrid systems offer solutions to the challenges present in environment and water management as mutual advantages of individual technologies can be used to maximize the benefits. Although still young field, the ability to use MFC and CW derived hybrid technologies for wastewater treatment is promising. However, a number of challenges should be overcome to achieve the advantages of the hybrid processes.

The high cost, low power density and difficult scale-up constitute the primary barriers of the implementation of MFCs and MFC-based hybrid systems (Li et al., 2014). The main goal of the hybrid systems involving MFCs is increased in power output simultaneously with producing high quality effluent. With the aim of improvement in electricity production, various aspects for future research that should be considered are as follows: exploration of cost-effective anode materials and their modification to significantly reduce the activation loss within the system; development of efficient and cost-effective ion-exchange membranes or separators to promote transfer of protons from anode to cathode; surface modification to increase the reaction rates on cathode; optimizing the operation parameters or modifying the substrates to prevent increase in internal resistance in hybrid systems leading to ohmic losses (Xu et al., 2016).

The hybrid systems can increase complexity in the treatment processes, therefore creating more challenges for process optimization and maintenance. Difficulty would arise in balancing the energy gains, environmental impacts, process stability and economics in MFC and different treatment units (Li et al., 2014). Researchers have identified that filamentous algae or macroalgae are advantageous over microalgae in terms of uptake of nutrients from wastewater and harvesting (Liu et al., 2020). These advantages can be utilized if the reactor designs support the larger area requirements that have not been researched and require in-depth exploration. Commercial applications should be implemented after establishing the optimal design and in-depth analysis of each component of the hybrid system.

In cold climates, the CW performance can be retained. Therefore, to obtain stable efficiency of hybrid systems including CW throughout the year, implementation of design strategies for longer retention time can be performed. For high removal efficiency of pollutants from CW-MFC systems, various factors are important such as variation of material separators, material electrodes, plant species, reactor configuration and temperature (Kataki et al., 2021). Significant research using the hybrid systems has been conducted at lab-scale with synthetic substrates, although working with real wastewater is lacking and requires immediate focus. The development of biofilms and identification of microbial communities and their relationship with the plants need to be researched to increase the pollutant removal and generate electricity from CW-MFC systems. Also, future work should be focused on the development of cost-effective materials like graphite, activated carbon and coal (Kataki et al., 2021). However, issues may arise with their performance such as catalytic rate problems, poor treatment efficiency or potential of corrosion in some cases; therefore, more research in these areas to address these issues is needed.

Scale-up studies of the hybrid systems are a challenge that needs to be overcome. Most of the studies discussed and available in literature were performed at lab-scale. To understand the complex mechanisms of the hybrid systems, stability and long-term practical applications, the larger scale studies are needed. Life cycle assessment studies to understand the impacts of various configurations of CW-MFC, algal-MFC, algal-CW and other hybrid systems are needed. Also, techno-economic analysis after implementation of these hybrid systems in individual houses or community-based level is lacking and requires urgent attention. Hybrid technologies have potential to offer a sustainable solution to decentralized water and wastewater treatment, and for their success, the continuous cooperation among the interdisciplinary field of environmental engineers, microbiologists, material scientists, government and local public is needed.

7.7 CONCLUSIONS AND PERSPECTIVES

With the limited results in energy harvesting from MFCs due to technical limitation, a more sustainable approach would be the integration of bioelectrical systems with the natural systems to improve their treatment efficiency and decrease the environment footprint. This chapter presents state-of-art systematic knowledge on various hybrid technologies (CW-MFC, algae-MFC, algae-CW) in water and wastewater treatment. Hybrid systems for sustainable, decentralized wastewater treatment and water reuse are a subject of ongoing research. Most of them are currently in the infancy stages, however, with the rapidly evolving technologies surrounding these hybrid systems; they can be integrated to the existing decentralized processes to improve treatment efficiency and recover energy.

The integration of individual technologies with each other for decentralized wastewater treatment has been proposed to increase the efficiency of the hybrid technologies and provides the advantages from both parent technologies. Most of the current studies in this field address the system components and configuration to improve the performance of the hybrid technology for its use to remove pollutants such as pesticides, dyes, heavy metals, pharmaceuticals and personal care products. Hybrid systems provide significant water reclamation and reuse opportunities that are important to meet the growing needs to the population. Since they provide with such opportunities, the net per capita water use can be reduced significantly, thus freeing up the water supplies for other uses.

The major challenges to implement hybrid systems at full scale are the long-term operation performance of the systems, which includes maintaining optimum operating conditions (organic load, physicochemical and environmental factors), preventing electrodes from fouling or deterioration, maintaining the vegetation vitality and efficient wastewater treatment. Potential options for implementation should be first screened to identify the system features and issues of the process. Then, the system should be screened for affordability in comparison to the water treatment, water supply and management options for the existing area. Then, a sustainability analysis should be done for the viable options. Good planning and social acceptance can accelerate the implementation of hybrid systems.

REFERENCES

Araneda, I., Tapia, N.F., Lizama Allende, K., Vargas, I.T. 2018. Constructed wetland-microbial fuel cells for sustainable greywater treatment. *Water*, **10**(7), 940.

Bieker, S., Cornel, P., Wagner, M. 2010. Semicentralised supply and treatment systems: Integrated infrastructure solutions for fast growing urban areas. *Water Science and Technology*, **61**(11), 2905–2913.

Buller, L., Sganzerla, W., Berni, M., Brignoli, S., Forster-Carneiro, T. 2022. Design and techno-economic analysis of a hybrid system for energy supply in a wastewater treatment plant: A decentralized energy strategy. *Journal of Environmental Management*, **305**, 114389.

Capodaglio, A.G., Bolognesi, S., Cecconet, D. 2021. Sustainable, decentralized sanitation and reuse with hybrid nature-based systems. *Water*, **13**(11), 1583.

Chauhan, A., Prajapati, S.K. 2022. Algae Coupled Constructed Wetland System for Wastewater Treatment. in: *Integrated Environmental Technologies for Wastewater Treatment and Sustainable Development*, Elsevier, pp. 191–216.

Chavan, R., Mutnuri, S. 2021. Domestic wastewater treatment by constructed wetland and microalgal treatment system for the production of value-added products. *Environmental Technology*, **42**(21), 3304–3317.

Cheng, S., Huai, J., Zhong, F., Wu, J., Yu, S. 2022. Enhancing denitrification in constructed wetland with algae addition. *Environmental Science and Pollution Research*, **29**, 1949–1960.

Corbella, C., Puigagut, J., Garfí, M. 2017. Life cycle assessment of constructed wetland systems for wastewater treatment coupled with microbial fuel cells. *Science of the Total Environment*, **584**, 355–362.

Cui, Y., Rashid, N., Hu, N., Rehman, M.S.U., Han, J.-I. 2014. Electricity generation and microalgae cultivation in microbial fuel cell using microalgae-enriched anode and bio-cathode. *Energy Conversion and Management*, **79**, 674–680.

Daigger, G.T. 2009. Evolving urban water and residuals management paradigms: Water reclamation and reuse, decentralization, and resource recovery. *Water Environment Research*, **81**(8), 809–823.

Daigger, G.T., Crawford, G.V. 2007. Enhancing water system security and sustainability by incorporating centralized and decentralized water reclamation and reuse into urban water management systems. *Journal of Environmental Engineering and Management*, **17**(1), 1.

Ding, Y., Wang, W., Liu, X., Song, X., Wang, Y., Ullman, J.L. 2016. Intensified nitrogen removal of constructed wetland by novel integration of high rate algal pond biotechnology. *Bioresource Technology*, **219**, 757–761.

Doherty, L., Zhao, Y., Zhao, X., Hu, Y., Hao, X., Xu, L., Liu, R. 2015. A review of a recently emerged technology: Constructed wetland–microbial fuel cells. *Water Research*, **85**, 38–45.

Elmaadawy, K., Hu, J., Guo, S., Hou, H., Xu, J., Wang, D., Liang, T., Yang, J., Xiao, K., Liu, B. 2020. Enhanced treatment of landfill leachate with cathodic algal biofilm and oxygen-consuming unit in a hybrid microbial fuel cell system. *Bioresource Technology*, **310**, 123420.

Han, Z., Miao, Y., Dong, J., Shen, Z., Zhou, Y., Liu, S., Yang, C. 2019. Enhanced nitrogen removal and microbial analysis in partially saturated constructed wetland for treating anaerobically digested swine wastewater. *Frontiers of Environmental Science & Engineering*, **13**, 1–11.

Hume, N.P., Fleming, M.S., Horne, A.J. 2002. Denitrification potential and carbon quality of four aquatic plants in wetland microcosms. *Soil Science Society of America Journal*, **66**(5), 1706–1712.

Kannan, N., Donnellan, P. 2021. Algae-assisted microbial fuel cells: A practical overview. *Bioresource Technology Reports*, **15**, 100747.

Kataki, S., Chatterjee, S., Vairale, M., Sharma, S., Dwivedi, S., Gupta, D. 2021. Constructed wetland, an eco-technology for wastewater treatment: A review on various aspects of microbial fuel cell integration, low temperature strategies and life cycle impact of the technology. *Renewable and Sustainable Energy Reviews*, **148**, 111261.

Kumar, V.K., Manangath, S.P., Gajalakshmi, S. 2023. Innovative pilot-scale constructed wetland-microbial fuel cell system for enhanced wastewater treatment and bioelectricity production. *Chemical Engineering Journal*, **460**, 141686.

Li, H., Chi, Z., Yan, B., Cheng, L., Li, J. 2017. An innovative wood-chip-framework substrate used as slow-release carbon source to treat high-strength nitrogen wastewater. *Journal of Environmental Sciences*, **51**, 275–283.

Li, W.-W., Yu, H.-Q., He, Z. 2014. Towards sustainable wastewater treatment by using microbial fuel cells-centered technologies. *Energy & Environmental Science*, **7**(3), 911–924.

Liu, J., Pemberton, B., Lewis, J., Scales, P.J., Martin, G.J. 2020. Wastewater treatment using filamentous algae–a review. *Bioresource Technology*, **298**, 122556.

Milledge, J.J., Nielsen, B.V., Maneein, S., Harvey, P.J. 2019. A brief review of anaerobic digestion of algae for bioenergy. *Energies*, **12**(6), 1166.

Rashid, N., Rehman, M.S.U., Han, J.-I. 2013. Recycling and reuse of spent microalgal biomass for sustainable biofuels. *Biochemical Engineering Journal*, **75**, 101–107.

Shukla, M., Kumar, S. 2018. Algal growth in photosynthetic algal microbial fuel cell and its subsequent utilization for biofuels. *Renewable and Sustainable Energy Reviews*, **82**, 402–414.

Thakur, S., Das, S., Das, B. 2022. Application of upflow constructed wetland microbial fuel cell for treating sewage spiked with reverse osmosis concentrate with the concomitant generation of green electricity. *Biomass Conversion and Biorefinery*, **12**(9), 1–11.

Velasquez-Orta, S.B., Curtis, T.P., Logan, B.E. 2009. Energy from algae using microbial fuel cells. *Biotechnology and Bioengineering*, **103**(6), 1068–1076.

Vymazal, J. 2007. Removal of nutrients in various types of constructed wetlands. *Science of the Total Environment*, **380**(1–3), 48–65.

Weber, B., Cornel, P., Wagner, M. 2007. Semi-centralised supply and treatment systems for (fast growing) urban areas. *Water Science and Technology*, **55**(1–2), 349–356.

Xu, L., Zhao, Y., Doherty, L., Hu, Y., Hao, X. 2016. The integrated processes for wastewater treatment based on the principle of microbial fuel cells: A review. *Critical Reviews in Environmental Science and Technology*, **46**(1), 60–91.

8 Decentralized Anaerobic Treatment of Rural Domestic Sewage

Suyun Xu[1], *Ting Song*[1], *Liwen Luo*[2], *and Jonathan W.C. Wong*[2]
[1] University of Shanghai for Science and Technology, Shanghai, China
[2] Dongguan University of Technology, Dongguan, China

8.1 INTRODUCTION

In recent years, rural areas around the world have undergone economic and lifestyle transformations, which pose serious challenges to the treatment of increasing amounts of wastewater. Domestic sewage containing large amounts of nitrogen (N), phosphorus (P), organic matter and other pollutants is discharged directly into rivers and lakes due to a lack of environmental technicians and energy shortages in most rural areas (Robles et al., 2018; Zha et al., 2018). Compared with the treatment rate of urban sewage, the treatment rate of sewage from rural areas is much lower. For example, in China, the treatment ratios of municipal sewage and rural sewage are 91.97 and 13–34%, respectively, for most provinces. In addition, more than 96% of villages in China lack drainage channels and sewage treatment systems, which are usually discharged directly or through open channels (Zhang et al., 2019). Rural sewage cannot be effectively treated due to its relative scattering, less water volume and large water quality fluctuations (Rong et al., 2023). Scattered population, underdeveloped economy and limited input costs make the governance of the rural environment unable to keep up with the development and construction of rural areas (Song et al., 2020). Therefore, more and more attention has been paid to rural environmental problems, especially the treatment of rural decentralized sewage.

Decentralized wastewater treatment solutions are built on the concepts of dispersion, simplicity and product reuse (Geetha Varma et al., 2022). Decentralization makes increasing sense for communities living in rural areas where it is economically unfeasible to build large collection systems (Badeti et al., 2022). Simplicity is achieved by the on-site handling without consuming chemicals, reducing equipment and energy requirements and easy maintenance, which requires less operation and maintenance skills. The expected advantages of decentralized sewage treatment facilities are their reliability, robustness, low energy consumption and ability to bear shock loads (Garrido-Baserba et al., 2022). Furthermore, treated sewage can be reused near the community, reducing water usage (Buller et al., 2022). Thus, the decentralized sewage treatment can be considered as a strategy that takes into account the unique economic and social circumstances of the community. For instance, about 25% of the population in the United States is served by small-scale, decentralized sewage treatment facilities (Capodaglio et al., 2017).

Quantity and quality of wastewater, legal discharge regulations, cost, local temperature, availability of land, cultural acceptance and social conditions and the final treatment of effluent (discharge or reuse) are all factors that influence the technical configuration of decentralized wastewater treatment solutions (Rattier et al., 2022). Anaerobic reactor (AR) and constructed wetland (CW) are commonly used treatment units in sewage decentralized treatment system, which are effective units of sewage treatment, especially suitable for small-sized and medium-sized communities and residential areas (Arias et al., 2020; Fernández del Castillo et al., 2022). The investment costs (including construction costs and maintenance costs) are highly different for different solutions such as sewage

DOI: 10.1201/9781003381419-10

treatment plant (STP), ARs and CWs. As per government report, the cost of building the capacity to treat 1 m^3 of sewage per day is estimated to be \$513, \$100 and \$24, respectively, for the STP, CWs and ARs system. Materials, energy, labor and land are the main inputs in the construction process. Meanwhile, the operating and maintenance costs of STP, CWs and anaerobic digestion systems generated for the government are estimated at US \$0.08, \$0.01 and \$0.001/m^3 of sewage, respectively. The operation of STP requires significant inputs of chemicals, electricity and highly specialized labor, resulting in much higher inputs of STP than CWs and anaerobic digestion systems. This means that STP is capital-intensive while the other two technologies are capital efficient.

Anaerobic treatments are essential in the decentralized wastewater treatment approach, which offers the capacity for water and nutrient recovery and represents the most cost-effective technologies for domestic wastewater (Chen et al., 2020; Santiago-Díaz et al., 2021). By applying anaerobic biotechnologies, various kinds of organic solid wastes generated from nearby community can be treated cooperatively (Le et al., 2022). Thus, depending on the technical setup, more biogas can be produced and used for cooking, lighting or electricity generation (Tian et al., 2021). Upflow anaerobic sludge blanket (UASB) reactor is the most popular anaerobic system, which is considered sustainable and suitable for decentralized treatment (Rattier et al., 2022). As an anaerobic membrane bioreactor (AnMBR) can effectively utilize organic matter in wastewater to generate biogas for subsequent energy production, remove a series of emerging pollutants associated with water reuse and realize the nutrient removal and recycling, as a result, the AnMBR for sustainable development and energy saving of decentralized wastewater treatment has great potential (Chen et al., 2020). Currently, AnMBRs is beginning to evolve to the next level of sustainable and energy-efficient operation.

8.2 DECENTRALIZED ANAEROBIC TREATMENT PROCESSES

Currently, there are four main types of anaerobic treatment processes applied in the decentralized STP, i.e., UASB reactor, anaerobic baffled reactor (ABR), anaerobic filter (AF) and AnMBR. Their performance and characteristics are compared and summarized in Table 8.1. UASB is the most popular anaerobic system, which is considered sustainable and suitable for decentralized treatment (Rattier et al., 2022). As AnMBR can effectively utilize organic matter in wastewater to generate biogas for subsequent energy production, remove a series of emerging pollutants associated with water reuse and realize the nutrient removal and recycling, as a result, the AnMBR for sustainable development and energy saving of decentralized wastewater treatment has great potential

TABLE 8.1
Comparison and Analysis of Anaerobic Bioreactors

No.	Reactor	Advantages	Constraints
1	UASB	Effective separation of mud and water; No microbial adsorption carrier required; High load and short hydraulic retention time	The structure of the three-phase separator is complex.
2	AF	Biocarrier is effective to prevent sludge loss; Strong resistance to impact load	Biocarrier is easy to age and the cost is high.
3	ABR	Effective sludge retention; Strong anti-shock load capacity; Distribution of functional bacteria in the compartment; Good effluent quality	Lack of agitation, prone to dead zones.
4	AnMBR	Effective sludge retention, good effluent quality; Low sludge yield and small footprint	Membrane fouling, high membrane cost.

(Chen et al., 2020). Currently, AnMBRs is beginning to evolve to the next level of sustainable and energy-efficient operation.

8.2.1 Upflow Anaerobic Sludge Blanket (UASB)

The upflow anaerobic sludge bed was developed in the 1970s by Lettinga et al. for industrial wastewater treatment in ARs, and the structure includes a water distribution system, a reaction zone, a three-phase separator, an outlet water and an air chamber area. As shown in Figure 8.1, the sewage is uniformly distributed through the water distribution system at the bottom of the reactor, through the sludge bed and suspended sludge layer of the reaction area, so that the sewage is in full contact with the microorganisms (Mainardis et al., 2020). Organic matter in sewage is converted into energy biogas, the water-gas-sludge is separated by a three-phase separator, where the biogas is collected through the collector swamp, the sludge returns to the sludge bed, and the effluent is discharged from the reactor through the sedimentation area.

The mean removal efficiencies reported for UASBs are 65 ± 12%, 68 ± 10% and 67 ± 10% in terms of total suspended solids (TSS), chemical oxygen demand (COD) and 5-day biochemical oxygen demand (BOD_5), respectively (Fernández del Castillo et al., 2022). The removal of phosphorus (P) is not commonly reported for UASB (Yang et al., 2021). However, due to the formation of ammonia during the anaerobic digestion, the negative removal efficiencies of nitrogen (N) have been reported for some UASB. The advantages of UASB are the absence of agitation equipment, the disturbance of the generated biogas, the enhancement of the slurry mixing efficiency and the addition of a three-phase separator to effectively prevent sludge loss and maintain a high biomass (Santiago-Díaz et al., 2021).

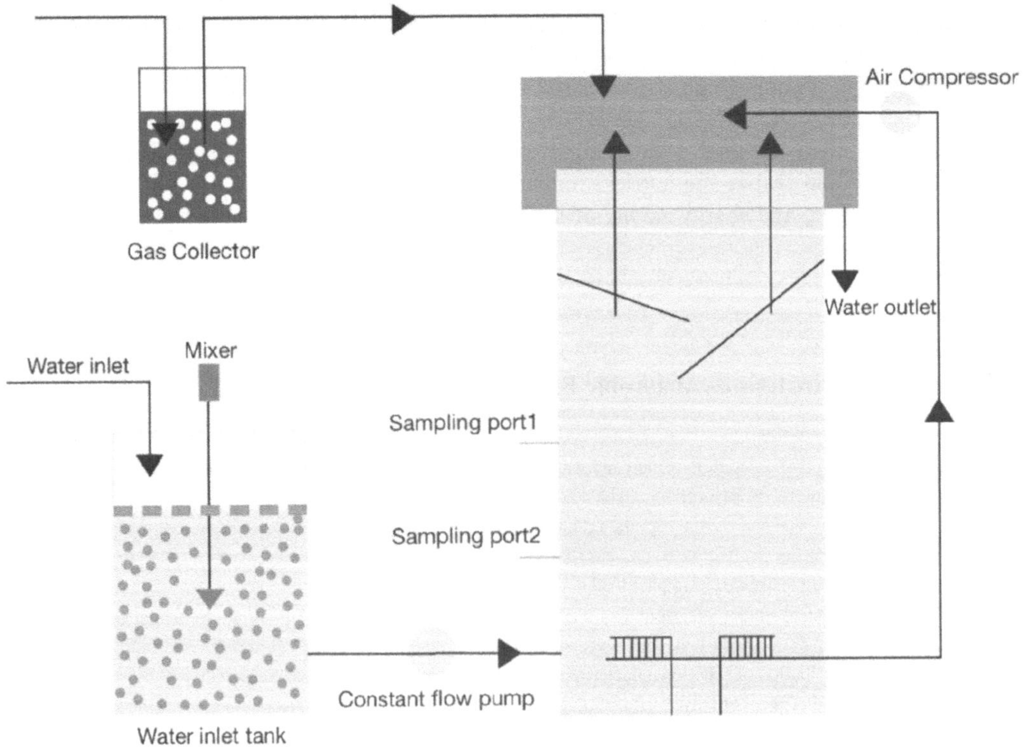

FIGURE 8.1 Configurations of upflow anaerobic sludge blanket (UASB).

To maximize biogas production and recovery of resources from sewage through anaerobic digestion, a combination approach of source separation and resource recovery has been proposed as an alternative that can improve the sustainability, reliability and resilience of sewage treatment systems (Garrido-Baserba et al., 2022). By using the corresponding toilet types (i.e., vacuum, urine-diverting and conventional toilets), black, gray, yellow, brown and combined wastewater can be separately obtained, which is more favorable for resource recovery. UASB can bear a higher organic loading rate as compared to other types of AR, which is suitable for the treatment of concentrated black (toilet) water. In the study of (De Graaff et al., 2010), the average removal rate of COD was 78% at a short HRT of 8.7 days at 25°C.

8.2.2 Anaerobic Baffled Reactor (ABR)

The ABR is the addition of vertical baffles in the reactor based on the theory of staged multiphase anaerobic process, which can also be considered as a series of connected simplified UASBs or serial septic tanks (Singh et al., 2009). As presented in Figure 8.2, within the ABR system, wastewater flows longitudinally through compartments separated by the vertical baffles, passes downward into the baffles and up through the fluidized anaerobic sludge zone in each compartment (Figure 8.2). So that the organic matters in the sewage can be fully contacted with anaerobic microorganisms, and then the organic matters are intercepted and adsorbed to degrade, and the inorganic suspended solids can also be settled down (Santiago-Díaz et al., 2021).

Typically, ABRs remove 30–50% of BOD_5 and 60–80% of TSS. Similarly, the removal rates of BOD_5 and COD in ABRs are in the range of 50%, while the TSS removal rate is about ~70% (Mladenov et al., 2017).

In summary, the main improvements on the ABR reactor are: (i) increased volume ratio of the upflow and downflow compartments, thereby reducing the flow rate of the upflow compartment; (ii) different types of fillers are add in the compartment to increase the sludge residence time and

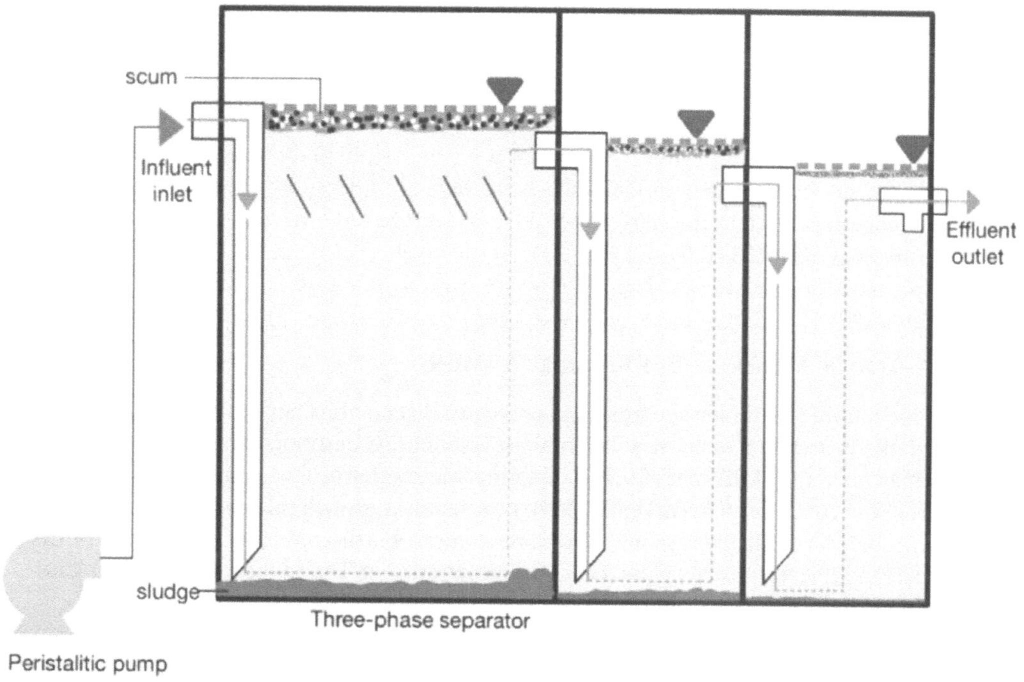

FIGURE 8.2 Configuration of anaerobic baffled reactor (ABR).

prevent sludge loss; (iii) letting the lower part of the baffle form a 45-degree angle with the horizontal plane; (iv) increasing the sedimentation chamber in the last compartment and returning the sludge to the first compartment (Liu et al., 2022).

The purpose of all these improvement measures is to increase the residence time of the sludge in the reactor, reduce the sludge loss and ensure that the bacterial cells are free from pollutants, thereby improving the processing efficiency of the reactor. For example, researchers widened the upward flow chamber of the ABR reactor and narrowed the downward flow chamber, which is more conducive to the contact between microorganisms and water (Wang et al., 2004). It will provide a high efficiency for extreme environmental conditions and inhibitory compounds, such as palm oil mill effluent.

Another approach to improve effluent quality is to combine an ABR system with a post-treatment polishing step, such as an AF to remove soluble COD or a CW to remove nutrients (e.g., NH_4^+-N). In fact, the packing materials can also be directly applied in ABR configurations to strengthen the contact between the organic matter in the sewage and the sludge and improve the treatment rate of the reactor (Sabzali et al., 2012). For example, Faisal and Unno (2001) improved the ABR reactor into a carrier ABR (CABR), which integrates the advantages of ABR reactor and the biofilm reactor, which is a kind of reactor suitable for rural areas. Biofilm growth on the carrier can lead to the adsorption of soluble compounds to the biofilm and the degradation of compounds by the microbial community in the biofilm. This kind of reactor can also be referred to as AF or anaerobic moving bed biofilm reactor (AMBBR).

8.2.3 Anaerobic Filter (AF)

The AF, also referred to as anaerobic fixed-film/bed reactor, is a process developed by Young and McCarty that uses filler as a carrier for microbial attachment and growth (Fernández del Castillo et al., 2022). Organic matters in sewage are intercepted, adsorbed, decomposed and converted into CH_4 and CO_2 by the attached microorganisms through the packing space. Biocarriers are an important component of AFs. Both natural fillers and synthetic fillers can be adopted as the biocarriers, such as gravel, brick, slag, sponge, Pall ring and other applications of a wide range (Sabzali et al., 2012).

According to the way of inlet water, AFs can be divided into upflow AFs and downflow AFs. Upflow AF is more commonly used as compared to downflow AFs. Upflow AFs have been reported to withstand impact loads and to be highly tolerant to high NH_4^+-N concentrations and elevated pH (Calli et al., 2006). The advantages of AF reactor are low cost, high treatment efficiency and easy regulation. The suspended solids in the domestic sewage are filtered by the filler, which prevents the loss of sludge and shortens the start-up period. The average removal rates of ARs for TSS, COD and BOD_5 are $86 \pm 9\%$, $79 \pm 10\%$ and $76 \pm 23\%$, respectively (Fernández del Castillo et al., 2022). However, the nutrient removal (i.e., N and P) by AFs is rarely reported.

8.2.4 Anaerobic Membrane Bio-Reactor (AnMBR)

AnMBRs are referred to the sewage treatment processes that combine anaerobic biological reaction and membrane separation to achieve solid-liquid separation. As described previously, some serious problems may arise in UASB reactors when treating wastewater of low strength, especially under the psychrophilic conditions (Arias et al., 2020). Low biomass growth rate and high hydraulic loading rate will increase sludge wash-out at low temperatures. Furthermore, the application of UASB in wastewater treatment is restricted by the high concentration of TSS in the influent (Lettinga et al., 2001).

Compared with traditional anaerobic bioreactors, AnMBR has a significant advantage i.e., it can retain a large amount of sludge, promote the anaerobic reaction, improve effluent quality, reduce sludge yield and have a small footprint (Chen et al., 2020). Membrane barrier leads to the retention

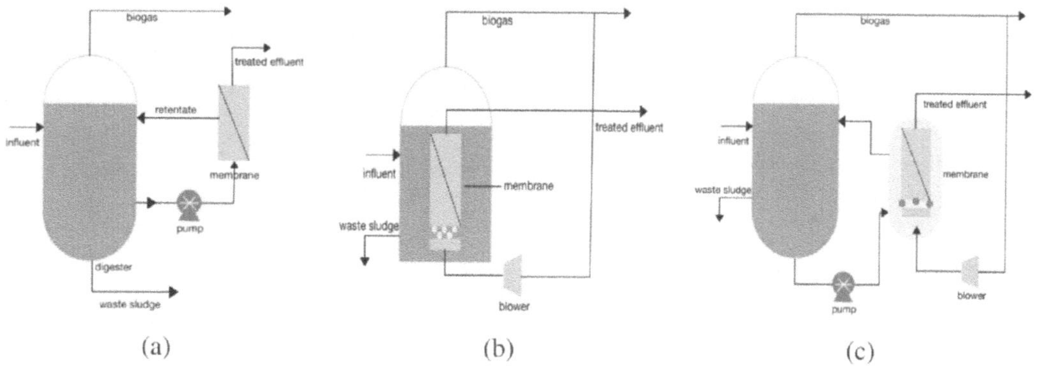

FIGURE 8.3 Three kinds of anaerobic membrane bioreactor (AnMBR): (a) side-stream membrane bioreactor; (b) immersed membrane bioreactor; (c) combination of side-stream approach and immersed AnMBR. (Revised from Robles et al., 2018.)

of biomass, suspended solids and dissolved/colloidal organic matters, which greatly increased the total COD removal rate to 92.3%, and about 84.7% of total COD was converted to CH_4 (Sanchez et al., 2022).

According to the position of the membrane module, AnMBR can be divided into a side stream membrane bioreactor and an immersed (submerged) membrane bioreactor (Figure 8.3). Among them, the immersed AnMBR sets the membrane module in the reactor and uses the biogas cycle generated in the reactor to control membrane fouling with low energy consumption, but the relative membrane flux is low (Robles et al., 2018).

Despite the promising nature of submerged AnMBR, membrane fouling is the major challenge that limits its widespread application, reducing its operating performance and decreasing the permeate flux (Olubukola et al., 2022). Various technical approaches have been proposed such as ultrasound irradiation, back-feeding, sonochemical control and *in situ* biogas sparging (Hu et al., 2018). Nevertheless, the high operational costs of ultrasound and sonochemical controls limit their application. Biogas sparging is deemed as one of the most effective approaches for membrane cake fouling control. As long as the biogas is produced, it is provided enough flow rate required for sparging operation, so the energy consumption is largely reduced. It has been reported that the gas injection energy and cost of the submerged membrane are approximately three times lower than that of the side-flow AnDMBR at a given flow rate (Jeison & van Lier, 2008).

Apart from biogas sparging, other emerging technologies have also been proposed to tackle the biofouling problem. For example, as seen in Figure 8.4a, membrane vibration in AnMBR improves the recovery of dissolved methane and energy-saving pollution mitigation (Wang et al., 2022b). Dynamic membranes (DMs) developed on supporting materials (filters) are also a good choice for controlling fouling problems (Sanchez et al., 2022). Once the DM is formed, the solids accumulate on the membrane, which experiences the formation and re-formation of the biofilm, while the mesh acts only as a support. Due to the characteristics of large pores, DM is widely adopted before the conventional MF/UF-MBRs (Hu et al., 2018). An electroconductive moving bed membrane bioreactor (EcMB-MBR) was proposed, in which electrodes and free-floating conductive media were introduced to a submerged MBR to reduce biofouling (Figure 8.4b). As a result, the simultaneous removal of organics (~97.1% of COD) and nutrients (~88.8 and 99.0% of TN and TP) was achieved from domestic wastewater; thus, 24% TN removal was enhanced through the hybrid bio-electrochemical denitrification (Udomkittayachai et al., 2021).

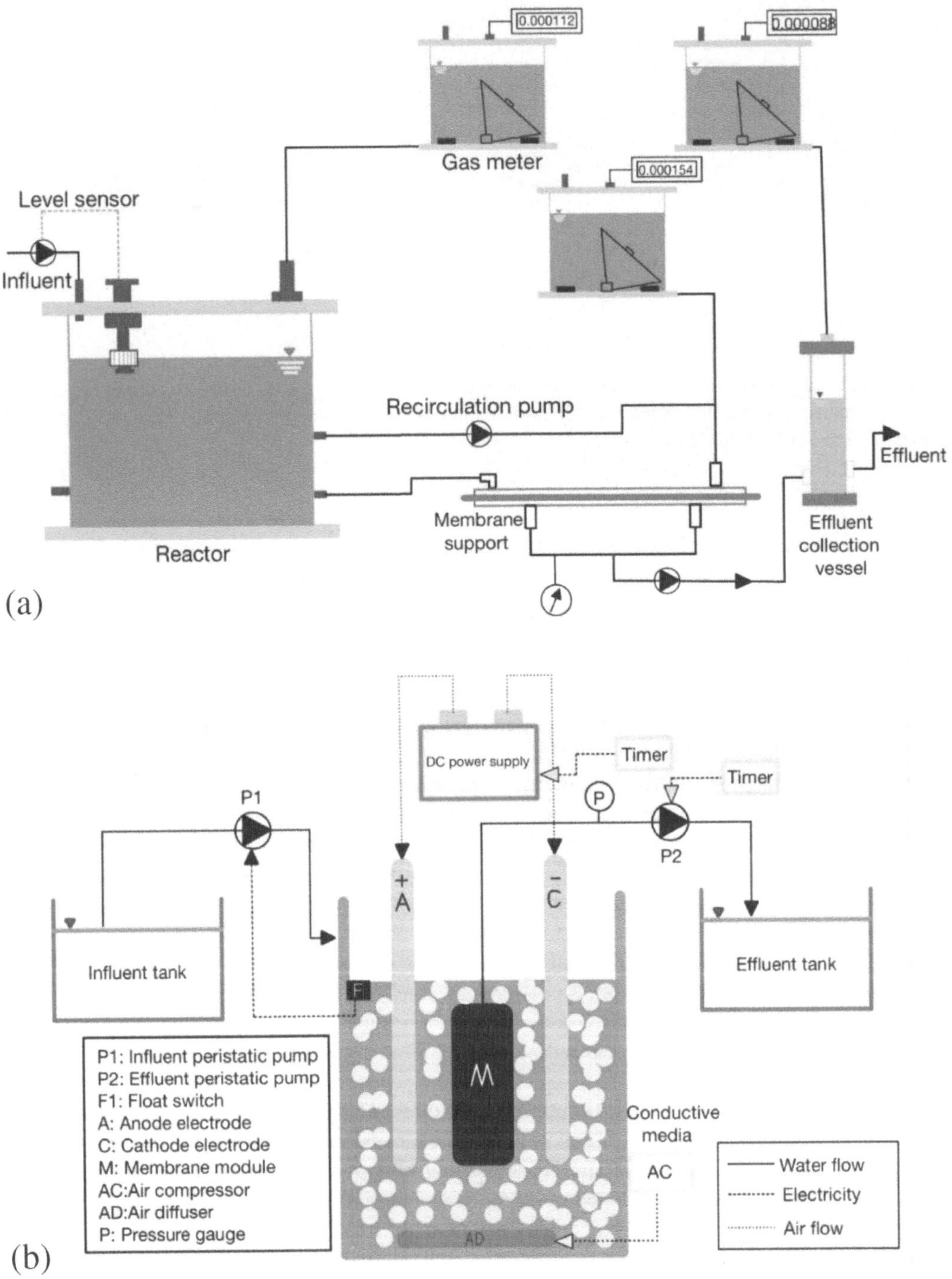

FIGURE 8.4 Schematic diagrams for (a) anaerobic dynamic membrane bioreactor (ADMBR) and (b) electroconductive moving bed membrane bioreactor (EcMB-MBR). (Revised from Alibardi et al., 2016 and Udomkittayachai et al., 2021.)

8.3 CONFIGURATION IMPROVEMENT OF DECENTRALIZED ANAEROBIC DIGESTER

8.3.1 Integrated Anaerobic Reactors with Constructed Wetlands (CWs) Systems

Although high removal efficiencies of TSS and COD can be achieved in various ARs, the low or even negative efficiencies have been reported for nutrient removal (Liu et al., 2013). In recent years, CWs have gained widespread popularity due to their relatively simple construction, operation and maintenance, as well as their cost-effectiveness. At the same time, this nature-based system exhibits high robustness and stability against organic and hydraulic load fluctuations (Gikas & Tchobanoglous, 2009; Li et al., 2009). The critical evaluation of the use of decentralized technologies in developing countries revealed widespread adoption of CWs (Muzioreva et al., 2022). CWs systems have shown the robust capacity to efficiently remove various nutrients, such as TN, NH_4^+-N and TP (Gikas & Tchobanoglous, 2009; Kamilya et al., 2022).

8.3.1.1 Constructed Wetlands (CWs)

CWs are a kind of nature-like system with engineering design, which is a combination of vegetation (emergent, floating or submerged), biological carrier (rock, soil, synthetic materials etc.) and the microbial community in a controlled environment. The wastewater flows through this system, and the organic and inorganic pollutants are removed via the combined physical-chemical-biological effects (Lemley et al., 2022; Verduzo Garibay et al., 2021).

The removal efficiency of pollutants in CW systems depends on various mechanisms and pathways. The following three mechanisms are characterized in the systems (Oliveira et al., 2021; Tan et al., 2021): (i) physical removal processes (sedimentation, volatilization, adsorption); (ii) chemical processes (photolysis, chemical precipitation, oxidation-reduction,); and (iii) biological processes (microbial degradation, nitrification-denitrification, phytoremediation, rhizosphere filtration).

However, CWs systems also have some limitations that can limit their optimal operation and long-term stability. For example, matrix clogging is one of the most common operational problems, especially when treating wastewater with high TSS and organic matter loads, which can also be harmful to macrophytes in CWs (Vymazal, 2013). Therefore, pre-treatment stages are required to reduce the organic load into the CWs. In addition, when used alone, CWs tend to exhibit low nutrient removal efficiency, hampered by low oxygen transfer capacity, and each type of CWs has some disadvantages (Wu et al., 2014).

8.3.1.2 Combination of Anaerobic Reactors with CWs

The incorporation of ARs with CW systems could buildup synergistic effects for sustainable wastewater treatment, in which the organic matters are degraded in the AR and the nutrients are removed in the CW systems (see Figure 8.5). For example, adding UASB and ABR systems before CWs has been reported to improve the removal efficiencies of both COD and nutrients (TN and TP), i.e., 81, 82 and 67%, respectively (Fernández del Castillo et al., 2022). The removal rates of COD and TSS can be elevated to 77 and 91% in the process of coupling ABR with horizontal flow and vertical flow CWs (Singh et al., 2009).

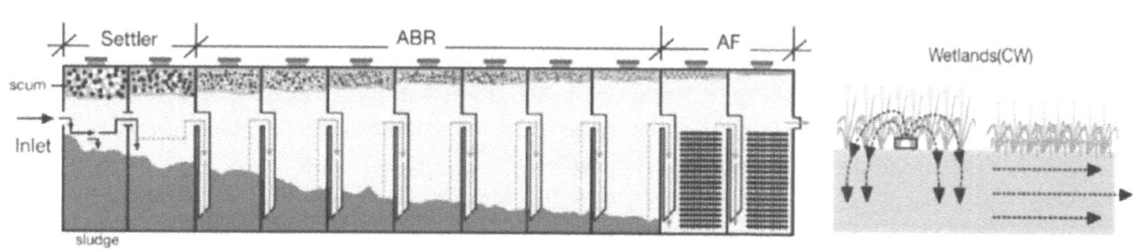

FIGURE 8.5 Configuration of anaerobic baffled reactor (ABR) and aerobic filter (AF) followed with the constructed wetlands (CW).

Combining a DM bioreactor with a CW system can achieve 93% and 90.4% TN and TP removal rates, respectively (Kamilya et al., 2022). The combined system of submerged membrane bioreactor and CWs can handle wastewater with high concentrations of COD (1008.08 mg/L), TN (95.22 mg/L) and TP (5.76 mg/L), with removal rates of 98, 96 and 80%, respectively (Xiao et al., 2010). Moreover, the coliform removal has also been reported in the AR + CW configuration, achieving removal efficiencies of approximately 97–100% (Fernández del Castillo et al., 2022; Raboni et al., 2014).

Dissolved oxygen (DO) levels in CWs play an important role in nutrient removal, especially for nitrogen removal. Atmospheric reoxygenation accounts for DO accumulation in CWs, and 99.8% of DO is consumed by microorganisms during the decomposition and nitrification of organic matter (Ye & Li, 2009). With the increase of DO concentration, the overall nitrification rate increased. At high DO levels, NH_4^+ can be completely oxidized to NO_3^-, whereas at a limiting DO level, the conversion of NH_4^+ to NO_3^- is inhibited. A stepwise plunge pool is designed to connect the ABR and the CWs; in this combined process, the stepwise passive aeration approach has the largest contribution to the removal of NH_4^+-N and TP (John et al., 2020). The water quality of the effluent from the combined process was 159.1–209.9 mg/L for COD, 54.3–79.7 mg/L for NH_4^+-N, 4.4–5.9 mg/L for TP and 102–153 mg/L for SS, which reached the secondary discharge standard of Agricultural Irrigation Water Quality Standard (GB5084) (Gui et al., 2018).

8.3.2 Integrated Anaerobic Reactors with Microbial Electrolysis

Microbial electrochemical technology has been declared as promising for limiting greenhouse gas emissions and promoting sustainable energy conversion. In particular, microbial electrolytic cell (MEC) has great potential for anaerobic sewage treatment (Park et al., 2020). Integrating MEC with AD process (MEC-AD) has been proposed as a practical solution to regulate the microbial community structures and electron transfer pathways, thereby improving overall system stability and energy efficiency (Wang et al., 2022a).

The modified anaerobic-anoxic-aerobic processes (A^2/O) are commonly used to treat mixed blackwater, but denitrification efficiency is usually limited due to the competition for carbon sources. Qin et al. (2022) have built a microbial fuel cell (MFC) and a MEC embedded in the anaerobic-two stage anoxic/aerobic ($A\text{-}(A/O)^2$) system treating mixed black water, where the MFC grades organic matter and generates electricity and then the electroactive bacteria in MEC catalyze denitrification at low carbon levels. The overall removal rates of TN and TP were up to 91.3% and 98.3%, respectively, at a stimulation voltage of 0.1 V.

8.4 MODELING AND CONTROLLING THE BIOTECHNOLOGICAL PROCESSES

The development of the decentralized sewage treatment model has just started, and the basic research on the treatment technology has just begun. In order to study the sensitivity of the decentralized wastewater treatment process to various operating parameters and optimize the reactor design, a mathematical model with a simple concept and close to the actual situation of the reactor is needed. In the past decades, there have been many studies on modeling UASB, AF and AnMBR reactors, as summarized by Saravanan and Sreekrishnan (2006).

The parameters that affect the performance of AR are as follows: (i) effect of hydrodynamics/flow pattern on reactor performance; (ii) mass transfer in suspended sludge and biofilms; (iii) kinetic effects; (iv) structure and composition of biogranules.

8.4.1 Anaerobic Digestion Models

Modeling of anaerobic processes is a mature and well-developed field that is largely guided by mechanistic model structures defined by the understanding of the underlying processes. Probably,

IWA ADM1 is the most widely used model, which has been effectively applied for different purposes (Robles et al., 2018). To obtain an accurate prediction of wastewater treatment plants treating low-intensity sewage, Biological Nutrient Removal Model No. 2 (BNRM2) was developed by Barat et al. (2013), which extended the components and processes required to simulate nitrogen removal via nitrite oxidation and ammonia oxidation as well as the formation of the solids.

8.4.2 Fluid Dynamics Model

Hydraulics and hydrodynamics are two of the most important factors in the design and operation of various sewage treatment systems (Hassanvand Jamadi & Alighardashi, 2017). The hydraulic characteristics of sewage crossflows are important factors in determining the specifications and dimensions of different units in biological treatment processes. The fluid characteristic parameters in the reactor include hydraulic retention time (HRT), material distribution, hydraulic dead zone, whether there is a short circuit in the reactor, and the type of flow regime depends on the flow hydraulic power and reactor volumetric efficiency (Sarathai et al., 2010).

Dynamic similitude is used to model hydraulic structures so that the calculation results obtained from the model reactors can be applied to the prototype reactor. Computational fluid dynamics (CFD) can determine important factors that are difficult or impossible to measure, including internal flow distribution and membrane shear rates. CFD has been applied to the modeling of different systems for the production of biomethane and biomethane by biochemical conversion (Wu, 2013).

The CFD simulations are not only able to predict mixing time but also the formation of a vortex using the volume of fluid (VOF) method and the suspension of solids using the algebraic slip model (ASM), which gives access to the necessary information for reactor scale-up. Trad et al. (2015) have developed the hydrodynamics modeling and optimized the mixing strategy in an AnMBR by using the single-phase and two-fluid CFD models. As a result, the optimum rotation speed and mixer design are consistent with those obtained from experiments and simulations.

8.5 CHALLENGES AND OPPORTUNITIES FOR DECENTRALIZED ANAEROBIC TREATMENT PROCESS

With the continuous development of the rural economy, the demand for water resources in rural areas is increasing, and the subsequent sewage discharge is also increasing. Therefore, the rural sewage decentralized treatment technology is still facing huge challenges. Wherein, the remarkable problems appeared are the uncertain effluent quality (resulting from the low removal efficiency of nutrients), the requirements of post-treatment to some extent, the increasing demand for resource recovery and water reuse etc.

8.5.1 Strive to Meet the Effluent Standards

The scale of rural decentralized sewage treatment technology is generally small, and it is currently only suitable for domestic sewage treatment in a small area, such as a restaurant or hotel, and the rural decentralized sewage treatment model is not particularly perfect. The quality of infrastructure construction of sewage treatment technology is generally not high, the safety performance cannot be guaranteed and it is easy to induce safety accidents.

Decentralized wastewater treatment plants are also regulated to comply with environmental protection standards. For example, the European Commission's Water Framework Directive 2000/60/EC sets limits for discharges from WTP in communities with a population of 2000 or more, that is, 25 mg/L BOD_5, 125 mg/L COD and 60 mg/L TSS and N/P are not included (Brooke, 2012). In China, more than 70 biogas standards have been issued at the national and industry levels. Most of these standards are organized and implemented by the Ministry of Agriculture due to its leading role in promoting rural biogas construction.

Although there is no unified standard for rural sewage discharge, some local standards have been issued by local governments in some developing areas such as DB31/T 1163-2019 issued by the Shanghai Ecology and Environment Department, DB11/1612-2019 issued by Beijing Municipal Ecological Environment Bureau (Xie et al., 2021). Therefore, in areas with existing standards, the sewage discharge system should be designed according to the standard. Meanwhile, it needs to consider the capacity of the water body. For example, in the grade I standard of the "Discharge Standard of Pollutants for Municipal Wastewater Treatment Plant" (GB 18918-2002), the limit value of COD required for Grade I-A is 50 mg/L. However, Beijing still proposes a higher threshold of 30 mg/L in the rural standard (DB11/1612-2019), due to the characteristics of urban development. The treatment effect of rural sewage can only be achieved if the discharge standard requirements for effluent water quality are met.

8.5.2 Onsite Water Quality Monitoring to Ensure the Stability of Decentralized Sewage Treatment System

Real-time and continuous monitoring of wastewater quality at decentralized onsite facilities is critical to ensure the system's operation, as the fluctuations of influent can limit system stability. On-site monitoring also allows early capture of effluent violations (Gu et al., 2017). To obtain the ideal long-term "maintenance-free" monitoring in decentralized wastewater treatment facilities, uninterrupted transmission of sensor readings may be a solution (Xu et al., 2016). However, frequent data transfers (e.g., once per second, once per minute) require external energy. Existing solar, thermal and wind harvesting systems are generally expensive. Therefore, it is challenging to find a reliable, low-cost *in situ* sensor with an energy harvesting and supply system to facilitate uninterrupted reading and transmission.

Conventional wastewater biosensors require coated enzymes and/or pure bacterial cells and an external power supply (Woznica et al., 2010). Microbial fuel cells (MFCs) generate electricity and promise self-sustaining monitoring, recording and data transmission, which greatly simplifies device setup and extends service life. As shown in Figure 8.6a, bacteria are used as biocatalysts to oxidize organic and inorganic matter and generate electrical currents in MFC devices. The microbes in the anodic chamber act as biometric elements, while the proton exchange membrane and electrodes act as transducers and can therefore be considered as a biosensor (Peixoto et al., 2011). Bacteria are sensitive to the sudden presence or change in the level of a target analyte (such as organic carbon, ammonia, nitrate or even heavy metals) and thereby respond to the output current (Do et al., 2022). The volume of MFC biosensors can be reduced to less than 10 mL (so-called miniature microbial fuel cell, MiMFCs), which has a great potential as a mini-power supply for water sensors (Fan et al., 2021). The MiMFCs can be deployed at multiple locations throughout the reactor to support a large number of water sensors harvesting power on-site from wastewater.

Yu et al. (2019) also found that the conductivity (EC) of the rural domestic sewage treatment facilities is in good correlation with the concentrations of TN ($R^2 = 0.7068$, $p < 0.01$), NH_4^+-N ($R^2 = 0.6718$, $p < 0.01$) and TP ($R^2 = 0.5944$, $p < 0.01$). The results show that EC can reflect the changes in TN, NH_4^+-N and TP concentrations in rural domestic sewage, so as to dynamically adjust the treatment load of decentralized treatment facilities (Yu et al., 2019).

Furthermore, to reduce the cost of online sensor technologies and the independently developed control system for decentralized wastewater treatment plants, the distributed data collection system architecture has been developed recently (such as iTWat) (O'Donovan et al., 2015). Figure 8.6b illustrates the iTWat architecture of a distributed data collection system, with each interacting component described in the context of its physical location, working together to provide an automatic data collection mechanism for numerical measurements and binaries (such as images or CSV) from

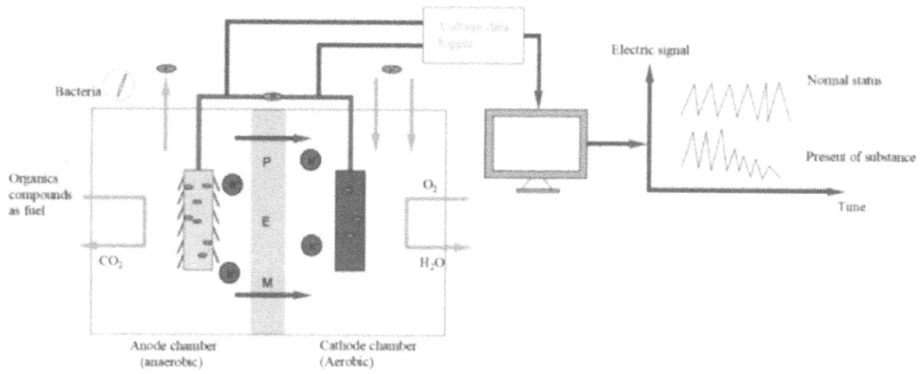

(a)

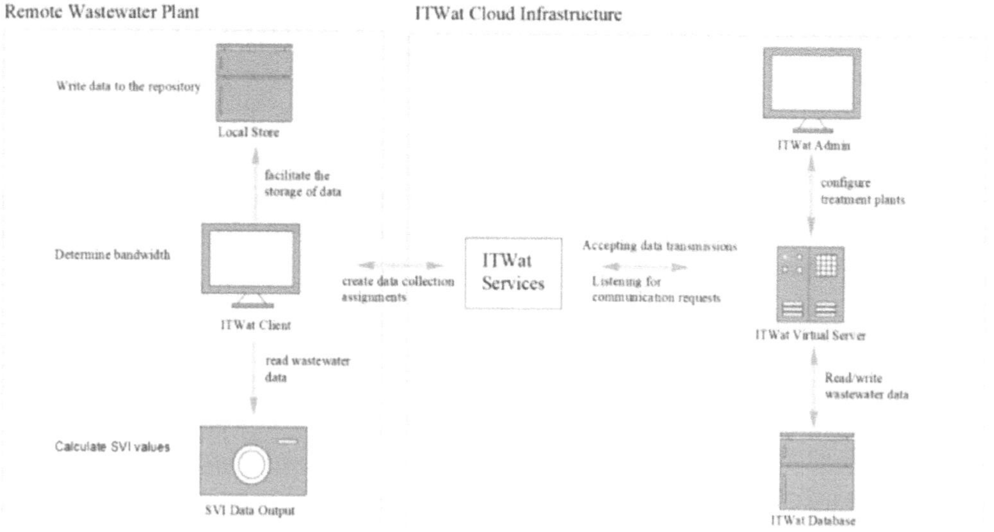

(b)

FIGURE 8.6 (a) Microbial fuel cell-based biosensor for online monitoring wastewater quality (revised from Do et al., 2020); (b) distributed data collection system architecture and components. (Revised from O'Donovan et al., 2015.)

a decentralized processing system. Simply put, this iTWat system provides subject matter experts with timely access to data for monitoring and analysis without the logistical and technical costs associated with manual data collection. It can help engineers or wastewater treatment plant managers process and interpret data quickly.

8.5.3 Technologies for Resource Recovery and Reuse Demand

Nutrient recovery is one of the main goals of sustainable development and circular economy. The best strategy for wastewater management is to shift the linear production path to a closed cycle, thereby minimizing losses and waste reuse (Lieder & Rashid, 2016). The concept of circular economy is based on reuse, reduction, recycling, recovery and recycling. Resource recovery involves a variety of physical, chemical and biological processes, such as precipitation, adsorption, nutrient or microbial assimilation (Chen et al., 2021).

8.5.3.1 Agricultural Irrigation

Since irrigation water is the main source of total water withdrawal, the use of sewage discharge for irrigation helps ensure the sustainability of water resources (Chen et al., 2021). In China, the level of water saving in agriculture is still very low, and water consumption needs to be controlled. Its development would benefit the achievement of water conservation targets in the agricultural sector if effluent from decentralized STPs could be reused in agriculture (Mannina et al., 2022).

As pH, electrical conductivity (EC), oil/grease etc. can interfere with nutrient uptake by the plant; thus, their levels should be pretreated and adjusted to a favorable range to meet the requirements for plant growth (Hussain et al., 2019). Among these parameters, EC significantly affects crop productivity. Table 8.2 summarizes irrigation water quality standards and/or recommendations for a number of different countries (Chen et al., 2021; Mannina et al., 2022). Generally, pH should be in the range of 6.0–8.0, while EC shows a wide range of 300–7000 μS/cm. In some Asian countries, SS is not allowed to exceed 100 mg/L. The standard of oil and grease for irrigation water is about 5.0 mg/L. Some Asian countries have more stringent standards for BOD_5, COD and TN. For example, in China, according to the standard of GB 5084-2005, parameters for irrigation water are COD 200, BOD_5 100, SS 100, coliforms 4000/100 mL, parasitic ovum 2/L and pH 5.5–8.5.

Furthermore, rural sewage is also accompanied by a large number of pharmaceuticals and personal care products (PPCPs) and other organic pollutants with environmental migration, bioaccumulation and ecotoxicity, which have become an emerging focus in the field of wastewater treatment and agricultural environment (Wu et al., 2023). Although the potential risks of PPCPs discharge in urban areas were greater than that in rural areas, its impact on aquaculture in rural areas was noteworthy. The development of economical and efficient degradation technology for organic pollutants is the current research hotspot and application frontier (Meyer et al., 2019).

8.5.3.2 Nutrient Recovery

As reported, both artificial and natural adsorbent materials can be used as substrates to recover nitrogen from ARs wastewater. In particular aluminosilicate material, such as zeolite, can effectively

TABLE 8.2
Irrigation Water Quality Standards and Recommendations

Item	China (GB5084-2021)	India (Porush Kumar et al., 2022)	Japan (Matsuno, 2008)	USEPA long term	USEPA short term	Egypt-advanced (Abu-Zeid, 1998)	Israel (Jeong et al., 2016)	Italy (Angelakis & Durham,2008)
Temp.	35	25	–	–	–	–	–	–
pH	5.5–8.5	8.5	6.0–7.5	6.0	6.0	–	6.5–8.5	6.0–9.5
EC	–	–	300	–	–	–	1400	3000
SS	80	–	100	–	–	20	10	10
DO	–	6.0	>5.0	–	–	–	–	–
Oil	–	–	–	–	–	5.0	–	–
BOD	60	2.0	–	–	–	20	10	20
COD	150	2.5	6.0	–	–	40	100	100
TN	–	–	–	–	–	–	25	15
Cl^-	350	250	–	–	–	350	–	–
F^-	2.0	1.0	–	1.0	15.0	–	2.0	1.5
SO_4^{2-}	–	200	–	–	–	–	–	–
SAR	–	–	–	–	–	20	–	–
RSC	–	–	–	–	–	–	–	–
Escherichia coli	40,000 mpn/L	500 mpn/L					≤100 cfu/L	≤1000 cfu/L

adsorb NH_4^+-N from wastewater (Xu et al., 2015). For example, zeolite Na-P1 synthesized from fly ash by alkaline hydrothermal reaction has good cation exchange performance. When the concentration of Z-P1 was 20 g/L and the concentration of Mg^{2+} was 4 mM, the removal rates of NH_4^+-N and phosphate reached 65.2 and 92.3% after 30 min, respectively (He et al., 2017). The adsorption capacity of NO_3^--N by modified sugarcane bagasse biochars was also reported to be up to 28.21 mg/g (Divband Hafshejani et al., 2016).

As for phosphorus recovery, normally, it requires to recover its inorganic form by forming struvite or calcium phosphate. However, recovery of phosphorus from the liquid phase is less efficient than that from the solid phase. Natural or industrial byproducts materials with high content of Ca or CaO exhibit significant phosphorus retention capacity (Kamilya et al., 2022). Adsorption or precipitation mechanisms are involved during phosphorus recovery, so the immobilization capacity of the material depends on its chemical, physical and hydrological properties. For example, the adsorption capacities of calcined egg shells, scallop shells, dolomite and crab shells were 23, 3, 52.91 and 108.9 mg/g, respectively (Köse & Kıvanç, 2011).

8.6 NON-TECHNICAL BOTTLENECKS AND MANAGEMENT RECOMMENDATIONS

8.6.1 Poor Management

For the decentralized domestic sewage treatment, it is still in a development stage. Till now, some general sewage treatment management modes are applied to a majority of the decentralized domestic sewage treatments, which would greatly reduce the management efficiency and increase the consumption of resources and manpower. First, the management model did not take into account the actual situation of rural areas when applied to decentralized sewage treatment, and the consideration of environmental protection issues in rural areas was not comprehensive enough. Second, due to various practical reasons, the current government has not been able to invest sufficient manpower and material resources to support and supervise the treatment of domestic sewage in rural areas. Therefore, the decentralized treatment mode in rural areas has not yet been fully supervised. Without supervision, the work will be slow or even without quality assurance, making rural domestic sewage decentralized. There is no way for the processing mode to develop well enough to take advantage of it.

8.6.2 Lack of Awareness and Standardization

Governments all over the world attach great importance to rural sewage treatment and have introduced a series of policies. For example, China has issued the "Three-year Action Plan for Improving the Rural Human Settlements Environment (2018)" and the "Five-Year Action Plan for Improving the Rural Human Settlements Environment (2021–2025)". Each explicitly calls for faster treatment of rural domestic sewage. However, the situation of rural sewage treatment is not optimistic, and there are many technical and attention problems (Xu et al., 2022). At present, the rural sewage treatment technology mainly follows the urban sewage treatment mode. However, centralized sewage treatment facilities are not suitable for rural areas with low population density and dispersed population, and they often lack professional facility of operation and maintenance (Buller et al., 2022). At present, the infrastructure construction of decentralized sewage treatment projects generally uses simple building materials. The construction process lacks effective supervision, and the strength and safety performance of infrastructure cannot be guaranteed. In the future, the infrastructure construction of decentralized sewage treatment projects in rural areas should be strengthened. Supervision and standards to ensure project quality are needed.

The publicity and popularization of environmental protection knowledge is not as comprehensive as that of cities, resulting in weak awareness of environmental protection among villagers. There are extensive discharges of domestic sewage. In the technical aspect, applications of decentralized sewage treatment in rural areas are less; in the awareness aspect, people living in rural areas lack of enough knowledge regarding the decentralized sewage treatment. Through continuous exploration, it is necessary to expand the scale of the decentralized treatment project of domestic sewage and increase the treatment capacity of the decentralized treatment project of domestic sewage.

8.7 CONCLUSIONS AND PERSPECTIVES

To sum up, sewage in rural areas is characterized by a large variation in water quantity and quality, and a wide pollution area due to the residential dispersion, leading to difficulties in collection and treatment. In the selection of treatment technology, environmental benefits, economic benefits and social benefits should be comprehensively considered along with the local conditions (such as the regional discharge standard). The anaerobic treatment system has a strong buffering capacity for organic load and a low yield of residual sludge, which can be well connected with other advanced purification units or CW. Many of the decentralized sewage treatment systems rely on source separation, i.e., treating each individual wastewater stream according to its composition, which allows a higher potential for resource recovery.

The topics of interest discussed in this field include effluent reuse, nutrient recovery and biogas reclamation from decentralized wastewater systems. Due to the economic reasons, the most commonly applied effluent reuse activity was found to be farmland irrigation and landscaping. Recycling nutrients from wastewater as soil fertilizer is still a major challenge for the future circular economy, in which the reduced energy input, improved recovery rate and product concentration are generally concerned in the pilot-scale plant. Apart from it, the potential contamination of the recovered nutrient with toxic compounds should be avoided.

Furthermore, the principles of understanding and modeling simulation are necessary to lay a good foundation for improving the sustainable performance of a decentralized anaerobic treatment system. In the small-scale high-strength decentralized facilities, higher concentrations of suspended solids are maintained in the process by increasing either biological growth or the recycling ratio. Under the condition of high sludge concentration, an in-depth understanding of residence time optimization, external circuit input and other parameters affecting the synergistic mechanism between microbial communities is conducive to improving the removal efficiency of organic and inorganic matters from wastewater. To reduce the capital and operational costs involved in the settling and pumping steps needed to recirculate the sludge, the optimization of reactor configuration is required by using fluid modeling. Thus, not only the mathematical model and fluid model but also the deep learning-based intelligent management mechanism are of great interest for accurate prediction. Last but not least, more pilot-scale practices are needed in the future to test the feasibility of the whole circular chain of source separation, decentralized wastewater treatment systems, nutrient recovery and reuse of fertilizer in agriculture.

REFERENCES

Abu-Zeid, M., Shady, A.M., Coulomb, R. 1998. The world water wouncil in three years—1994–1997. *Water International*, **23**(2), 97–106.

Alibardi, L., Bernava, N., Cossu, R., Spagni, A. 2016. Anaerobic dynamic membrane bioreactor for wastewater treatment at ambient temperature. *Chemical Engineering Journal*, **284**, 130–138.

Angelakis, A.N., Durham, B. 2008. Water recycling and reuse in EUREAU countries: Trends and challenges. *Desalination*, **218**, 3–12

Arias, A., Feijoo, G., Moreira, M.T. 2020. 11 – Environmental profile of decentralized wastewater treatment strategies based on membrane technologies. in: *Current Developments in Biotechnology and Bioengineering*, (Eds.) G. Mannina, A. Pandey, C. Larroche, H.Y. Ng, H.H. Ngo, Elsevier, pp. 259–287.

Badeti, U., Jiang, J., Almuntashiri, A., Pathak, N., Dorji, U., Volpin, F., Freguia, S., Ang, W.L., Chanan, A., Kumarasingham, S., Shon, H.K., Phuntsho, S. 2022. Impact of source-separation of urine on treatment capacity, process design, and capital expenditure of a decentralised wastewater treatment plant. *Chemosphere*, **300**, 134489.

Barat, R., Serralta, J., Ruano, M.V., Jimenez, E., Ribes, J., Seco, A., Ferrer, J. 2013. Biological nutrient removal model no. 2 (BNRM2): A general model for wastewater treatment plants. *Water Science and Technology*, **67**(7), 1481–1489.

Brooke, J. 2012. Briefing: Recent developments in EU water framework directive implementation. *Proceedings of the Institution of Civil Engineers – Maritime Engineering*, **165**(1), 3–5.

Buller, L.S., Sganzerla, W.G., Berni, M.D., Brignoli, S.C., Forster-Carneiro, T. 2022. Design and techno-economic analysis of a hybrid system for energy supply in a wastewater treatment plant: A decentralized energy strategy. *Journal of Environmental Management*, **305**, 114389.

Calli, B., Mertoglu, B., Roest, K., Inanc, B. 2006. Comparison of long-term performances and final microbial compositions of anaerobic reactors treating landfill leachate. *Bioresource Technology*, **97**(4), 641–647.

Capodaglio, A.G., Callegari, A., Cecconet, D., Molognoni, D. 2017. Sustainability of decentralized wastewater treatment technologies. *Water Practice and Technology*, **12**(2), 463–477.

Chen, C., Liu, Z., Huang, X. 2020. 14 – Anaerobic membrane bioreactors for sustainable and energy-efficient municipal wastewater treatment. in: *Current Developments in Biotechnology and Bioengineering*, (Eds.) H.H. Ngo, W. Guo, H.Y. Ng, G. Mannina, A. Pandey, Elsevier, pp. 335–366.

Chen, C.-Y., Wang, S.-W., Kim, H., Pan, S.-Y., Fan, C., Lin, Y.J. 2021. Non-conventional water reuse in agriculture: A circular water economy. *Water Research*, **199**, 117193.

De Graaff, M.S., Temmink, H., Zeeman, G., Buisman, C.J.N. 2010. Anaerobic treatment of concentrated black water in a UASB reactor at a short HRT. *Water*, **2**(1), 101–119.

Divband Hafshejani, L., Hooshmand, A., Naseri, A.A., Mohammadi, A.S., Abbasi, F., Bhatnagar, A. 2016. Removal of nitrate from aqueous solution by modified sugarcane bagasse biochar. *Ecological Engineering*, **95**, 101–111.

Do, M.H., Ngo, H.H., Guo, W., Chang, S.W., Nguyen, D.D., Liu, Y., Varjani, S., Kumar, M. 2020. Microbial fuel cell-based biosensor for online monitoring wastewater quality: A critical review. *Science of The Total Environment*, **712**, 135612.

Do, M.H., Ngo, H.H., Guo, W., Chang, S.W., Nguyen, D.D., Pandey, A., Sharma, P., Varjani, S., Nguyen, T.A.H., Hoang, N.B. 2022. A dual chamber microbial fuel cell based biosensor for monitoring copper and arsenic in municipal wastewater. *Science of The Total Environment*, **811**, 152261.

Faisal, M., Unno, H. 2001. Kinetic analysis of palm oil mill wastewater treatment by a modified anaerobic baffled reactor. *Biochemical Engineering Journal*, **9**(1), 25–31.

Fan, Y., Qian, F., Huang, Y., Sifat, I., Zhang, C., Depasquale, A., Wang, L., Li, B. 2021. Miniature microbial fuel cells integrated with triggered power management systems to power wastewater sensors in an uninterrupted mode. *Applied Energy*, **302**, 117556.

Fernández del Castillo, A., Garibay, M.V., Senés-Guerrero, C., Orozco-Nunnelly, D.A., de Anda, J., Gradilla-Hernández, M.S. 2022. A review of the sustainability of anaerobic reactors combined with constructed wetlands for decentralized wastewater treatment. *Journal of Cleaner Production*, **371**, 133428.

Garrido-Baserba, M., Barnosell, I., Molinos-Senante, M., Sedlak, D.L., Rabaey, K., Schraa, O., Verdaguer, M., Rosso, D., Poch, M. 2022. The third route: A techno-economic evaluation of extreme water and wastewater decentralization. *Water Research*, **218**, 118408.

Geetha Varma, V., Jha, S., Himesh Karthik Raju, L., Lalith Kishore, R., Ranjith, V. 2022. A review on decentralized wastewater treatment systems in India. *Chemosphere*, **300**, 134462.

Gikas, P., Tchobanoglous, G. 2009. The role of satellite and decentralized strategies in water resources management. *Journal of Environmental Management*, **90**(1), 144–152.

Gui, L., Ji-Ping, M.A., Yao, J. 2018. Study on rural domestic sewage treatment with combined process of anaerobic pool, collecting tank, stepwise drop aeration tower and artificial wetland. *Journal of Anhui Agricultural Sciences*, **46**, 179–181.

Gu, Y., Li, Y., Li, X., Luo, P., Wang, H., Robinson, Z.P., Wang, X., Wu, J., Li, F. 2017. The feasibility and challenges of energy self-sufficient wastewater treatment plants. *Applied Energy*, **204**, 1463–1475.

Hassanvand Jamadi, M., Alighardashi, A. 2017. Application of Froude dynamic similitude in anaerobic baffled reactors to prediction of hydrodynamic characteristics of a prototype reactor using a model reactor. *Water Science and Engineering*, **10**(1), 53–58.

He, H.F., Xu, S.Y., Han, R.Q., Wang, Q.C. 2017. Nutrient sequestration from wastewater by using zeolite Na-P1 synthesized from coal fly ash. *Environmental Technology*, **38**(8), 1022–1029.

Hussain, M.I., Muscolo, A., Farooq, M., Ahmad, W. 2019. Sustainable use and management of non-conventional water resources for rehabilitation of marginal lands in arid and semiarid environments. *Agricultural Water Management*, **221**, 462–476.

Hu, Y., Wang, X.C., Ngo, H.H., Sun, Q., Yang, Y. 2018. Anaerobic dynamic membrane bioreactor (AnDMBR) for wastewater treatment: A review. *Bioresource Technology*, **247**, 1107–1118.

Jeison, D., van Lier, J.B. 2008. Feasibility of thermophilic anaerobic submerged membrane bioreactors (AnSMBR) for wastewater treatment. *Desalination*, **231**(1), 227–235.

Jeong, H., Kim, H., Jang, T. 2016. Irrigation Water Quality Standards for Indirect Wastewater Reuse in Agriculture: A Contribution toward Sustainable Wastewater Reuse in South Korea. in: *Water*, **8**, 169

John, Y., Langergraber, G., Adyel, T.M., Emery David, V. 2020. Aeration intensity simulation in a saturated vertical up-flow constructed wetland. *Science of The Total Environment*, **708**, 134793.

Kamilya, T., Majumder, A., Yadav, M.K., Ayoob, S., Tripathy, S., Gupta, A.K. 2022. Nutrient pollution and its remediation using constructed wetlands: Insights into removal and recovery mechanisms, modifications and sustainable aspects. *Journal of Environmental Chemical Engineering*, **10**(3), 107444.

Köse, T.E., Kıvanç, B. 2011. Adsorption of phosphate from aqueous solutions using calcined waste eggshell. *Chemical Engineering Journal*, **178**, 34–39.

Kuldeep, K., Kumar, P., Mathur, A.K., Kamboj, P. 2022. Analysis of Irrigation Water Quality Indices for Chambal River at Kota, Rajasthan, India. *ECS Transactions*, **107**(1), 10797

Lemley, D.A., Lakane, C.P., Taljaard, S., Adams, J.B. 2022. Inorganic nutrient removal efficiency of a constructed wetland before discharging into an urban eutrophic estuary. *Marine Pollution Bulletin*, **179**, 113727.

Le, T.-S., Nguyen, P.-D., Ngo, H.H., Bui, X.-T., Dang, B.-T., Diels, L., Bui, H.-H., Nguyen, M.-T., Le Quang, D.-T. 2022. Two-stage anaerobic membrane bioreactor for co-treatment of food waste and kitchen wastewater for biogas production and nutrients recovery. *Chemosphere*, **309**, 136537.

Lettinga, G., Rebac, S., Zeeman, G. 2001. Challenge of psychrophilic anaerobic wastewater treatment. *Trends in Biotechnology*, **19**(9), 363–370.

Lieder, M., Rashid, A. 2016. Towards circular economy implementation: A comprehensive review in context of manufacturing industry. *Journal of Cleaner Production*, **115**, 36–51.

Li, F., Wichmann, K., Otterpohl, R. 2009. Review of the technological approaches for grey water treatment and reuses. *Science of The Total Environment*, **407**(11), 3439–3449.

Liu, J., Lu, Z., Zhang, J., Xing, M., Yang, J. 2013. Phylogenetic characterization of microbial communities in a full-scale vermifilter treating rural domestic sewage. *Ecological Engineering*, **61**, 100–109.

Liu, J., Zang, N., Gao, L., Liu, X., Tian, H., Yue, P., Li, T. 2022. A modified packed anaerobic baffled reactor based on phase separation for the treatment of decentralized wastewater: Performance and microbial communities. *Biochemical Engineering Journal*, 183, 108455.

Mainardis, M., Buttazzoni, M., Goi, D. 2020. Up-flow anaerobic sludge blanket (UASB) technology for energy recovery: A review on state-of-the-art and recent technological advances. *Bioengineering*, **7**(2), 43.

Mannina, G., Gulhan, H., Ni, B.-J. 2022. Water reuse from wastewater treatment: The transition towards circular economy in the water sector. *Bioresource Technology*, **363**, 127951.

Matsuno, Y., Hatcho, N., Shindo, S. 2007. Water transfer from agriculture to urban domestic users: a case study of the Tone River Basin, Japan. *Paddy and Water Environment*, **5**, 239–246

Meyer, M.F., Powers, S.M., Hampton, S.E. 2019. An evidence synthesis of pharmaceuticals and personal care products (PPCPs) in the environment: Imbalances among compounds, sewage treatment techniques, and ecosystem types. *Environmental Science & Technology*, **53**(22), 12961–12973.

Mladenov, N., Bigelow, A., Pietruschka, B., Palomo, M., Buckley, C. 2017. Using submersible fluorescence sensors to track the removal of organic matter in decentralized wastewater treatment systems (DEWATS) in real time. *Water Science and Technology*, **77**(3), 819–828.

Muzioreva, H., Gumbo, T., Kavishe, N., Moyo, T., Musonda, I. 2022. Decentralized wastewater system practices in developing countries: A systematic review. *Utilities Policy*, **79**, 101442.

O'Donovan, P., Coburn, D., Jones, E., Hannon, L., Glavin, M., Mullins, D., Clifford, E. 2015. A cloud-based distributed data collection system for decentralised wastewater treatment plants. *Procedia Engineering*, **119**, 464–469.

Oliveira, G.A., Colares, G.S., Lutterbeck, C.A., Dell'Osbel, N., Machado, ÊL., Rodrigues, L.R. 2021. Floating treatment wetlands in domestic wastewater treatment as a decentralized sanitation alternative. *Science of The Total Environment*, **773**, 145609.

Olubukola, A., Gautam, R.K., Kamilya, T., Muthukumaran, S., Navaratna, D. 2022. Development of a dynamic model for effective mitigation of membrane fouling through biogas sparging in submerged anaerobic membrane bioreactors (SAnMBRs). *Journal of Environmental Management*, **323**, 116151.

Park, J.-G., Jiang, D., Lee, B., Jun, H.-B. 2020. Towards the practical application of bioelectrochemical anaerobic digestion (BEAD): Insights into electrode materials, reactor configurations, and process designs. *Water Research*, **184**, 116214.

Peixoto, L., Min, B., Martins, G., Brito, A.G., Kroff, P., Parpot, P., Angelidaki, I., Nogueira, R. 2011. In situ microbial fuel cell-based biosensor for organic carbon. *Bioelectrochemistry*, **81**(2), 99–103.

Qin, S., Liu, H., Meng, Q., Zhou, Y., Xu, S., Lichtfouse, E., Chen, Z. 2022. Enhanced nutrient removal from mixed black water by a microbial ultra-low weak electrical stimulated anaerobic-two stage anoxic/aerobic process. *Chemical Engineering Journal*, **434**, 134615.

Raboni, M., Gavasci, R., Urbini, G. 2014. UASB followed by sub-surface horizontal flow phytodepuration for the treatment of the sewage generated by a small rural community. *Sustainability*, **6**(10), 6998–7012.

Rattier, M., Jimenez, J.A., Miller, M.W., Dhanasekar, A., Willis, J., Keller, J., Batstone, D. 2022. Long-term comparison of pilot UASB and AnMBR systems treating domestic sewage at ambient temperatures. *Journal of Environmental Chemical Engineering*, **10**(6), 108489.

Robles, Á, Ruano, M.V., Charfi, A., Lesage, G., Heran, M., Harmand, J., Seco, A., Steyer, J.-P., Batstone, D.J., Kim, J., Ferrer, J. 2018. A review on anaerobic membrane bioreactors (AnMBRs) focused on modelling and control aspects. *Bioresource Technology*, **270**, 612–626.

Rong, Y., Zhang, Y., Sun, Z., Liu, Z., Jin, X., Jin, P. 2023. Characteristics of rural sewage discharge and a case study on optimal operation of rural sewage treatment plant in Shaanxi, China. *Chemical Engineering Journal*, **453**, 139853.

Sabzali, A., Nikaeen, M., Bina, B. 2012. Performance evaluation of cigarette filter rods as a biofilm carrier in an anaerobic moving bed biofilm reactor. *Environmental Technology*, **33**(15), 1803–1810.

Sanchez, L., Carrier, M., Cartier, J., Charmette, C., Heran, M., Steyer, J.-P., Lesage, G. 2022. Enhanced organic degradation and biogas production of domestic wastewater at psychrophilic temperature through submerged granular anaerobic membrane bioreactor for energy-positive treatment. *Bioresource Technology*, **353**, 127145.

Santiago-Díaz, ÁL., Mugica-Álvarez, V., de los Cobos-Vasconcelos, D., Vaca-Mier, M., Salazar-Peláez, M.L. 2021. Performance evaluation and kinetic modeling of an upflow anaerobic sludge blanket septic tank for domestic wastewater treatment. *Environmental Science and Pollution Research*, **28**(47), 67414–67428.

Sarathai, Y., Koottatep, T., Morel, A. 2010. Hydraulic characteristics of an anaerobic baffled reactor as onsite wastewater treatment system. *Journal of Environmental Sciences*, **22**(9), 1319–1326.

Saravanan, V., Sreekrishnan, T.R. 2006. Modelling anaerobic biofilm reactors—A review. *Journal of Environmental Management*, **81**(1), 1–18.

Singh, S., Haberl, R., Moog, O., Shrestha, R.R., Shrestha, P., Shrestha, R. 2009. Performance of an anaerobic baffled reactor and hybrid constructed wetland treating high-strength wastewater in Nepal—A model for DEWATS. *Ecological Engineering*, **35**(5), 654–660.

Song, X., Liu, R., Yu, Q., Sheng, X., Chen, L. 2020. Management mode construction for operation and supervision of rural sewage treatment facilities: Towards the information-to-intelligence strategy. *Bioresource Technology Reports*, **11**, 100481.

Tan, X., Yang, Y.-L., Li, X., Gao, Y.-X., Fan, X.-Y. 2021. Multi-metabolism regulation insights into nutrients removal performance with adding heterotrophic nitrification-aerobic denitrification bacteria in tidal flow constructed wetlands. *Science of The Total Environment*, **796**, 149023.

Tian, H., Wang, X., Lim, E.Y., Lee, J.T.E., Ee, A.W.L., Zhang, J., Tong, Y.W. 2021. Life cycle assessment of food waste to energy and resources: Centralized and decentralized anaerobic digestion with different downstream biogas utilization. *Renewable and Sustainable Energy Reviews*, **150**, 111489.

Trad, Z., Vial, C., Fontaine, J.-P., Larroche, C. 2015. Modeling of hydrodynamics and mixing in a submerged membrane bioreactor. *Chemical Engineering Journal*, **282**, 77–90.

Udomkittayachai, N., Xue, W., Xiao, K., Visvanathan, C., Tabucanon, A.S. 2021. Electroconductive moving bed membrane bioreactor (EcMB-MBR) for single-step decentralized wastewater treatment: Performance, mechanisms, and cost. *Water Research*, **188**, 116547.

Verduzo Garibay, M., Fernández del Castillo, A., de Anda, J., Senés-Guerrero, C., Gradilla-Hernández, M.S. 2022. Structure and activity of microbial communities in response to environmental, operational, and design factors in constructed wetlands. *International Journal of Environmental Science and Technology*, **19**(11), 11587–11612.

Vymazal, J. 2013. The use of hybrid constructed wetlands for wastewater treatment with special attention to nitrogen removal: A review of a recent development. *Water Research*, **47**(14), 4795–4811.

Wang, C., Ding, M., Albert Ng, T.C., Yong Ng, H. 2022b. Enhanced dissolved methane recovery and energy-efficient fouling mitigation via membrane vibration in anaerobic membrane bioreactor. *Resources, Conservation and Recycling*, **184**, 106404.

Wang, J., Huang, Y., Zhao, X. 2004. Performance and characteristics of an anaerobic baffled reactor. *Bioresource Technology*, **93**(2), 205–208.

Wang, B., Liu, W., Liang, B., Jiang, J., Wang, A. 2022a. Microbial fingerprints of methanation in a hybrid electric-biological anaerobic digestion. *Water Research*, **226**, 119270.

Woznica, A., Nowak, A., Karczewski, J., Klis, C., Bernas, T. 2010. Automatic biodetector of water toxicity (ABTOW) as a tool for examination of phenol and cyanide contaminated water. *Chemosphere*, **81**(6), 767–772.

Wu, B. 2013. Advances in the use of CFD to characterize, design and optimize bioenergy systems. *Computers and Electronics in Agriculture*, **93**, 195–208.

Wu, S., Kuschk, P., Brix, H., Vymazal, J., Dong, R. 2014. Development of constructed wetlands in performance intensifications for wastewater treatment: A nitrogen and organic matter targeted review. *Water Research*, **57**, 40–55.

Wu, Y., Song, S., Chen, X., Shi, Y., Cui, H., Liu, Y., Yang, S. 2023. Source-specific ecological risks and critical source identification of PPCPs in surface water: Comparing urban and rural areas. *Science of The Total Environment*, **854**, 158792.

Xiao, E.-R., Liang, W., He, F., Cheng, S.-P., Zhen-Bin, W. 2010. Performance of the combined SMBR–IVCW system for wastewater treatment. *Desalination*, **250**(2), 781–786.

Xie, Y.D., Zhang, Q.H., Dzakpasu, M., Zheng, Y.C., Tian, Y., Jin, P.K., Yang, S.J., Wang, X.C. 2021. Towards the formulation of rural sewage discharge standards in China. *Science of The Total Environment*, **759**, 143533.

Xu, Y., Li, H., Li, Y., Zheng, X., Zhang, C., Gao, Y., Chen, P., Li, Q., Tan, L. 2022. Systematically assess the advancing and limiting factors of using the multi-soil-layering system for treating rural sewage in China: From the economic, social, and environmental perspectives. *Journal of Environmental Management*, **312**, 114912.

Xu, Z., Liu, Y., Williams, I., Li, Y., Qian, F., Zhang, H., Cai, D., Wang, L., Li, B. 2016. Disposable self-support paper-based multi-anode microbial fuel cell (PMMFC) integrated with power management system (PMS) as the real time "shock" biosensor for wastewater. *Biosensors and Bioelectronics*, **85**, 232–239.

Xu, S.Y., Luo, L.W., He, H.F., Liu, H.B., Cui, L.F. 2015. Nitrogen and phosphate recovery from source-separated urine by losing with magnesite and zeolite. *Polish Journal of Environmental Studies*, **24**(5), 2269–2275.

Yang, F., Zhang, H., Zhang, X., Zhang, Y., Li, J., Jin, F., Zhou, B. 2021. Performance analysis and evaluation of the 146 rural decentralized wastewater treatment facilities surrounding the Erhai Lake. *Journal of Cleaner Production*, **315**, 128159.

Ye, F., Li, Y. 2009. Enhancement of nitrogen removal in towery hybrid constructed wetland to treat domestic wastewater for small rural communities. *Ecological Engineering*, **35**(7), 1043–1050.

Yu, Q., Liu, R., Chen, J., Chen, L. 2019. Electrical conductivity in rural domestic sewage: An indication for comprehensive concentrations of influent pollutants and the effectiveness of treatment facilities. *International Biodeterioration & Biodegradation*, **143**, 104719.

Zha, X., Ma, J., Lu, X. 2018. Performance of a coupling device combined energy-efficient rotating biological contactors with anoxic filter for low-strength rural wastewater treatment. *Journal of Cleaner Production*, **196**, 1106–1115.

Zhang, P., Huang, G., An, C., Fu, H., Gao, P., Yao, Y., Chen, X. 2019. An integrated gravity-driven ecological bed for wastewater treatment in subtropical regions: Process design, performance analysis, and greenhouse gas emissions assessment. *Journal of Cleaner Production*, **212**, 1143–1153.

9 Development of Decentralized Drinking Water Treatment Plant

Song Yan[1], *Patrick Drogui*[1], *Rajeshwar D. Tyagi*[2,3], *and Jonathan W.C. Wong*[2]
[1] Université du Québec, Québec, Canada
[2] Dongguan University of Technology, Dongguan, China
[3] BOSK Bioproducts, Québec, Canada

9.1 INTRODUCTION

A total of 2.1 billion people worldwide lack access to safe, readily available water at home (WHO 2017, 2022). More than 2 billion people live in countries where water scarcity is a pressing issue, a problem that is expected to worsen in certain regions due to climate change and population growth. At a global level, at least 2 billion individuals rely on a drinking water source that is contaminated with fecal matter. Among the 2.1 billion people who lack safe access to potable water, 844 million do not even have access to basic drinking water services. This includes 263 million people who must spend over 30 minutes per trip to collect water from sources outside their homes, and 159 million who still drink untreated water from surface sources like streams or lakes. While arsenic, fluoride, and nitrate are the most significant chemical hazards in drinking water, emerging contaminants such as pharmaceuticals, pesticides, per- and polyfluoroalkyl substances (PFASs), and microplastics are also causing public concern (WHO 2017, 2022).

In many cities all over the world, a centralized water treatment system (WTS), also known as conventional treatment, is used; it treats water with a combined process of coagulation, flocculation, filtration, and disinfection in a central location and then distributes water via dedicated distribution networks. This system is well developed and can effectively remove practically any range of raw water turbidity along with harmful pathogens, including bacteria, virus, and *protozoa*. However, centralized systems are often deficient or non-existent, unavailable or unaffordable in certain area or developing and transition countries. Therefore, decentralized drinking WTSs are an important alternative option (Estévez et al. 2022; Geetha Varma et al. 2022; Hafeez et al. 2021; Lakho et al. 2021, 2022; Mladenov et al. 2022; Ngoma et al. 2020; Nguyen et al. 2020; Otter et al. 2020).

Decentralized water systems are used not only in developing regions lacking centralized infrastructure but also in economically developed countries as a low-cost and efficient way to provide good quality drinking water locally. Compared to centralized systems, which can have significant capital, operating, and maintenance costs, decentralized systems can provide supplementary water resources at a more affordable cost (Hafeez et al. 2021; Lakho et al. 2021; Otter et al. 2020).

This chapter provides a detailed overview of drinking water problems worldwide and explores solutions through the use of decentralized water systems. The focus is on pressure-driven membrane processes, which are a promising solution for decentralized drinking water systems. Other decentralized water treatment methods, such as physical removal, heat or UV-based treatments, and chemical treatments, are also discussed. Solutions for limited water supply, such as rainwater harvesting and groundwater wells, are illustrated as ways to transform local water sources into drinking water. Decentralized drinking WTSs are adaptable and can be implemented in extreme conditions, including remote areas, rural areas, and cold climates. The decentralized approach

DOI: 10.1201/9781003381419-11

offers sustainable benefits as an alternative to centralized systems. Lastly, the challenges and perspectives of these processes are presented.

9.2 DRINKING WATER PROBLEMS EXISTING GLOBALLY AND THEIR SOLUTIONS USING DECENTRALIZED WATER SYSTEMS

9.2.1 Drinking Water Problems Existing Globally

Most industrialized countries use a centralized water supply approach to control, treat, and distribute water among their populations. These first-generation systems consist of separate centralized drinking water system and wastewater system (Arora et al. 2015; De Graaf 2009). While with more attention to the public health concerns, the centralized drinking water system is being questioned. Even this has proved an effective way of reaching the various regions in nations where a developed infrastructure is in place. However, there are problems with centralized supplies, including leakages (which amount to 45 million cubic meters a day), bacteria that can develop when water stagnates, and uneven water quality across different areas (Arora et al. 2015; De Graaf 2009). The traditional way of providing water from a central location to a large area may no longer be effective or safe due to aging infrastructure and new environmental regulations (Arora et al. 2015; De Graaf 2009). As a result, decentralized water supply systems are gaining popularity because they are more flexible, optimize utility, and provide greater independence for communities.

The traditional drinking water treatment process is not suitable for decentralized drinking water supply. In particular, it is difficult to achieve centralized drinking water treatments in remote rural areas, which raises questions about safe drinking water supply (Estévez et al. 2022; Geetha Varma et al. 2022; Gora et al. 2020; Lakho et al. 2021, 2022; Maryati, Firman and Humaira 2022; Ngoma et al. 2020).

9.2.2 Solutions Using Decentralized Water Systems

Decentralized water systems use smaller scale purification methods to turn local water supplies into drinking water. Communities in several developing countries have made successful use of decentralized drinking water systems, and the decentralized process can also be used as an easy and sustainable alternative even in the places with a centralized water supply.

Decentralized water supplies can improve microbial water quality and reduce contamination between treatment and use because the treatment is done much closer to the point of access. Decentralized treatment filtration systems are designed to bring water treatment closer to where it's needed. This reduces the distance the treated water has to travel, resulting in a safer and better quality product. Decentralized water supply systems can offer many benefits. They can be low cost, as they reduce the need for expensive groundwater pumping procedures. Using sustainable methods to source water locally can also have environmental benefits. Decentralized systems can help preserve water by reducing waste, and they use less energy than centralized systems, which means they can save on gas and electricity costs. In cases where a centralized community treatment system is unavailable or too expensive, a decentralized WTS can be used to obtain potable water (Schowalter 2006).

Decentralized water supply system has been already applied in many parts of developing and transition countries (Maryati, Firman and Humaira 2022; Peter-Varbanets et al. 2009). Decentralized treatment is a crucial solution to address the challenges of global water scarcity and climate resilience. Typically, single households or communities are the most commonly used decentralized systems (Peter-Varbanets et al. 2009).

Although the advantages of decentralized treatment have long been recognized, technological challenges have limited its widespread application. Therefore, it is important to pay attention to new advances in technology to improve the efficiency and effectiveness of decentralized water systems. Since most water quality problems are due to hygiene factors and pathogens, various type of decentralized WTS/technologies (Table 9.1) is available to solve these problems, including membrane processes, heat and/or UV treatment, and physical and chemical treatment of water.

TABLE 9.1
Various Decentralized Water Treatment Technologies

SN	Water	Technology	Treatment efficacy	Reference
1	Tap water Parameter value color (HU) 0 Turbidity (NTU) 0.09–1.21 Free chlorine (mg/L) 0.02–0.63 pH 6.86–7.51	Microfiltration (M), 1. M Unitary membrane process 2. M + GAC Hybrid process (membrane and plain GAC) 3. M + GAC/Cu05 Hybrid process (membrane and GAC impregnated with 0.5 wt.% Cu) 4. M + GAC/Cu10 Hybrid process (membrane and GAC impregnated with 1.0 wt.% Cu)	• Color 0 HU • The GAC ensured higher chlorine removals • Copper with 1.0 wt.% concentration was the only filtration system that exhibited bactericidal efficiency of 99.99%, in the initial assay • Hybrid filtration system with activated carbon significantly increased chlorine removal, higher than 90% in all collected samples	Yamaguchi et al. (2017)
2	Water collected from lake	1. Ultrafiltration (UF) 2. Ultrafiltration (UF) units with polyaluminum chloride (PACl) coagulation	• PACl suppressed both total fouling and hydraulically irreversible fouling • PACl improved the permeate quality by enhancing the removal of particulates and dissolved organics • Compared with the conventional UF system, the hybrid coagulation-UF system contributed to a 21% increase in the flux rate, allowing for a 27% reduction in membrane area and thus, providing cost benefits in terms of both capital and operating costs	Arhin et al. (2019)
3	Water	Nanofiltration	Excellent performance of removing DOC, TDS, hardness, fluoride, and pathogens in groundwater, and the permeate water of the NF plant has been well-accepted by the stakeholders of the society	
4	Groundwater	End-of-life reverse osmosis (EoL-RO) Reverse osmosis (RO) at the point-of-use	• The increase in permeability from 4.5 to 50.4 L/m^{2}h.bar and decrease in salinity rejection from 99.4 to 11.6% suggest that the RO was converted to membranes similar to UF • A steady permeate flux (21.2 ± 0.46 L/m^{2}h) and filtration resistance (6.37 ± 0.16 1014 1/m), producing a total permeate volume (3,000 L) comparable with commercial filtration systems and in compliance with the 109 potability standards analyzed	Moreira et al. (2022)
5	Water	Physical methods Biosand filtration (BSF)	Removes bacteria consistently if not completely, on average by 81–100%, and protozoa by 99.98–100%	Lantagne et al. (2006).
6	Drinking water	Chemical treatment Chlorination	Reductions of diarrheal disease incidence among users ranging from 22 to 84%	CDC (2022)

(*Continued*)

TABLE 9.1 (*Continued*)
Various Decentralized Water Treatment Technologies

SN	Water	Technology	Treatment efficacy	Reference
7	Water	UV treatment Combination of inline electrolytic chlorine generation (ECl_2) with low-pressure UV reactors (UV/ECl_2)	• From a free available chlorine (FAC) concentration range between 1 and 18 mg/L produced by ECl_2 up to 84% can be photolyzed to form • Hydroxyl radicals (OH) with a UV energy input of 0.48 kWh/m^3. This ratio could be increased to 97% by doubling the UV energy input to 0.96 kWh/m^3 and was constant throughout the tested FAC range. Whereas with ECl_2 alone only 5% of benzotriazoles could be degraded, the combination with UV improved the degradation to 89%	Otter et al. (2020)

Biosand filtration (BSF); End-of-life reverse osmosis (EoL-RO); Electrolytic chlorine generation (ECl_2); Free Available Chlorine (FAC); Microfiltration (M); Nanofiltration (NF); point-of-use (POU); Polyaluminum chloride (PACl); Reverse osmosis (RO); Ultrafiltration (UF).

9.3 PRESSURE-DRIVEN MEMBRANE PROCESSES

Membrane technology is widely accepted as a means of producing various qualities of water from surface water, well water, brackish water, and seawater. Pressure-driven membrane processes such as microfiltration (MF), ultrafiltration (UF), nanofiltration (NF), and reverse osmosis (RO) can be used at different stages: MF in macroscopic pretreatment, UF in the separation of macro- and micro-molecules, and NF and RO in purification/concentration. All these processes allow the concentration and separation to be carried out without using heat; another advantage is that the equipment requires little space, running costs and energy consumption are low, and products and co-products are high in quality. All these factors are very important for the recovery of by-products. On the other hand, a disadvantage of membrane processes is related to the sensitivity to concentration polarization and membrane fouling due to chemical interaction with feed components, for which a pretreatment is necessary (Montenegro-Landívar et al. 2022; Sobana and Panda 2014; Tang et al. 2016, 2018).

Membrane technology is used in the water industry to improve the quality of water for use, reuse, or discharge to the environment. Membranes range from finely porous structures to nonporous and can remove contaminants such as bacteria and protozoa down to ions. The advantages of membrane technology include its modular nature, allowing application at very large or small scale, the quality of the product water, the relatively small footprint, and, in some cases, the lower energy usage. Membrane processes could be a solution for an advanced physical treatment of water for drinking purposes other than for wastewater recovery. Normally MF and UF are low-pressure processes that effectively remove microorganisms and suspended solids (MF) and colloids (UF). NF and RO work at relatively high transmembrane pressures (1–10 MPa). RO and NF are very effective at removing dissolved ions and organic solutes (Montenegro-Landívar et al. 2022; Peter-Varbanets, Gujer and Pronk 2012; Yue et al. 2021).

9.3.1 MICROFILTRATION

A study (Yamaguchi et al. 2017) evaluated the application of a polymeric MF membrane in a gravitational filtration module and its combination with granular activated carbon (GAC) impregnated with copper, resulting in a hybrid process. The proposed system would be used to improve the quality of water for human consumption in developing countries. The GAC ensured higher chlorine removals, as well as higher permeate flux. Furthermore, the GAC impregnated with copper oxide nanoparticles exhibited higher *Escherichia coli* removal (Table 9.1). Therefore, the hybrid gravitational membrane system applying GAC impregnated with copper oxide could be considered a potential alternative point-of-use (POU) treatment to improve the quality of water for human consumption (Yamaguchi et al. 2017).

9.3.2 Ultrafiltration

A comparative assessment of two decentralized UF units with and without polyaluminum chloride (PACl) coagulation was carried out (Arhin et al. 2019) to evaluate the influence of coagulation on the fouling, water quality, and cost nexus. A membrane module had an effective area of 4.5 m^2, a molecular weight cut-off (MWCO) of 100 kDa, a fiber outer diameter of 1.6 mm, and an inner diameter of 1.0 mm. The results showed that PACl suppressed both total fouling and hydraulically irreversible fouling. It also revealed that PACl improved the permeate quality by enhancing the removal of particulates and dissolved organics. Compared with the conventional UF system, the hybrid coagulation-UF system contributed a 21% increase in the flux rate, allowing for a 27% reduction in membrane area, and thus, providing cost benefits in terms of both capital and operating costs. These results suggested that PACl coagulation was potentially a cost-effective antifouling method for decentralized membrane-based water systems (Arhin et al. 2019).

9.3.3 Nanofiltration

NF is a relatively newly introduced pressure-driven filtration membrane process. During transport through the membrane, both diffusive and convective fluxes take place due to the porous type of membrane and the transmembrane pressure difference. Obviously, the same transport flux may occur in the feed side external boundary layer as well. It showed the concentration distribution, the expression of outlet concentration, the effect of the fluid mass transfer coefficient, and permeation flux on the rejection coefficient. Additionally, it gave transport expressions for solvent-resistant NF, the Spiegler-Kedem model, and the NF of ionic components (Nagy 2019).

In Sri Lanka, over 3.6 million villagers face acute water quality problems (Cooray et al. 2019); installation of decentralized water treatment plants is an ideal option to supply safe drinking water for rural communities. Most of the drinking water in these villages is unpalatable due to high hardness and salinity. As an interim measure, RO water treatment plants are introduced to provide safe water. RO-treated water may not taste good to some people because it does not have enough important minerals called electrolytes; hence, people refuse it after prolonged use. The operation, maintenance, and management of RO plants are other major problems. Aimed at providing safe drinking water to the rural sector in a cost-effective manner, Cooray et al. (2019) fabricated an automated drinking water purification system based on NF membrane technology, which could remove divalent cations, dissolved organic carbon (DOC) and pathogens efficiently, and monovalent ions partially, and thus keep electrolytes to some degree. Ten commercial NF membranes were tested in a laboratory, for solute and DOC removal efficiency and robustness. The DF-90 membrane showed the highest removal of DOC and hardness, and it was therefore selected, to design a pilot NF drinking water treatment plant. The adhered DOC by the membrane can be cleaned with NaOH solution (pH = 12). The pilot NF drinking water treatment plant has been in use since September 2018, and it showed excellent performance of removing DOC, TDS, hardness, fluoride, and pathogens in groundwater, and the permeate water of the NF plant has been well-accepted by the stakeholders of the society. The dominant genus of source water, and throughout the two processes (NF and RO), is *Pseudomonas*, and their difference is significant in the concentrates of the NF and RO processes (Cooray et al. 2019).

Many of the more recent membranes are not only polymeric but contain also an inorganic phase. Tailoring innovative materials with organic and inorganic phases coexisting in a nanoscale with multi-functionalization is an appealing approach to control at the same time diffusivity and gas solubility. Other advanced materials, which are now being considered for membrane development, are organic or organic-inorganic self-assemblies, metal-organic frameworks, and different forms of carbon fillers (Trinh 2018).

9.3.4 Reverse Osmosis

Removal of organic contaminants in water may sometimes be more easily achieved with a pair rather than with a single unit operation. Ollis (2003) investigated the combination of an ambient temperature, chemical oxidation photocatalysis (PC) process with physical separation via a membrane operation, MF, UF, or RO using both conceptual statements. Four configurations are noted and discussed here: (1) PC + MF for catalyst slurry recycle, (2) PC + UF for catalyst slurry and (polymer) reactant recycle, (3) immobilized PC and UF/RO for reactant recycle, and (4) immobilized PC on UF/RO membrane for membrane self-cleaning. The paucity of examples indicates a need for substantial effort to fully exploit the suggested possibilities for process development (Ollis 2003).

Decentralized systems applied at the POU could be an alternative for a safe drinking water supply in isolated regions and developing countries. Based on that, Moreira et al. (2022) investigated the recycling of end-of-life RO (EoL-RO) household modules as an alternative for groundwater treatment at the POU. Spiral-wound EoL-RO membranes were Dow® FilmTec TW30, 0.5 m^2. The membranes were recycled through chemical oxidation using NaOCl (440,000 ppm h) and were validated in a long-term operation. The increase in permeability from 4.5 to 50.4 $L/m^2h.bar$ and decrease in salinity rejection from 99.4 to 11.6% suggest that the EoL-RO was effectively converted to UF. The monitoring data demonstrated a steady permeate flux (21.2 ± 0.46 L/m^2h) and filtration resistance ($6.37 \pm 0.16 \times 10^{14}$ 1/m), producing a total permeate volume (3,000 L) comparable with commercial filtration systems and in compliance with the 109 potability standards analyzed. From a broader perspective, the advantages of using recycled membranes could be extended to environmental aspects, preventing the production of waste from spent membranes and focusing on cleaner production of technical processes (Moreira et al. 2022).

9.4 CASE STUDIES OF WELL-KNOWN MEMBRANE-BASED DECENTRALIZED SYSTEMS

A hybrid process (Sartor et al. 2008) was used to produce drinking and service water from surface water in Thailand. The process consisted of ozonation and membrane filtration, followed by activated carbon filtration. The study focused on evaluating the entire process, with a particular emphasis on the membrane reactor technology's performance under tropical conditions. To this end a demonstration plant with a capacity of about 5 m^3/h in continuous operation was designed and built up in a 40 ft sea container. Submerged ceramic MF and UF membranes with a cut-off of 200 and 80 nm were used for the separation process inside the ozone reactor. Pilot tests with different kinds of surface water were carried out in Thailand, a typical tropical country. The tests have shown that during ozonation, the permeate fluxes of the submerged membranes have constant fluxes on a high level for UF in a range of 60–80% and for MF in a range of 35% of the pure water flux. The low amount of concentrate generated by this process contains a high amount of mineralized elements and is harmless for discharging.

The comparison of the different raw water qualities with the qualities of the produced freshwater has shown that the process is suitable for the decentralized production of drinking and service water in tropical countries (Sartor et al. 2008).

A work (Arnal et al. 2010) has been performed in Mozambique, where 43% of the population can only access to improved water sources and 32% of the population lives with access to improved sanitation. Main reasons that explain lack of drinking water in developing countries are as follows: absence of proper infrastructures and chemicals, deteriorated distribution network or interrupted service that encourages stagnancy of water and growth of pathogens (Arnal et al. 2010).

The AQUAPOT project was developed at Polytechnic University of Valencia (Spain), and it was intended for the installation of decentralized drinking water systems based on membrane technologies as an alternative to the current water management in developing areas. This project was being satisfactorily applied in rural areas of Ecuador (South America) and Mozambique (Africa). This work described the evaluation of drinking water system, main characteristics of the UF-purification plant placed at Ressano Garcia (Mozambique), and assessed the main results obtained since the plant was working from the point of view of membrane performance and drinking water quality (Arnal et al. 2010).

Solutions to alleviate water crisis in developing countries have to be oriented to develop innovative, effective, sustainable, low-cost, and socially accepted drinking water treatment technologies that allow to improve sanitary situation and living conditions.

9.5 OTHER PROCESSES FOR TREATMENT

9.5.1 Physical Removal Methods

Several types of media can be used, including granular media of various grain sizes. Such mechanical filters may be an attractive option for household treatment because they can be produced on the spot with locally available materials; they are mostly simple, easy to use, and potentially long-lived. Slow sand filtration has been adapted for use in the home and is known as Biosand filtration (BSF). Biosand filters are containers filled with sand in which a bioactive layer is allowed to grow as a means of eliminating disease-causing organisms. Laboratory and field tests showed that BSF removes bacteria consistently if not completely, on average by 81–100%, and protozoa by 99.98–100%. However, these filters have limited virus-removal efficiency (Lantagne, Quick and Mintz 2006). Furthermore, paper, fiber, or fabric filters may be applied at household level. They can be effective in the removal of larger water-borne pathogens such as free-swimming larval forms (cercariae) of *schistosomes* and *Fasciola* species, guinea worm larvae within their intermediate crustacean host (Cyclops), and bacterial pathogens associated with relatively large copepods and other zooplankton in water, such as the bacterium *Vibrio cholerae*. However, these filters are not recommended for the general treatment of household water because their pores are too large to significantly retain viruses, bacteria, and smaller protozoan parasites. Activated carbon filters, often in the form of pressed blocks, followed by UV disinfection or silver (Ag) pre-coating, are being used as table-top units for additional tap water treatment. Membrane processes, including ceramic membranes, can also be considered filtration processes (Peter-Varbanets et al. 2009).

9.5.2 Heat or UV-Based

Boiling with fuel effectively destroys all classes of water-borne pathogens. However, a major disadvantage of boiling is its consumption of energy in relation to the availability, cost, and sustainability of fuel. Therefore, boiling household water is considered unrealistic and inaccessible under the scarcity and high cost of fuels conditions or region. Another problem of boiling is that it provides no residual protection: water can easily be recontaminated after cooling and is also associated with the risk of scalding, especially among infants.

UV irradiation with lamps has received renewed interest in recent years because of its well-documented ability to extensively (>99.9%) inactivate two water-borne, chlorine-resistant protozoans, Cryptosporidium parvum oocysts and Giardia lamblia cysts, at relatively low doses. However,

UV lamp disinfection has some disadvantages for use as a drinking water disinfectant at household level. Particulates, turbidity, and certain dissolved constituents can interfere with or reduce the efficiency of microbial inactivation. A reliable and affordable source of electricity is required to power the UV lamps.

These lamps require periodic cleaning, especially in the case of submerged lamps; they have a finite lifespan and must be periodically replaced. To make the cleaning and replacement possible, an efficient infrastructure is needed, which may not always be possible. This increases the operational costs of UV-based systems and impacts their environmental sustainability (Peter-Varbanets et al. 2009).

A key component of a decentralized WTS should be an ultraviolet (UV) disinfection unit. Water disinfection (and disinfection in general) by short-wavelength UV radiation is well understood. Mercury lamps emit the bulk of their UV radiation at a wavelength of 254 nm. The availability of UV LEDs would allow some applications that are not achievable with mercury lamps, for example, a disinfection unit in a cool water dispenser such as a refrigerator, air and water purification inside of airplanes, and portable/emergency water purification systems. Utilization of POU or entry units based on UV-LEDs would also translate into energy savings because of the switchable nature of LEDs, as the light up of the disinfection unit and its output power could be synchronized with the water flow (Schowalter 2006).

9.5.3 Chemical Treatment Methods

Chemical treatment is widely used for disinfection purposes. Among the drinking water disinfectants, free chlorine is the simplest, most widely used and most affordable. It is highly effective against nearly all water-borne pathogens, with the notable exception of Cryptosporidium parvum oocysts and the mycobacteria species.

To treat trace organic compounds in water, various advanced oxidation processes (AOPs) have been studied. One study (Otter et al. 2020) explored using a combination of inline electrolytic chlorine generation (ECl_2) and low-pressure UV reactors (UV/ECl_2) to create a chlorine-based AOP without adding additional chlorine. This process resulted in up to 84% of free available chlorine (FAC) concentration being photolyzed to create hydroxyl radicals (OH) with a UV energy input of 0.48 kWh/m^3. Doubling the UV energy input to 0.96 kWh/m^3 increased the ratio to 97%, which remained constant for the tested FAC range. When only ECl_2 was used, only 5% of benzotriazoles could be degraded, but when combined with UV, the degradation increased to 89%.

To treat trace organic compounds in water, various AOPs have been studied. Otter et al. (2020) explored using a combination of inline electrolytic chlorine generation (ECl_2) and low-pressure UV reactors (UV/ECl_2) to create a chlorine-based AOP without adding additional chlorine. This process resulted in up to 84% of FAC concentration being photolyzed to create hydroxyl radicals (OH) with a UV energy input of 0.48 kWh/m^3. Doubling the UV energy input to 0.96 kWh/m^3 increased the ratio to 97%, which remained constant for the tested FAC range. When only ECl_2 was used, only 5% of benzotriazoles could be degraded, but when combined with UV, the degradation increased to 89% (Otter et al. 2020).

AOPs consist of the usage of oxidizing agents to oxidize contaminants. Among the strongest oxidizing radicals used is the hydroxyl radical (•OH). Sulfate radical (SO_4•–) has also been widely studied for the removal of organic contaminants from water. AOPs are environmentally friendly chemical techniques that can degrade organic contaminants into harmless products that neither transfer contaminants from one phase to another nor produce massive amounts of sludge. Furthermore, this technique has various advantages, including rapid reaction rates leading to less retention time compared to other conventional treatment techniques, and it does not require a large area for processing the needed flow rate for the system.

Chemical water treatment consists of a variety of chemical reactions that help in hydrolyzing contaminants into safer chemicals. The main chemical methods are coagulation and AOPs, including ozonation and Fenton treatment. Chemical treatments are usually used in combination with photocatalysis and/or membrane techniques. Chlorination is a method of treating water with chemicals to limit the growth of bacteria in order to control taste and odor. This method is advantageous because it is inexpensive and easy to use.

9.6 SOLUTIONS IN CASE OF LIMITED WATER SUPPLY

9.6.1 Rainwater Harvesting

Rainwater is a natural resource that can be used for various purposes in water-scarce areas. Rainwater harvesting involves collecting water from a building's roof surface. Rain that falls on the ground is usually called stormwater and is not as clean as rainwater. There are three important things to consider when harvesting rainwater. First, the water collected should be clean enough for its intended use. Second, it should be available during the right seasons and reliably enough. Third, the area where the water is collected and the amount collected should be enough to meet the intended water demand (Council 2011).

Harvesting rainwater can be used for non-potable and potable purposes, but it's essential to ensure the water collected is clean, available seasonally with sufficient dependability, and enough catchment area and volume to meet the water demands for which the system is intended to serve.

Rainwater is often treated before use to ensure it is safe for consumption. When used for drinking purposes, the treatment process is more extensive to remove potential pathogens and harmful chemicals. A series of filters are used to remove particles, with increasingly finer filters used to remove smaller particles. After filtration, sterilization is required to kill any remaining microorganisms. UV treatment is a chemical-free method of sterilization that uses UV lamps but requires energy. Carbon filtration can also be used to remove chemicals and organic compounds. Regular monitoring of the system is necessary to ensure proper functioning and maintenance, including periodic cleanout of sediment buildup and water quality testing. A well-maintained rainwater harvesting system can provide quality water for decades before replacement is necessary (Council 2011).

Designing and planning a rainwater harvesting system involves various considerations, including choosing appropriate materials for cisterns, ensuring the water collected is clean, and monitoring the system regularly. Rainwater harvesting can reduce water bills and may result in savings in sewerage charges in some municipalities. Proper planning and design can also integrate cistern overflow into the overall site stormwater management plan, protecting properties from flooding. Overall, rainwater harvesting is a valuable tool for water conservation and management, with benefits that extend beyond the individual site to the broader community and environment (Council 2011).

In some dry areas of the world, people have been using rainwater harvesting for centuries. For example, in Tanzania, half of the area relies completely on rainwater for survival. Rainwater harvesting is a method of collecting water where it is needed, and families have control over their own systems, reducing operation and maintenance problems. There are also community systems where water is collected from roads or fields. However, there are some disadvantages, such as seasonal variability in supply, uncertainty of rainfall, and unreliable water quality due to contamination during storage (Peter-Varbanets et al. 2009).

Urban storm water runoff can not only pollute water sources but it can also be a useful resource. Zhang et al. (2015) compared storm water management in Singapore and Berlin, Germany. The two cities have different strategies for managing storm water due to factors like geography, infrastructure, regulations, and incentives. Singapore focuses on managing storm water through a centralized network of drains and canals to control flooding and provide potable water, since water scarcity is a serious issue. On the other hand, Berlin aims to protect surface water quality and provide non-potable water supply by using a low-impact solution based on source control (Zhang et al. 2015).

Decentralized rainwater harvesting in Berlin is more economically efficient than the centralized drainage system in Singapore. Green roofs can save energy by mitigating the effects of urban heat islands and global warming. Combining conventional and decentralized urban stormwater management is expected to be the most practical solution for most cities in the future (Zhang et al. 2015).

Naddeo, Scannapieco, and Belgiorno (2013) developed an innovative decentralized system to collect and treat rainwater for potable use in a single household or small community. The system is called Filtration Adsorption Disinfection (FAD) and combines a filtration process with an adsorption step on GAC and a UV disinfection phase. The FAD technology is effective in removing pathogens and organic contaminants, and reducing turbidity, thus improving the water quality. Full-scale tests were conducted to examine the efficiency of FAD technology in terms of pH, turbidity, COD, TOC, DOC, *E. coli*, and total coliforms. FAD technology has low costs compared to other water treatment technologies and could become a viable option for developing countries (Naddeo, Scannapieco and Belgiorno, 2013).

Gravity-driven membrane (GDM) filtration is a technology that can treat rainwater without the need for backwashing, flushing, or chemicals. A lab-scale GDM system was tested for two months using stored rainwater and compared to stored tap water. GDM was found to be effective in removing bacteria and turbidity, but not DOC due to low rejection of low molecular weight fulvic. The permeate flux reached stable levels of 6–6.5 L/m^2h during 60 days of operation in the rainwater system, and bio-fouling with lower extracellular polymeric substance (EPS) contents resulted in higher flux. These results suggest that GDM is suitable for decentralized water supply as it requires low maintenance (Ding et al. 2017).

9.6.2 Groundwater Wells

Groundwater is a major source of drinking water worldwide. Groundwater is recharged from precipitation, which is complemented by natural infiltration by surface water or by artificial recharge. On a global scale, 20% of the irrigation water and 40% of the water used in industry are derived from groundwater (Connor 2015).

In most inhabited parts of the world, there is a large amount of groundwater, and despite that the abstracted volumes are huge, they are often readily supplemented. Furthermore, the upper soil layers act as a filter against physical, chemical, and biological deterioration which is effective both in terms of quality and cost. Finally, groundwater use often brings great economic benefits per unit volume compared to surface water because of ready local availability, high drought reliability, and a generally good quality requiring only minimal treatment (Katsanou and Karapanagioti 2017).

If drinking water originates from groundwater, it is usually treated on site of the point it is collected, e.g., on the well (Katsanou and Karapanagioti 2017).

Conventional treatment for groundwater originating from boreholes or wells, the most common treatment scheme is a combination of softening, aeration, and disinfection. Softening is a chemical process used for the removal of hardness that is due to calcium and magnesium carbonate minerals dissolved in groundwater. Hardness is associated with scaling in the pipelines or in machines that heat water. Aeration is also a chemical process used for oxidation and, thus, formation of less soluble species and precipitation of iron and manganese. It is also possible to use physical, chemical method.

If water is polluted, it may require additional treatment before it can be consumed. Toxic metals present in groundwater due to pollution or to natural minerals can be removed by sorbent materials that will strongly bind them. However, in many countries, it is considered preferable to change water supply than try to treat a polluted one. In places where water supplies are scarce, in-house water filters are commonly used employing ceramic pots and sorbent filtering media (e.g., activated carbon, zeolites, biosorbents, and bone chars) (Katsanou and Karapanagioti 2017).

When a centralized water supply is not available or the quantity is limited, poor households have to rely on water obtained from rivers or shallow wells. Wells are often the preferred solution, as the distance to them may be shorter, access easier, and the water considered to be less polluted. In many cases, households build private shallow wells at their own cost, obtaining water for their family needs or sharing it with neighbors. Wells operated by hand pumps and tube wells with motor pumps are also common. Contamination can occur if the wells are placed too close to sources of contamination or if the wells are too shallow. Furthermore, depletion can occur due to overdraft, and salinization can occur in case of inadequate drainage. If groundwater quality is unsatisfactory, additional treatment is necessary (Peter-Varbanets et al. 2009).

9.7 DECENTRALIZED DRINKING WATER TREATMENT SYSTEMS IN EXTREME CONDITIONS

Decentralized systems for treating drinking water can work well in challenging conditions such as remote areas, rural areas, and places with extreme weather events like floods and droughts. These conditions can affect the operation of regular water and sewer systems, as well as wastewater treatment plants, which can have an impact on public health (WHO 2011).

Huttinger et al. (2015) studied the performance of ten WTSs that were given to rural healthcare facilities in Rwanda. The systems had pre-filtration, membrane UF, and chlorine disinfection. The study lasted 22 months, during which the systems worked properly for 74% of the time. The most common reasons for downtime were water scarcity (8%) and chlorine mechanism failure (7%). When the systems were functioning, 98% of the water samples collected met the World Health Organization (WHO) standards for safe drinking water. Water quality decreased when there were interruptions or when water was stored in containers. The systems' effectiveness depended on the ability of the healthcare facility technician to maintain and repair them, as well as on environmental factors such as water and power availability, and obtaining replacement parts and chlorine in Rwanda (Huttinger et al. 2015).

WTSs that use membrane UF are a good way to treat water on-site at healthcare facilities. These systems can produce a lot of high-quality water, but they are only practical in areas with reliable access to water and electricity. When water needs to be stored after treatment, it is important to add disinfectant to maintain water quality. Maintaining the membrane UF component of the WTS is easy and does not require much time or technical knowledge, but the chlorination component often needs repairs that are beyond the operator's ability. The study found that the low level of chlorine was not enough to keep the water safe from contamination in the health center's plumbing network and in storage containers (Huttinger et al. 2015).

A study conducted by Gora et al. (2020) investigated the quality of water in a decentralized drinking water system in Pond Inlet, Nunavut, located at the northernmost tip of Baffin Island in the Canadian Arctic. The community has roughly 1,600 inhabitants, predominantly Inuit. The researchers gathered drinking water samples from various sources, including the water source, water delivery truck, domestic water storage tanks, and the POU in the decentralized system.

The findings showed that the quality of drinking water in Pond Inlet did not meet territorial and federal guidelines due to issues within the system and buildings, which made it susceptible to harmful microbes. The water had high levels of turbidity and color, and low levels of free chlorine. The physical characteristics of the water remained fairly constant throughout the system, as only chlorination was used, which does not affect pH, turbidity, or color. Other factors, such as conductivity and chlorine levels, varied across different locations. Additionally, the study showed that the bacteria present in the water changed as it moved from the source to the community buildings, with more harmful bacteria present in the buildings (Gora et al. 2020).

9.8 CHALLENGES IN THE PROCESS

Many countries lack data on the quality of water and sanitation services. Therefore, it is necessary to carry out the systematic research on this, and the guidelines should be established based on each region or country.

In some countries, water quality problems are caused by things like germs, dirty conditions, or problems with sewage systems. In areas with decentralized water systems, there are often issues with pollution in the sources of water because there aren't enough protections in place. Also, there may not be a centralized sewage system for homes and buildings, especially in rural areas that are still being developed. Sometimes, wells are used to get water from natural sources like rivers or lakes, but there aren't enough technologies to treat the water properly, so it doesn't meet accepted standards. Additionally, some natural sources of water have high levels of minerals like iron, manganese, and magnesium, which can affect water quality (Bayanova 2019; WHO 2022). Decentralized WTSs have a lot of potential for improvement in terms of technology. We can choose the best technologies to make them more efficient. Additionally, these systems require more participation and management from the community (Ma et al. 2015).

Developing WTSs that are specifically designed for different types of water sources and use energy efficiently can help decrease the expenses of operating water refill stations, while promoting environmental sustainability for future generations (Sima and Elimelech 2013).

Various studies have shown that drinking water systems in many developing countries cannot rely on only one formal treatment system. Thus, drinking water in developing countries in the future would comprise a combination of various types of centralized and decentralized WTSs based on its local conditions.

9.9 CONCLUSIONS AND PERSPECTIVES

Decentralized drinking WTSs offer a practical and viable alternative in areas where centralized systems are inadequate, unavailable, or unaffordable, particularly in developing and transition countries. The advantages of decentralized systems, such as cost-effectiveness, ease of use, environmental sustainability, reduced maintenance, and independence from energy sources, make them valuable solutions to address water scarcity and access challenges. A diverse array of technologies can be applied to these systems, showcasing high performance in providing safe drinking water.

Pressure-driven membrane processes, including MF, UF, NF, and RO, have demonstrated their efficacy at different stages of decentralized water treatment. These processes effectively remove contaminants, ensuring the delivery of clean and potable water to communities. Additionally, physical removal methods, heat or UV-based treatments, and chemical treatment processes have proven to be effective solutions in various settings, contributing to the versatility of decentralized water treatment options.

In areas with limited water supply, innovative solutions such as rainwater harvesting and groundwater wells can be employed to convert local water sources into safe drinking water. These sustainable approaches empower communities to become self-reliant and resilient in the face of water scarcity, particularly in regions with extreme conditions.

However, despite their promising benefits, decentralized drinking WTSs encounter several challenges that must be addressed for long-term success. Guidelines, regulations, and effective management practices must be established and integrated into the system's design, including the implementation of appropriate drainage systems. Ensuring proper implementation, operation, and maintenance of these systems requires skilled personnel and regular training to uphold their performance and safety standards.

Community participation and education play a pivotal role in the sustainability of decentralized WTSs. Empowering local communities through active involvement in the decision-making process fosters a sense of ownership and responsibility, making them more likely to embrace and support these initiatives. Education on proper water usage and hygiene practices is equally vital to maximize the health benefits of clean drinking water and prevent water-borne diseases.

As look to the future, investing in research, development, and capacity building for decentralized drinking WTSs will be crucial. Tailoring these systems to suit specific regional needs, including extreme conditions, will optimize their effectiveness and impact. Governments, NGOs, and international organizations must collaborate to create sustainable funding mechanisms and infrastructure support to ensure the widespread adoption of decentralized solutions in areas where they are most needed.

To achieve long-term success, comprehensive planning, proper management, and investment in capacity building are imperative. By embracing decentralized drinking water treatment solutions, it can enhance water security, empower communities, and contribute to a healthier and more sustainable future for all.

REFERENCES

Arhin, S. G., N. Banadda, A. J. Komakech, W. Pronk, and S. J. Marks. 2019. "Application of hybrid coagulation–ultrafiltration for decentralized drinking water treatment: Impact on flux, water quality and costs." *Water Supply* 19 (7):2163–2171. doi: 10.2166/ws.2019.097

Arnal, J. M., B. García-Fayos, M. Sancho, G. Verdú, and J. Lora. 2010. "Design and installation of a decentralized drinking water system based on ultrafiltration in Mozambique." *Desalination* 250 (2):613–617. doi: 10.1016/j.desal.2009.09.035

Arora, Meenakshi, Hector Malano, Brian Davidson, Rebecca Nelson, and Biju George. 2015. "Interactions between centralized and decentralized water systems in urban context: A review." *Wiley Interdisciplinary Reviews: Water* 2 (6):623–634.

Bayanova, A. A. 2019. Monitoring the quality of drinking water of the regional decentralized water supply. In *IOP conference series: Earth and environmental science.*

Connor, Richard. 2015. *The United Nations world water development report 2015: Water for a sustainable world*, volumes 1. UNESCO Publishing.

Cooray, Titus, Yuansong Wei, Junya Zhang, Libing Zheng, Hui Zhong, Sujithra K. Weragoda, and Rohan Weerasooriya. 2019. "Drinking-water supply for CKDu affected areas of Sri Lanka, using nanofiltration membrane technology: From laboratory to practice." *Water* 11 (12):2512.

Council, Cascadia Green Building. 2011. *Toward net zero water: Best management practices for decentralized sourcing and treatment.* Cascadia Green Building Council.

De Graaf, Rutger Ewout. 2009. *Innovations in urban water management to reduce the vulnerability of cities: Feasibility, case studies and governance.*

Ding, An, Jinlong Wang, Dachao Lin, Xiaobin Tang, Xiaoxiang Cheng, Hui Wang, Langming Bai, Guibai Li, and Heng Liang. 2017. "A low pressure gravity-driven membrane filtration (GDM) system for rainwater recycling: Flux stabilization and removal performance." *Chemosphere* 172:21–28. doi: 10.1016/j.chemosphere.2016.12.111

Estévez, Sofía, Sara González-García, Gumersindo Feijoo, and María Teresa Moreira. 2022. "How decentralized treatment can contribute to the symbiosis between environmental protection and resource recovery." *Science of the Total Environment* 812:151485. doi: 10.1016/j.scitotenv.2021.151485

Geetha Varma, V., Swetti Jha, L. Himesh Karthik Raju, R. Lalith Kishore, and V. Ranjith. 2022. "A review on decentralized wastewater treatment systems in India." *Chemosphere* 300:134462. doi: 10.1016/j.chemosphere.2022.134462

Gora, Stephanie L., Tim Anaviapik Soucie, Nicole E. McCormick, C. Carolina Ontiveros, Vincent L'Hérault, Megan Gavin, Benjamin F. Trueman, Jessica Campbell, Amina K. Stoddart, and Graham A. Gagnon. 2020. "Microbiological water quality in a decentralized arctic drinking water system." *Environmental Science: Water Research & Technology* 6 (7):1855–1868.

Hafeez, Ainy, Zufishan Shamair, Nasir Shezad, Fahed Javed, Tahir Fazal, Saifur Rehman, Aqeel Ahmed Bazmi, and Fahad Rehman. 2021. "Solar powered decentralized water systems: A cleaner solution of the industrial wastewater treatment and clean drinking water supply challenges." *Journal of Cleaner Production* 289:125717. doi: 10.1016/j.jclepro.2020.125717

Huttinger, Alexandra, Robert Dreibelbis, Kristin Roha, Fidel Ngabo, Felix Kayigamba, Leodomir Mfura, and Christine Moe. 2015. "Evaluation of membrane ultrafiltration and residual chlorination as a decentralized water treatment strategy for ten rural healthcare facilities in Rwanda." *International Journal of Environmental Research and Public Health* 12 (10):13602–13623.

Katsanou, Konstantina, and Hrissi K Karapanagioti. 2017. Surface water and groundwater sources for drinking water. In *Applications of advanced oxidation processes (AOPs) in drinking water treatment*, pp. 1–19.

Lakho, Fida Hussain, Hong Quan Le, Flor Mattheeuws, Wouter Igodt, Veerle Depuydt, Joachim Desloover, Diederik P. L. Rousseau, and Stijn W. H. Van Hulle. 2021. "Decentralized grey and black water reuse by combining a vertical flow constructed wetland and membrane based potable water system: Full scale demonstration." *Journal of Environmental Chemical Engineering* 9 (1):104688. doi: 10.1016/j.jece.2020.104688

Lakho, Fida Hussain, Asif Qureshi, Wouter Igodt, Hong Quan Le, Veerle Depuydt, Diederik P. L. Rousseau, and Stijn W. H. Van Hulle. 2022. "Life cycle assessment of two decentralized water treatment systems combining a constructed wetland and a membrane based drinking water production system." *Resources, Conservation and Recycling* 178:106104. doi: 10.1016/j.resconrec.2021.106104

Lantagne, Daniele S., Robert Quick, and Eric D. Mintz. 2006. "Household water treatment and safe: Storage options in developing countries." *Navig* 99:17–38.

Ma, Xin, Xiaobo Xue, Alejandra González-Mejía, Jay Garland, and Jennifer Cashdollar. 2015. "Sustainable water systems for the city of tomorrow—A conceptual framework." *Sustainability* 7 (9):12071–12105.

Maryati, Sri, Tommy Firman, and An Nisaa Siti Humaira. 2022. "A sustainability assessment of decentralized water supply systems in Bandung City, Indonesia." *Utilities Policy* 76:101373. doi: 10.1016/j.jup.2022.101373

Mladenov, Natalie, Nathan G. Dodder, Lauren Steinberg, William Richardot, Jade Johnson, Bice S. Martincigh, Chris Buckley, Tolulope Lawrence, and Eunha Hoh. 2022. "Persistence and removal of trace organic compounds in centralized and decentralized wastewater treatment systems." *Chemosphere* 286:131621. doi: 10.1016/j.chemosphere.2021.131621

Montenegro-Landívar, María Fernanda, Paulina Tapia-Quirós, Xanel Vecino, Mónica Reig, Mercè Granados, Adriana Farran, José Luis Cortina, Javier Saurina, and César Valderrama. 2022. "Recovery of natural polyphenols from spinach and orange by-products by pressure-driven membrane processes." *Membranes* 12 (7):669.

Moreira, Victor Rezende, Yuri Abner Rocha Lebron, Lucilaine Valéria de Souza Santos, and Míriam Cristina Santos Amaral. 2022. "Low-cost recycled end-of-life reverse osmosis membranes for water treatment at the point-of-use." *Journal of Cleaner Production* 362:132495. doi: 10.1016/j.jclepro.2022.132495

Naddeo, Vincenzo, Davide Scannapieco, and Vincenzo Belgiorno. 2013. "Enhanced drinking water supply through harvested rainwater treatment." *Journal of Hydrology* 498:287–291. doi: 10.1016/j.jhydrol.2013.06.012

Nagy, Endre. 2019. "Chapter 15 – Nanofiltration." In *Basic equations of mass transport through a membrane layer* (Second Edition), edited by Endre Nagy, 417–428. Elsevier.

Ngoma, Wezzie, Zvikomborero Hoko, Shepherd Misi, and Russel C. G. Chidya. 2020. "Assessment of efficiency of a decentralized wastewater treatment plant at Mzuzu University, Mzuzu, Malawi." *Physics and Chemistry of the Earth, Parts A/B/C* 118–119:102903. doi: 10.1016/j.pce.2020.102903

Nguyen, Thi Hai, Hai Nguyen Tran, Hai Anh Vu, Minh Viet Trinh, Tien Vinh Nguyen, Paripurnanda Loganathan, Saravanamuthu Vigneswaran, Tuan Minh Nguyen, Van Tuyen Trinh, Duc Loi Vu, and Thi Hoang Ha Nguyen. 2020. "Laterite as a low-cost adsorbent in a sustainable decentralized filtration system to remove arsenic from groundwater in Vietnam." *Science of the Total Environment* 699:134267. doi: 10.1016/j.scitotenv.2019.134267

Ollis, David F. 2003. "Integrating photocatalysis and membrane technologies for water treatment." *Annals of the New York Academy of Sciences* 984 (1):65–84. doi: 10.1111/j.1749-6632.2003.tb05993.x

Otter, Philipp, Katharina Mette, Robert Wesch, Tobias Gerhardt, Frank-Marc Krüger, Alexander Goldmaier, Florian Benz, Pradyut Malakar, and Thomas Grischek. 2020. "Oxidation of selected trace organic compounds through the combination of inline electro-chlorination with UV radiation (UV/ECl_2) as alternative AOP for decentralized drinking water treatment." *Water* 12 (11):3275.

Peter-Varbanets, Maryna, Willi Gujer, and Wouter Pronk. 2012. "Intermittent operation of ultra-low pressure ultrafiltration for decentralized drinking water treatment." *Water Research* 46 (10):3272–3282. doi: 10.1016/j.watres.2012.03.020

Peter-Varbanets, Maryna, Chris Zurbrügg, Chris Swartz, and Wouter Pronk. 2009. "Decentralized systems for potable water and the potential of membrane technology." *Water Research* 43 (2):245–265. doi: 10.1016/j.watres.2008.10.030

Sartor, M., B. Schlichter, H. Gatjal, and V. Mavrov. 2008. "Demonstration of a new hybrid process for the decentralised drinking and service water production from surface water in Thailand." *Desalination* 222 (1):528–540. doi: 10.1016/j.desal.2007.03.013

Schowalter, L. 2006. *Development of UV-based water purification systems for point of use/point of entry*. Critical National Need Idea.

Sima, Laura C., and Menachem Elimelech. 2013. "More than a drop in the bucket: Decentralized membrane-based drinking water refill stations in Southeast Asia." *Environmental Science & Technology* 47 (14):7580–7588. doi: 10.1021/es304384n

Sobana, S., and Rames C. Panda. 2014. "Modeling and control of reverse osmosis desalination process using centralized and decentralized techniques." *Desalination* 344:243–251. doi: 10.1016/j.desal.2014.03.014

Tang, Xiaobin, Xiaoxiang Cheng, Xuewu Zhu, Binghan Xie, Yuanqing Guo, Jinlong Wang, An Ding, Guibai Li, and Heng Liang. 2018. "Ultra-low pressure membrane-based bio-purification process for decentralized drinking water supply: Improved permeability and removal performance." *Chemosphere* 211:784–793. doi: 10.1016/j.chemosphere.2018.07.183

Tang, Xiaobin, An Ding, Fangshu Qu, Ruibao Jia, Haiqing Chang, Xiaoxiang Cheng, Bin Liu, Guibai Li, and Heng Liang. 2016. "Effect of operation parameters on the flux stabilization of gravity-driven membrane (GDM) filtration system for decentralized water supply." *Environmental Science and Pollution Research* 23 (16):16771–16780. doi: 10.1007/s11356-016-6857-0

Trinh, Ngoc Dao. 2018. *Development of a decentralized drinking water treatment plant based on membrane technology for rural areas in Vietnam*, volumes 42. Kassel University Press GmbH.

WHO. 2011. *Guidance on water supply and sanitation in extreme weather events*. World Health Organization. Regional Office for Europe.

WHO. 2017. 2.1 billion people lack safe drinking water at home, more than twice as many lack safe sanitation, 12 July 2017, News release, GENEVA; https://www.who.int/news/item/12-07-2017-2-1-billion-people-lack-safe-drinking-water-at-home-more-than-twice-as-many-lack-safe-sanitation#:~:text=lack%20safe%20sanitation-,2.1%20billion%20people%20lack%20safe%20drinking%20water%20at%20home%2C%20more,as%20many%20lack%20safe%20sanitation&text=Some%203%20in%2010%20people,report%20by%20WHO%20and%20UNICEF

WHO. 2023. Drinking-water, 13 September,UN-Water. Summary progress update 2021: SDG 6 – water and sanitation for all. https://www.unwater.org/sites/default/files/app/uploads/2021/12/SDG-6-Summary-Progress-Update-2021_Version-July-2021a.pdf; https://www.who.int/news-room/fact-sheets/detail/drinking-water

Yamaguchi, Natália Ueda, Silvia Amélia Liwa Abe, Flávia Vieira da Silva Medeiros, Angélica Marquetotti Salcedo Vieira, and Rosangela Bergamasco. 2017. "Hybrid gravitational microfiltration system for drinking water purification." *Revista Ambiente & Água* 12:168–178.

Yue, Xiao, Yanling Yang, Xing Li, Jiawei Ren, Zhiwei Zhou, Yu Zhang, and Haikuan Yu. 2021. "Effect of Fe-based micro-flocculation combined with gravity-driven membrane ultrafiltration on removal of aluminum species during water treatment." *Journal of Environmental Chemical Engineering* 9 (6):106803. doi: 10.1016/j.jece.2021.106803

Zhang, Dong-Qing, Richard Gersberg, Jern Ng, and Soon Tan. 2015. "Conventional and decentralized urban stormwater management: A comparison through case studies of Singapore and Berlin, Germany." *Urban Water Journal* 14:1–12. doi: 10.1080/1573062X.2015.1076488

10 Septage as a Resource and Potential Strategies for Resource Recovery

Basem Haroun[1,2], *Amr Ismail*[3], *Farokh Laqa Kakar*[4], *Rania A. Hamza*[1], *Mohamed El-Qelish*[2], *and Elsayed Elbeshbishy*[1]

[1] Toronto Metropolitan University (formerly Ryerson University), Toronto, Canada

[2] National Research Centre, Cairo, Egypt

[3] Anaergia Inc., Burlington, Ontario, Canada

[4] Brown and Caldwell, Burnaby, BC, Canada

10.1 INTRODUCTION

According to the World Bank, globally 3.6 billion people lack access to safely managed sanitation services (World Bank, 2017). Sanitation coverage is typically higher in urban areas than in rural communities (World Health Organization [WHO], 2002). Lacking adequate sanitation services leads to immense challenges, including, but not limited to, diseases such as diarrhoeal, stunting, and undernutrition in addition to worsening quality of life, decreasing attendance of girls at school, holding back economic growth, and causing environmental pollution (World Bank, 2017).

Significant attention to the development of wastewater treatment plants was given to urban areas compared to rural areas which are lagging far behind. One of the major challenges to build and/or extend wastewater treatment plants is the high capital cost. Operation and maintenance costs increase the burden on the government's shoulders to increase access to sanitation systems. Besides lack of funding, lack of local expertise and restricted local budgets result in inadequate operation of wastewater treatment plants in developing countries (Paraskevas et al., 2002). Because of the aforementioned challenges, decentralized systems, which are quite simple and cost-effective, can serve small villages and low-density communities (U.S. EPA, 2005). The huge capital investment of collecting and pumping costs associated with centralized systems can be minimized, therefore increasing the affordability of wastewater management in decentralized systems.

Several treatment systems, such as trickling filters, activated sludge, septic tanks, membrane bioreactors, constructed wetlands, and ponds, have been applied for on-site wastewater treatment. Septic tanks are the most commonly known primary treatment system for decentralized wastewater treatment and they are more economical for areas where centralized wastewater treatment is too costly due to sparse population or difficult terrain. Approximately 10%–15% of the population globally uses septic tank systems or some form of household sewer treatment system (HSTS). In the USA, 20 million people use septic tanks and 60 million people use different onsite wastewater systems (Bradley et al., 2002). In Australia, about 12% of the population uses septic tanks for domestic wastewater treatment. Greece is of no difference, where 14% of the population is served by decentralized systems in rural areas (Tsagarakis et al., 2001). Septic tanks remove most settleable solids by settling and act as an anaerobic bioreactor that can achieve partial removal of organic matter. Accumulated sludge may cause an odour problem if left untreated for a long time. These septic tanks require periodic cleaning, which is typically between three and ten years, as suggested by the

DOI: 10.1201/9781003381419-12

US Environmental Protection Agency (USEPA). Typical septic tank size in the United States ranges from 3 to 6 m^3 depending on house area, number of occupants, and location.

During cleaning, the septic tank is stirred and the settled sludge, scum layer, and liquid contents are sucked into a septage hauling truck. Septage hauling companies usually charge $200–$500 to homeowners for septic tank cleaning, depending on the size, location, and complexity of the tank. This septage is then either treated in a nearby wastewater treatment facility or landfilled, depending on domestic regulations. The sludge found in septage contains much higher nutrient content than sewage sludge, for example 3.7% phosphorus by mass. Additionally, energy can be recovered by waste incineration, while the subsequent treatment of its ash can remove heavy metals (McGaughy and Reza, 2018).

Conventional decentralized wastewater treatment systems (DEWATS) are not effective in removing nitrogen and phosphorus compounds and in reducing pathogenic organisms. Septic tanks may be used alone or in combination with other processes to treat raw wastewater before it is discharged to a subsurface infiltration system. The tanks are always covered, buried, watertight concrete which provides primary treatment by creating quiescent conditions. In addition to primary treatment, the septic tank can reduce the sludge and scum volumes by as much as 40%. The outlet connections (e.g. a sanitary "tee" fitting) retain the sludge and scum layers in the tank and effluent can only be withdrawn from the clarified zone between the sludge and scum layers. Septage, which is a mixture of sludge, scum, and liquid, which is generated during septic tank cleaning, is a type of faecal sludge that is specifically removed from an individual septic tank (Nasr and Mikhaeil, 2013). Several studies exhibited the presence of Severe Acute Respiratory Syndrome Coronavirus 2 (SARS-CoV-2), which caused the deadliest pandemic of the last hundred years across the globe, in blackwater (Majumder et al., 2021).

Septic tank systems can remove total suspended solids (TSS) and total chemical oxygen demand (TCOD) up to 77%–78% depending on the tank configuration, hydraulic retention time, and temperature (Nasr and Mikhaeil, 2015). Septic tank configuration has been modified by applying baffled and inserting packing materials for better contact with the liquid and the sludge inside the tank. Figure 10.1 shows the schematic diagram of different types of septic tanks: (a) conventional septic tank, (b) single baffle septic tank, (c) two baffle septic tank, and (d) packed septic tank. Modified septic tank systems have been investigated in different countries. A baffled septic tank (Figures 10.1b and 10.1c) consists of several chambers in series in which the baffles force the incoming wastewater to flow under and over the baffles from the inlet to the outlet of the tank. This enhances the contact between anaerobic accumulated sludge and entering wastewater which, in turn, improves the removal of suspended solids due to enmeshment of dissolved anaerobically biodegradable organics. A baffled septic tank with packing materials or the so-called packed septic tank as shown in Figure 10.1d comprises several compartments in addition to packing materials, such as river stones, pieces of plastic electro ducts, and bamboo rings, in the last chamber. They are available in the market, are low cost, and have a good specific superficial area for bacterial adherence (Andrade, 2004).

10.2 POTENTIAL REUSE OF SEPTAGE

Septage, liquid, and solid material accumulated in septic tanks have been applied on agricultural lands without treatment or disposed of in landfills for decades. But recently, new standards and regulations have been in place requiring proper treatment and handling of septage. Both liquid and a solid portion of the septage can be utilized in different forms. For example, septage water is used for irrigation and carwash in many countries where septage is adopted and treated. Furthermore, the solids are treated differently. Given the richness of septage in organic carbon, the potential for reuse and recovery has been extensively deployed. Figure 10.2 shows the reuse options for septage solids (sludge).

For many reuse options, such as animal farming (helping digestion), construction, and land application for non-food crops, septage sludge can be utilized in its original form after dewatering. However, if the intention is the land application for food crops, especially crops meant to be eaten raw, further treatment is required. These treatments are mainly applied to reduce the number of

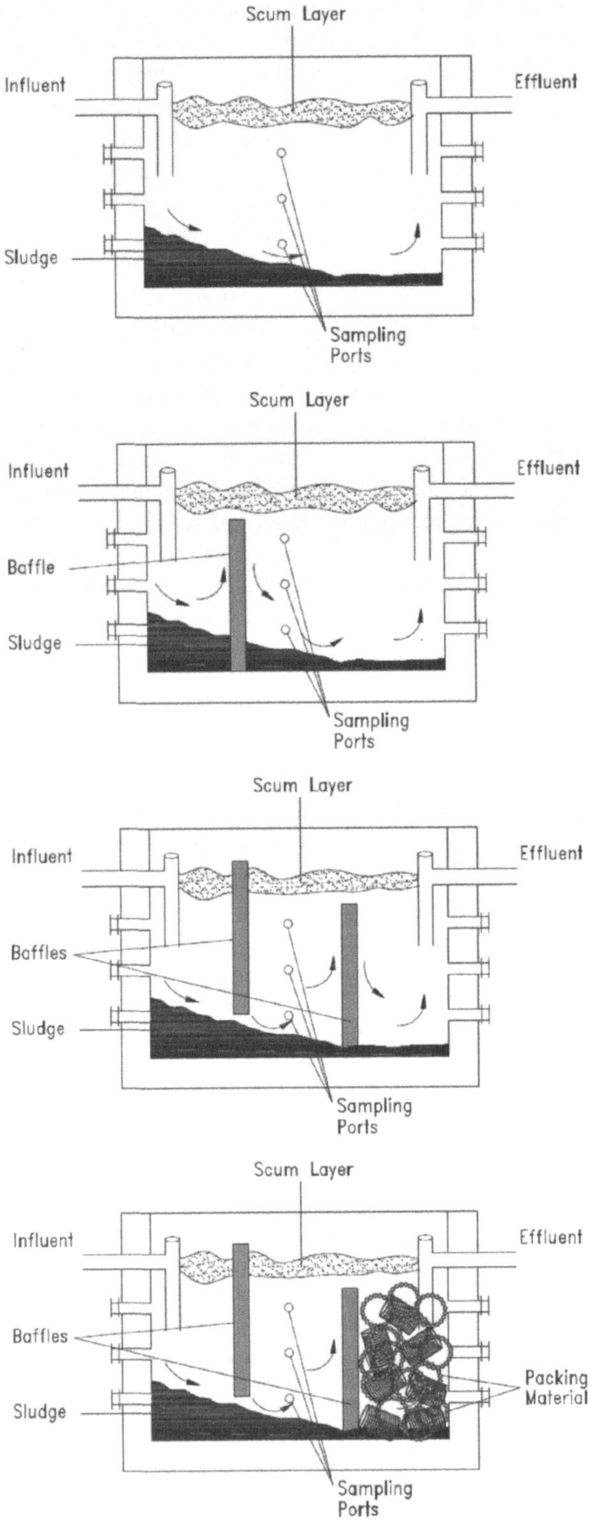

FIGURE 10.1 Schematic diagram of different types of septic tanks: (a) conventional septic tank, (b) single baffle septic tank, (c) two baffle septic tank, and (d) packed septic tank. (From Nasr and Mikhaeil, 2015.)

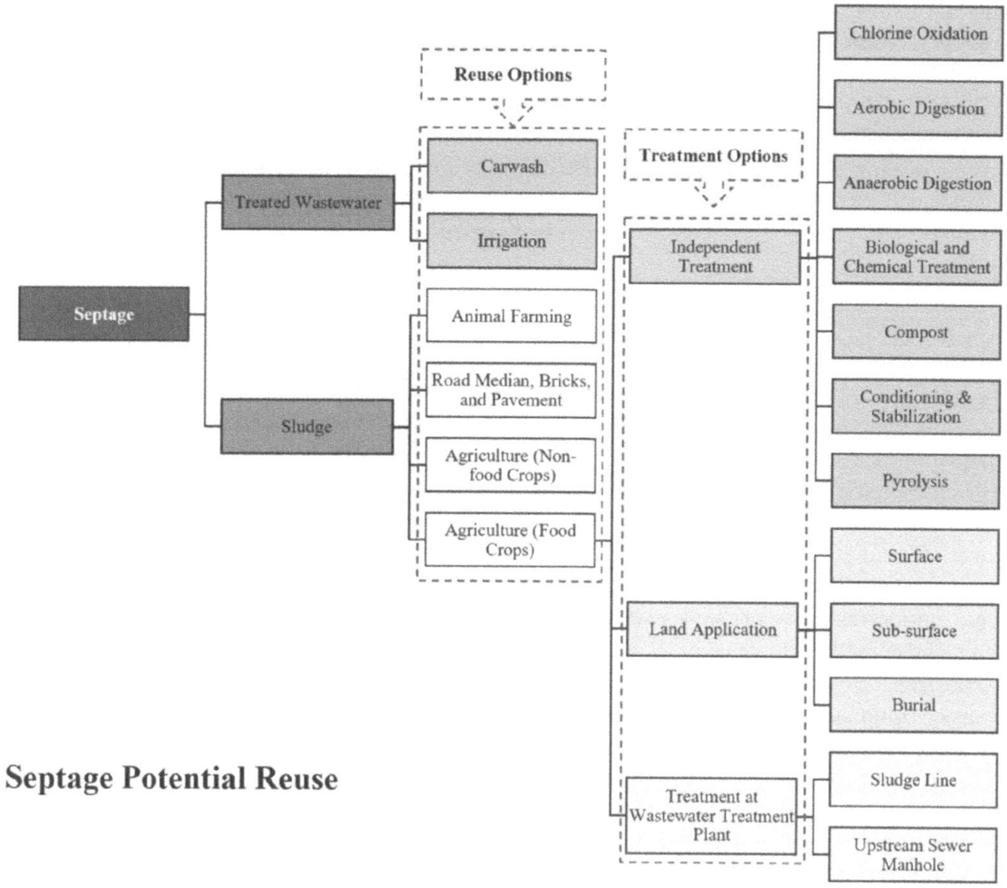

FIGURE 10.2 Septage potential reuse options and corresponding treatment approaches.

pathogens in the fertilizer or soil conditioner. It is noteworthy that each region has its regulation in terms of the level of the pathogen, or they would follow World Health Organization (WHO) Guidelines, which could be measured as the number of helminth eggs, with Nematodes being the indicator (U.S. EPA, 2000).

Septage could be either treated independently, with sewage at the wastewater treatment plant, or used in land application. Independent treatment options are similar to wastewater sludge treatment technologies, including anaerobic digestion (AD), aerobic digestion, thermal treatments, air-drying, stabilization, and more (Kinsley et al., 2017). When combined with sewer, septage could be added to the upstream sewer maintenance hole, to plant headworks, to the sludge handling process, and both liquid stream and sludge handling processes. Furthermore, land application of septage, which is currently the most commonly used disposal method, is relatively simple and cost-effective, uses low energy, and recycles organic material and nutrients to the land. The land application includes spreading septage from designed land application vehicles or tank wagons onto sites using spray irrigation, ridge and furrow irrigation, and overland flow (Kim et al., 2018).

10.3 RESOURCE/NUTRIENT RECOVERY DURING WASTEWATER TREATMENT

The wastewater treatment process consumes huge amounts of energy (23% of US public municipalities) and materials in the different stages of the process to produce final effluent complying with the increasingly stringent discharge requirements (Means, 2004). However, wastewater is rich in finite nutrients (e.g. phosphorus, nitrogen, and potassium) and energy-carrying molecules which if properly recovered during/after the treatment process would achieve a win-win situation (El-Qelish and Mahmoud, 2022). In addition, treated wastewater itself is a valuable resource and could be applied for different purposes in water-stressed regions. The goal of the wastewater treatment process should not be limited to wastewater purification, and a proper attention has to be paid to overall process sustainability including secondary waste minimization and resources/nutrients recovery (Lundin et al., 2000). Resource recovery from septage involves onsite energy generation in the form of biogas and nutrient recovery in the form of biofertilizers for land application (Mo and Zhang, 2013).

10.3.1 Nitrogen Recovery

The negative impacts of the uncontrolled discharge of wastewater rich in nitrogen (Doma et al., 2016) and phosphorus (such as eutrophication) have made them primary targets for the wastewater treatment processes (Xie et al., 2016). However, nitrogen was emphasized over phosphorus for its more challenging removal and remediation requirements (Zhou et al., 2021). Nitrogen removal from wastewater has traditionally proceeded via nitrification/denitrification or via an advanced anammox process (Zahid et al., 2023). For decades, the nitrogen removal process focused on the remediation of wastewater with less or no attention paid to resource recovery, sustainability, or energy consumption. Considering these principles would allow the achievement of the circular flow of nutrients, water, and energy resources (Qin et al., 2023).

Nitrogen in domestic wastewater is considered the second most abundant resource, and its recovery could be applied for many industries such as fertilizers, explosives, foods, and chemicals (Zhang et al., 2019) with a cost-benefit of 0.018–0.086 $\$/m^3$ (Errico et al., 2018). Nitrogen recovery from high-strength wastewater (nitrogen concentration 2–3 g/L) takes place through ammonia stripping and struvite precipitation, while recovery from low-strength wastewater (nitrogen concentration 20–50 mg/L) faces challenges of recovery technologies. The key factors which affect nitrogen recovery suitable for different scenarios from various waste streams are the removal efficiency and economic feasibility. This approach could be achieved following a particular strategy of nitrogen removal from the mainstream (to comply with the discharge standards) and recovery of nitrogen from the concentrate in the side stream (Qin et al., 2023). This aims to concentrate a targeted element from a diluted stream and eliminate the undesirable contaminants by specific methods.

Nitrogen removal from domestic wastewater takes place through ammonification, nitrification, and denitrification at different anaerobic/aerobic conditions. However, nitrogen contamination in the effluent of WWTP based on nitrification-denitrification process is not complying with the discharge standards of many countries (Tang and Chen, 2015). In addition, complying with the discharge requirements of nitrogen would face three main challenges including higher cost (more energy is needed for complete nitrification), more carbon emission (due to the energy required for the aeration accounting for 50%–60% of the total process energy required), and enlarged land occupancy (for building the anaerobic treatment units for the denitrification at long hydraulic retention time (HRT)) (Guven et al., 2019; Maurer et al., 2003; Zhou et al., 2021). The final products of denitrification include nitrous oxide which is a major participant in global warming for the depletion of the ozone layer, and it comprises 14%–16% of the total emissions of the WWTP (Ravishankara et al., 2009). Promising nitrogen removal processes have been developed such as the anammox process and SHARON, which couple many advantages over the traditional processes, such as lower energy consumption, lower sludge production, and smaller footprint, and applied in full-scale configuration (Khin and Annachhatre, 2004; Xiao et al., 2021).

Haber-Bosh process accounts for one-third of the global nitrogen production and consumes from 1% to 2% of the global energy. Energy consumption of the nitrogen removal from municipal wastewater treatment (12.2 kWh/kg N) is higher than the energy consumed in the nitrogen fixation in the

Haber-Bosh process (10.3 kWh/kg N), and this amount is expected to rise due to increasingly stringent discharge requirements (Beckinghausen et al., 2020; Cruz et al., 2019; Maurer et al., 2003). Therefore, a circular economy approach which couples nitrogen remediation-recovery is required for promoting sustainable production and consumption patterns (Stumpf et al., 2021). In addition, the Haber-Bosh process is energy intensive and would be challenging with the extremely vacillating oil prices (Smith et al., 2020). Thus, developing an environmentally friendly, sustainable, and economically feasible nitrogen remediation-recovery is an insistent need. However, nitrogen recovery from ammonia-rich wastewater has been successfully applied; merging this approach in a circular economy layout is a challenging issue. The circular economy aims to maximize the finite resources recycling within a closed cycle.

Several technologies have been experimented with and applied for nitrogen recovery from different kinds of wastewater which are crucial processes for water remediation and nutrient recovery. These technologies include ion exchange, membrane separation, electrochemical systems, and nutrient upconcentration from activated sludge. The selection of recovery technology depends on the nature of the wastewater and its feasibility. Herein the different recovery technologies will be reviewed as follows:

10.3.1.1 Upconcentration from Activated Sludge

Activated sludge process which is the most applied wastewater treatment process globally, is not efficient in the removal of nitrogen, instead nitrogen is partially accumulated inside the activated sludge cells. The nitrogenous compounds content of the sludge is accounted to be 40–60 g/kg dry matter (Huang et al., 2020; Shou et al., 2019), and this high nitrogen content could be recovered through adsorption unit following the traditional anaerobic/anoxic/aerobic process. However, this nitrogen content accumulated in the excess sludge is not high enough for upconcentration; therefore, direct land application as a fertilizer enables benefiting from the available nutrients and minerals in the excess activated sludge. However, the application of excess activated sludge for land application faces some challenges regarding the environmental risk, the existence of pathogens, high transportation cost, and lack of public awareness. Excess activated sludge which is called surplus sludge is pumped into the AD process which results in the solubilization of the nitrogenous compound and the release of free ammonia in the resultant reject water. However, the excess activated sludge is not an easily biodegradable substrate for the anaerobic treatment process; therefore, several pretreatment processes are applied to enhance the hydrolysis process such as alkali, acidic, thermal, and combined pretreatment (El-Qelish et al., 2020). Sludge reject water was applied for microbial nutrient recovery cells, and main nutrients including nitrogen, phosphorus, and potassium have been successfully concentrated as a liquid biofertilizer (El-Qelish and Mahmoud, 2022).

10.3.1.2 Ion Exchange

The ion exchange process can separate materials based on their electrostatic attraction between differently charged ions, where ammonia could be effectively swapped from domestic wastewater (Manto et al., 2018). Ammonia-nitrogen was exchanged with ion exchange anions, and ammonia was concentrated in a side stream solution (Gong et al., 2017). Similarly, nitrate was upconcentrated from municipal wastewater using natural zeolite and modified forms of zeolite (Malekian et al., 2011). Zeolite was applied for ammonia nitrogen upconcentration and achieved a removal efficiency of over 95.0% (Lahav and Green, 1998). Iranian zeolite removed ammonia nitrogen from wastewater and was applied as a biofertilizer and was able to release the loaded ammonia within 200 minutes (Malekian et al., 2011). Life cycle assessment of the application of the ion exchange process for nitrogen upconcentration from wastewater concluded that it is efficient and environmentally friendly. It saves up to 25% of the energy requirements and reduces the emission of greenhouse gases through eutrophication (Qin et al., 2023).

10.3.1.3 Membrane Separation

Reverse osmosis (RO) is a widely applied technology for the separation of different solutes from aqua streams including nitrogen upconcentration from municipal wastewater (Zhang and Liu, 2021). Membrane efficiency depends on membrane material and selectively, while the targeted molecules are upconcentrated in the concentrate chamber. A ten-month pilot-scale RO experiment was conducted to recover ammonia nitrogen from coagulated domestic wastewater in the form of ammonium sulphate (Bilstad, 1995). Although many opportunities are offered by RO process, it still faces many challenges including high energy consumption and membrane fouling (El-Qelish et al., 2022b). Membrane fouling results from the deposition of calcium ions and bicarbonate on the surface of the membrane. In addition, this process depends on many other parameters including pH (which affects the membrane scaling and salt diffusion coefficients) and acid washing (for removal of the deposited scales on the membrane surface). Nitrogen was successfully upconcentrated from domestic wastewater in the form of struvite (Gao et al., 2018).

10.3.1.4 Electrochemical Process

Electrochemical methods depend on electric current for driving specific elements through ion-selective membranes. This process possesses more advantages over the traditional membrane separation methods such as mitigating membrane fouling and shows more tolerance to organic contaminants and suspended solids (Walker et al., 2014). The existence of nitrogen in positively charged form (ammonium NH_4^+) facilitates its upconcentration by negative electric current.

Nitrogen, which is usually in the cationic ammonium form in wastewater (Chen et al., 2021), is more easily transferred and upconcentrated under the electric drive. Electrically driven processes can better mitigate membrane fouling compared to pressure-driven membrane processes because the passive flow of water, which is not filtered through the membranes, elevates the tolerance of electrical processes to organic matters and turgidities (Landsman et al., 2020; Walker et al., 2014), and the representatives include electrodialysis (ED), capacitive deionization (CDI), and bio-electrochemical processes.

Electrodialysis depends on the upconcentration of nitrogen in different forms of either negatively (NO_3^-) or positively (NH_4^+) charged nitrogen compounds forced by an externally applied voltage through ion exchange membrane (Al-Amshawee et al., 2020; Mohammadi et al., 2021). It has been reported that nitrate can be concentrated up to a factor of 16–24 at a nitrogen recovery efficiency of 80.0% (Mohammadi et al., 2021). Electrodialysis has been successfully applied for nitrogen recovery from high-strength nitrogen wastewater including urine, manure, and sludge reject water (Tarpeh et al., 2018). Electrodialysis produces a high-concentration and impurities-free solution which could be directly applied after an evaporation step to have the desired nitrogenous products. However, the many advantages offered by the electrodialysis and membrane fouling represent the main challenge of this process (Rodrigues et al., 2021; van Linden et al., 2019). Several technologies have been applied for minimizing the membrane fouling of the electrodialysis cell such as membrane cleaning by mechanical and chemical methods. In addition, changing the configuration of the electrodialysis cell showed a better performance with uninterruptible operation (Rodrigues et al., 2021).

Bioelectrochemical process is another electrically driven method for the upconcentration of nitrogenous compounds catalysed by electroactive microorganisms attached to the surface of the electrodes under a continuous flow of current. A microbial fuel cell is a bioelectrochemical system which is usually applied for nitrogen remediation and nitrogen upconcentration from different wastewater such as rejected water and livestock wastewater (El-Qelish and Mahmoud, 2022). El-Qelish and Mahmoud (2022) have successfully experimented with co-substrate treatment for waste remediation and nutrient recovery from sludge reject water and livestock wastewater and achieved the upconcentration factor of $\sim 0.43 \pm 0.15$ to produce a liquid biofertilizer instead of the chemical fertilizers. This process is influenced by many factors including current density, voltage, pH, dissolved oxygen, and existence of electroactive bacteria. Higher pH at the cathode results in conversion of NH_4^+ to NH_3, which enhances the mobility of the ammonium ion through the ion-selective membrane, while the increase in the dissolved oxygen inhibits the process as it leads to reduction of the oxygen at the cathode and consequently unstable potential (Zamora et al., 2017). However, the microbial activity of the bioelectroactive species is considered the core motive for the whole process as it is responsible for the conversion of the different forms of the nitrogen and link between the rate of conversion and the potential (Li et al., 2020).

10.3.2 Phosphorus Recovery

The world's ever-increasing population and the massive economic development are pressing on the natural resources. Annually, global fertilizer demand is growing by 4%, which should be covered by either non-renewable reserves or renewable resources such as waste streams (Xie et al., 2016). However, rock phosphate deposits (70 million tons in the whole world) are the main source of the global phosphorus market which is exclusive and available only in few countries (Morocco and Western Sahara, China, Algeria, Brazil, and South Africa) (Jasinski, 2017); phosphate demand could be retrieved from renewable resources such as waste streams (Rittmann et al., 2011). It is estimated that the global phosphorus rock reserves will be depleted within the next century (Cooper et al., 2011). In addition, food production cost is influenced by the prices of phosphate rocks used for fertilizer production; therefore, maintaining stable phosphate supplies is of great importance. Phosphorus (P) is one of the three major plant nutrients, along with nitrogen (N) and potassium (K), and its scarcity may become a driving force behind dramatic changes in agricultural practices.

Phosphorus recovery from renewable resources achieves a more efficient and sustainable production method and avoids the depletion of the already limited global phosphorus reserves. In addition, phosphorus recovery from waste streams couples with environment protection and resource-loss minimization. EU has assigned promotions for large-scale fertilizer production from different waste streams (domestic or industrial) as a result of the studies which concluded that more than 20% of the phosphorus consumed in the EU, ends up in the wastewater treatment plants. Two-thirds of this share is concentrated in the sludge of the wastewater treatment plant and the remaining one-third ends up in the hydrosphere itself (Ott and Rechberger, 2012). A considerable fraction of phosphorus exists in fertilizers, and sewage sludge used in agriculture is lost into surface waters by rain (Coelho et al., 2012). In 2013, more than 50% of the total amount of phosphorus consumed globally (37.9 million metric tons) was wasted in the seas as a result of washing away by rain (Chen and Graedel, 2016). This results in serious environmental issues such as eutrophication phenomena and its consequences including oxygen depletion and blockage of sunlight leading to destruction of the aquatic life (Conley et al., 2009; Dittmann and Wiegand, 2006).

Resource recovery and recycling from municipal wastewater in isolated areas has emerged in line with the concept of serving these communities with decentralized wastewater treatment plants such as soak pits and septic tanks. Septage is formed under anaerobic conditions and is therefore rich in reduced nitrogen (e.g. ammonia) and orthophosphates (Withers et al., 2011). Resource recovery from effluents of small-scale decentralized wastewater treatment plants such as septic tanks has received modest attention except for some studies on phosphorus recovery in the form of vivianite (Azam and Finneran, 2014; Liu et al., 2018).

Several methods have been applied for phosphorus recovery from waste streams such as crystallization, wet chemical extraction, and thermochemical processes, depending on the efficiency of recovery (amount and crystallinity) and waste nature, e.g. liquid stream, dried sludge, or sludge ash (Goel and Kansal, 2020). Different recovery methods produce phosphorus in different forms such as hydroxyapatite ($Ca_5(PO_4)_3(OH)$) and aluminium phosphate ($AlPO_4$), while struvite ($MgNH_4PO_4 \cdot 6H_2O$) is the most prominent form of recovered phosphorus. In addition, phosphorus in the form of struvite is preferable for plant consumption as it is non-toxic (Liu et al., 2013) and does not harm the plant roots (Kumar and Pal, 2015).

The main goal of wastewater treatment plants is to reduce the concentration of the main nutrients including phosphorus from the final effluent by concentrating it in the biomass or the sludge (El-Shafai et al., 2022). The wastewater treatment process depends on the mineralization of the organic compounds into simpler forms, including the release of P from the complex substances. However, there are many studies which reported phosphorus removal from waste streams, but few ones are implemented in full-scale applications (Di Capua et al., 2022). The recovery method is selected based on the matrix representing the P source. The phosphorus recovery methods from wastewater are discussed as follows:

10.3.2.1 Crystallization

Crystallization is one of the main methods which are applied for phosphorus recovery from liquid streams (Chrispim et al., 2019). Desmidt et al. (2015) developed a crystallization system using a fluidized-bed reactor, where suitable doses of precipitation (lime and Mg $(OH)_2$) reagents were added to the targeted medium resulting in increasing the size of the material and consequently precipitated in the bottom of the fluidized-bed reactor in the form of calcium phosphate, magnesium phosphate, or struvite (Di Capua et al., 2020). The efficiency of P recovery from wastewater through the crystallization process can reach up to 40% which is higher than other reported values (from 10% to 25%) (Egle et al., 2016). Fixed-bed reactor (FBR) and continuously stirred tank reactor (CSTR) have been also applied for P recovery from the supernatant digester. FBR dosed with calcium silicate hydrate achieved 10%–11% recovery efficiency (Green, 2015). CSTR fed with $MgCl_2$ has been successfully applied for phosphorus recovery in the form of struvite pellets. Recovering phosphorus in the form of hydroxyapatite ($Ca_5(PO_4)_3OH$) has been tested for synthetic wastewater and resulted in phosphorus removal efficiency of 87.5% compared to effluent concentration (Zou and Wang, 2016).

However, the main goal of all studies for P recovery from wastewater aimed to produce struvite to be used as biofertilizers, and other studies experimented P recovery for other purposes (detergents, food additives, phosphoric acid) in the form of hydroxyapatite or amorphous calcium phosphate (Chen and Graedel, 2016).

10.3.2.2 Adsorption

Adsorption is one of the most applied technologies for phosphorus recovery from wastewater and liquid phases. This process is motivated by the electrostatic forces between the charges of the adsorbent and the adsorbate (Ajmal et al., 2018; Kumar and Viswanathan, 2018). The adsorption process is influenced by operational conditions including pH, temperature, adsorbate dose, surface area, and contact time (Alsawy et al., 2022). Phosphorus has been successfully recovered from water and wastewater through the adsorption/precipitation process. The mechanism of the process involves ion exchange between the cations (Ca^{+2} and Mg^{+2}) and the phosphate anions in the liquid phase. This step is followed by the precipitation of phosphate at a high pH value (dos Reis et al., 2020).

Raw organic waste materials have been applied for the adsorption process such as agricultural waste (Alsawy et al., 2022), algae (Elgarahy et al., 2022), sludge (Nasr et al., 2021), and biochar-derived materials as low-cost, sustainable, and chemical-free materials. Modified and decorated adsorbents with $AlCl_3$, $MgCl_2$, and Fe_3O_4 have shown much higher adsorption capacities for phosphorus recovery from wastewater (Field et al., 2019; Zheng et al., 2020). Pyrolysed wood chips, supplemented with nanosized aluminium oxyhydroxide (AlOOH), achieved a promising result in P recovery and application as a biofertilizer (Zhou et al., 2019). A biochar filter was applied for the P recovery from real wastewater and achieved a 78% efficiency, which could be used as a guideline for full-scale application (Manyuchi et al., 2018).

10.3.2.3 Electrochemical Precipitation

Electrochemical precipitation is considered a sustainable method for P recovery from waste streams since it is chemical-free, cost-effective, and simple in operation. The mechanism of electrochemical precipitation is based on water electrolysis at the cathode where the calcium phosphate (e.g.

hydroxyapatite) is precipitated and harvested, excluding the separation step of the other recovery methods (Di Capua et al., 2022). Electrochemical precipitation has been effectively applied for P recovery from municipal wastewater and found that cathode size and Ca/P ratio are the driving parameters for this process and yielded hydroxyapatite of 1.67 mol/mol (Lei et al., 2018). However, electrochemical precipitation experimented for P recovery from raw domestic wastewater achieved only 50% efficiency, which is significantly lower than the recovery percentage of crystallization and absorption technologies. In addition, the yield of P from this method is in the form of hydroxyapatite which is considered of lower value than the other P forms from the other methods (Di Capua et al., 2022).

10.3.3 Biogas Production

Biogas production has been investigated from different kinds of waste streams including sewage sludge, agricultural waste residues, and industrial wastewater (El-Qelish et al., 2022a). Biogas production through AD of septage is a sustainable management approach at a low cost. AD process is a complex series of biochemical reactions, which takes place by different kinds of microorganisms (Mahmoud and El-Qelish, 2022). AD could be divided into four main steps to produce methane from complex organic compounds, namely hydrolysis (complex organics such as carbohydrates, proteins, and lipids are converted into simpler monomers such as C5 and C6 sugars, peptides, and amino acids and short-chain fatty acids), acidogenesis (simple organics such as hexose and pentose sugars are converted into volatile fatty acids and hydrogen by acidogens), acetogenesis (volatile fatty acids are converted into acetic acids by acetogens), and methanogenesis (volatile fatty acids are converted into methane) (Afroze et al., 2022). Septage collected from municipal wastewater treatment septic tanks and faecal treatment systems is rich in organic compounds which could be digested at anaerobic conditions yielding considerable amounts of methane, which could be subjected to electricity production (Bella and Venkateswara Rao, 2022). India is one of the biggest countries (in terms of number, 148 million septic tanks) depending on on-site sanitation systems such as septic tanks for municipal wastewater treatment in rural areas due to the lack of public sewer networks. It has been estimated by the United States Agency for International Development (USAID) that India has around 148 million septic tanks for household service (Wankhade, 2015). Indian national urban sanitation policy has forbidden the collection and transporting of septage, which results in the bad management of septage through open dumping, resulting in pollution of the surface waters and soils and severe health issues (Rohilla et al., 2017). Application of AD for biogas production could be a sustainable and cost-effective management route. However, the high ammonia content of septage inhibits the process, and this could be compromised through co-substrate digestion (Park and Li, 2012). In addition, septage digestion sometimes exhibits a slow digestion rate due to the existence of complex compounds and stabilized solids. Therefore, the C/N ratio is one of the critical parameters which affect the AD of septage and co-substrate treatment was proposed as one of the successful strategies. Co-substrate treatment couples many advantages including buffering capacity, dilution of toxic substances, and providing deficient nutrients (Kim et al., 2019).

10.3.4 Organic Fertilizer

Many attempts have been proposed for septage management, and because it is very rich in organics and nutrients such as nitrogen and phosphorus, it has been applied as a chemical-free biofertilizer suitable for direct land application (Wood and Cowie, 2004). Application of biofertilizer couples many advantages including saving 1%–2% of the global energy used for the production of chemical fertilizers, protecting the environment from pollution through uncontrolled discharge of the septage and conserving global reserves of phosphate and other elements mines.

Septage has been applied as a biofertilizer in the form of pellets, and it was found that this leads to a reduction of the monetary value of septage biofertilizer (Grau et al., 2017). Composting is an environmentally friendly technology to stabilize organic waste materials into bio-fertilizers (compost) (Cerda et al., 2018) through a series of biological aerobic/anaerobic reactions. Composting as a septage management approach has many advantages, e.g. reduction of landfills (including reduction of greenhouse gas emission and contamination of groundwater), sustainability, efficiency, and production of high-value bio-fertilizer which improves the soil characteristics (Schott et al., 2016). Production of high-quality compost requires controlling the process conditions (moisture content, temperature, aeration, and porosity) and the waste materials constituents (C/N ratio, impurities, and heterogeneity) (Swan et al., 2002). Improper control of the initial adjustment of the composting process and the existence of a high amount of non-biodegradable materials such as plastics and glass would negatively affect the process leading to the release of offensive odours and the production of low-quality compost (Cerda et al., 2018).

10.4 CASE STUDIES FOR NUTRIENTS RECOVERY

10.4.1 Phosphorus Recovery from Septic Tank Liquor in India

The growing interest in resource recovery and water recycling from wastewater in isolated areas has directed a great attention to the decentralized wastewater systems. Septic tank has been recognized as the dominant system in developing countries for sanitation-focused policies and interventions (Goel and Kansal, 2020). Global phosphorus reserves are limited; therefore, phosphorus recovery from waste streams has been considered one of the practical strategies to deal with such depletion. AD of septage converts phosphorus from the complex form into free soluble orthophosphates which facilitates its recovery.

A case study was reported by Goel and Kansal (2020), who investigated phosphorus recovery from septic tank liquor in India. In the aforementioned study, wastewater was collected from a two-chamber septic tank serving six households in Delhi, India, and working at an HRT of 12 days. Several factors were tested for phosphorus recovery from septage including pH, molar ratio of magnesium and phosphorus, velocity gradient, and mixing duration (Goel and Kansal, 2020).

The recovery of P was evaluated in terms of the extent of reduction in P in the liquor, and the presence of struvite crystals was verified using X-ray diffraction (XRD) and scanning electron microscopy. Based on the four factors studied, it has been found that the optimal operational conditions for phosphorus recovery were pH, 9; magnesium to phosphorus molar ratio of 2.2; velocity gradient of 343.2 s^{-1}; and mixing duration of 15 minutes. Different operational conditions showed independent interaction. pH showed the most significant influence on the recovery of phosphorus, which was around 80%. Velocity gradient showed a much less effect on the recovery process by 5%, followed by a molar ratio of 2%. Phosphorus was recovered in the form of struvite, which is considered the most suitable form as a fertilizer. Besides, calcium phosphate was also obtained more easily from the septage because of the existence of high concentrations of calcium in the septage. Therefore, phosphorus recovery from septage is achievable in a mixture of struvite and calcium phosphates (Goel and Kansal, 2020).

10.4.2 Nutrient Recovery from Wastewater by Algal Biofilm for Fertilizers Production in Remote Areas in Germany

Decentralized wastewater treatment plants are the cost-effective treatment option for small communities in isolated areas. Therefore, the on-site treatment systems such as septic tanks, soak pits and pit latrines, and twin pit latrines have been applied for wastewater treatment in small-scale agriculture and industries (Shukla et al., 2023).

A case study on the efficiency of algal biofilms using Algal Turf Scrubber (ATS) systems for nutrient recovery from different kinds of wastewater was carried out at ideal places in Germany for one year (Reinecke et al., 2023). Four ATS systems were operated and fed with municipal wastewater, fresh pig manure, fresh cattle manure, and biogas effluent. The four ATS units have successfully recovered the macro- and micronutrients from all kinds of the wastewater. The ATS biofilms have successfully recovered nitrogen, phosphorus, potassium, and carbon throughout the duration of the experiment. The ATS system was a low-maintenance and cost-effective wastewater treatment technology. Economic evaluation for fertilizer production indicated competitive costs of 2.22 €/kg in an optimized ATS of 1,760 m^2 (Reinecke et al., 2023).

10.5 WATER REUSE AND RECYCLING

Septage is a low-quality highly fluctuating organic waste that comprises large fractions of grease, grit, hair, and debris in addition to harbouring disease-causing bacteria, viruses, and parasites; hence, it cannot be discharged to the aquatic environment or reused without meeting the processing, reuse, and disposition regulations and guidelines governed by the U.S. Code of Federal Regulation (40 CFR) Part 503 (Gutterer et al., 2009; U.S. EPA, 1994b). Land application is one of the most common and economic means for the reuse of septage. However, the availability of suitable land infrequently visited by the public; the need for holding facilities during periods of frozen soil; and the possible enforcement of stabilization to reduce odours, pathogens, and vector attraction are limiting factors (U.S. EPA, 1994b). Reed beds and freezing beds or constructed wetlands are low-cost technologies that can dewater and treat septage while meeting reuse standards year-round (Kinsley et al., 2017). Reed beds are similar in design to drying beds, only planted with common reeds such as *Phragmites australis*, where septage is allowed to accumulate for six to ten years, significantly reducing operating costs (Kinsley et al., 2017). The common reeds play an important role in breaking the accumulated solids layers and permit continuous draining and improving the dewaterability of the septage through evapotranspiration. The advantages of using reed beds also materialize in their tolerance to low and high pH and salinity, and their ability to withstand different water and drought levels (de Maeseneer, 1997). The potential for reuse of the sludge cake will be governed by biosolids regulations. Nutrient content as nitrogen, phosphorus, and potassium in addition to organic matter content crucial economic drivers for beneficial reuse as dewatered sludge cake can substitute the manure, compost, and organic soil applied to agricultural land and reclamation sites (U.S. EPA, 1994a). The potential reuse of the filtrate will depend on wastewater reuse guidelines and regulations which include limits on specific water quality parameters such as TSS, biochemical oxygen demand (BOD), and salinity in addition to pathogen indicators (Alberta Environment, 2000; U.S. EPA, 1994a). Septage, in addition to excreta, wastewater sludge, and biomass slurry, can be used to fertilize fishponds, cultivate microalgae in high-rate algal ponds, and raise crustaceans such as shrimps and crayfish (Mara et al., 1989). The main purpose of using septage is to provide nutrient-rich fertilizer to the ponds to produce natural food for the fish (Eves et al., 1995). In developing countries where land is available, waste stabilization ponds used to treat septage and other organic wastes permit the wastes to be recycled into food (Edwards, 1980). Cultivating fish can be carried out successfully in the maturation ponds of a series of waste stabilization ponds with annual yields of 4 tons/ha, although the better managed ponds produce 7–8 tons/ha. However, fish muscle is susceptible to pathogenic invasion when raised in ponds that contain concentrations of faecal coliform and salmonella of $>10^4$ and $>10^5$, respectively. Even at lower concentrations, high pathogenic concentrations may be available in the intestines and the intraperitoneal fluid of the fish; hence, pretreatment of the waste is recommended (El-Shafai et al., 2004; Mara, 2013).

10.6 ADVANTAGES OF THE SYSTEM

Septic tanks are known to be part of DEWATS which are characterized by their suitability for diverse conditions and versatility in applications; reliability and efficiency; requiring very short planning and implementation periods; low capital, operational, and maintenance costs; and independence from external energy input and sophisticated technical equipment (Gutterer et al., 2009). Septic tanks provide a combination of physical treatment through sedimentation and biological treatment through sludge stabilization; consequently, no design considerations for biomass with poor settleability or ancillary sludge separation units are required. The design of septic tanks is fairly simple and durable considering no moving or mechanical parts are required; hence, they can be constructed from locally available materials by the local workforce (Barber and Stuckey, 1999; Gutterer et al., 2009), while occupying a small footprint since their construction is typically underground (Sasse, 1998). Operationally, septic tanks are resilient to intermittent operation, hydraulic fluctuations, and organic shock loads (Barber and Stuckey, 1999; Gutterer et al., 2009). Additionally, sludge accumulation rates in septic tanks steadily decline over time, with a significant plummet after 12 months due to biological degradation resulting in lower sludge generation and more stabilized sludges (Gray, 1995).

10.7 CHALLENGES IN THE PROCESS

Septic tanks have many advantages as decentralized systems for primary treatment; however, several challenges are associated with their usage. The settled sludge in the bottom of the tank is partially undigested sludge which needs further digestion for biogas production and stabilization to be used as fertilizers. The mesophilic AD process could be a feasible technology for septic sludge treatment and biomethane production (Lu et al., 2019). The septic tank itself acts as an anaerobic bioreactor; however, most of the tanks are not designed for biogas production. This could be due to the uncontrolled temperature around the buried tank. The optimum temperatures for mesophilic and thermophilic ADs are 37°C and 55°C, respectively, which is quite difficult to be controlled for the installed underground tanks. Septic systems are considered primary treatment, nutrient, colloidal and soluble organic removal rates were significantly low (Saeed et al., 2021; Sharma et al., 2014). Additionally, the maximum faecal coliform removal efficiency could be one log removal in the effluent (Nasr and Mikhaeil, 2015). Therefore, post-treatment is required to achieve effluent quality that meets the local legislations. Besides, improper disposal and treatment of septic tank effluent affect both water resources and ecosystems. Treated wastewater can provide valuable resources (water, nutrients, soil conditioner, briquettes, and energy) especially in urban areas and in water-scarce environments (World Bank, 2017). In 2022, some European countries including, but not limited to, France, Germany, England, and Italy have experienced the worst drought conditions over the past 500 years due to heatwaves. Consequently, water use such as watering lawns, filling pools, or cleaning cars was restricted in some parts of the aforementioned countries (Toreti et al., 2022). Therefore, treated domestic wastewater could be a feasible option for the abovementioned usages.

Furthermore, septic tanks that are generally poorly designed and maintained have resulted in pollution of surface and groundwater with serious health impacts (World Bank, 2015). For example, if the installed baffled and/or dip pipes have collapsed, either all the wastewater backs up into the house or an overflow of untreated wastewater to the surrounding locations could occur which cause severe health impact (Carroll et al., 2006). Additionally, discharging non-biodegradable material and hazardous chemicals in the tank can cause septic tank failure.

On the other hand, resource recovery from septage represents a major challenge for many reasons, including but not limited to the lack of knowledge and poor infrastructure, and directs all efforts towards the centralized wastewater treatment technologies. McConville et al. (2019) proposed a four-step scenario for septage management, starting with the surveying of the available treatment options, then the selection of the optimum treatment process, then investigating treatment option sustainability in terms of socio-cultural, economic, and environmental impact, and finally discussing the full process with on-site stakeholders. However, septage is rich in biosolids and nutrients; resource recovery in full-scale application is facing some constraints regarding the high cost involved in its implementation (containment, emptying, transport, solid-liquid separation) (Shukla et al., 2023).

10.8 CONCLUSIONS AND PERSPECTIVES

Septic tank systems are the most reported decentralized wastewater treatment processes in rural areas, small villages, and low-density communities with regard to alleviating the impact of wastewater disposal on the environment and public health. Septic tanks remove most settleable solids by settling and can achieve partial removal of soluble organic matter anaerobically. The performance of the system is quite simple, and it is inexpensive in terms of operation and maintenance. Desludging is required every three to five years depending on the volume of the tank and the number of people who are served using the tank. The sludge in the bottom of the tank is partially stabilized and mineralized which reduces its volume. Septic tanks are water-sealed to avoid underground water pollution and made up of several chambers for better contact of the upcoming wastewater with the inoculum (sludge in the bottom of the tank). They may have packing materials in the last chamber where microorganisms can be attached to their surface. Septic tank effluents are discharged to subsoil infiltration which could cause environmental pollution due to poor treatment. To avoid such pollution, post-treatment is required in order to enhance pathogen elimination. On the other hand, septic systems have some drawbacks such as leakage into the groundwater and into the adjacent lagoons if not properly constructed. Additionally, septic tank overflow could negatively impact coral reefs where flooding of untreated wastewater into the adjacent marine may cause macroalgal blooms.

Septage, which is the slurry taken out of the septic tank, contains much higher nutrient content than sewage sludge. Septage could be treated similar to wastewater sludge in AD, aerobic digestion, thermal treatments, air-drying, and stabilization. After proper treatment, septage can be used in land application by recycling organic material and nutrients to the land.

Nitrogen, which is the second abundant resource, can be recovered from high-strength wastewater (nitrogen concentration 2–3 g/L) through ammonia stripping and struvite precipitation; however, nitrogen recovery low strength wastewater (nitrogen concentration 20–50 mg/L) faces some challenges. Nitrogen can be used in many industries such as fertilizers, explosives, foods, and chemicals.

Septage, which is rich in nitrogen, can be used through ion exchange, membrane separation, electrochemical systems, and nutrient upconcentration for nitrogen recovery. Selection of recovery technology depends on the nature of the wastewater and its feasibility. Septage is also orthophosphates-rich. Phosphorus recovery from septage is feasible. Several methods have been applied for phosphorus recovery from waste streams such as crystallization, adsorption, and electrochemical precipitation. Struvite ($MgNH_4PO_4{\cdot}6H_2O$) is the most predominant form of recovered phosphorus while hydroxyapatite ($Ca_5(PO_4)_3(OH)$) and aluminium phosphate ($AlPO_4$) may be produced depending on the recovery methods.

Septage collected from septic tank systems is organic-rich compounds which can produce biomethane gas through a mesophilic digestion process. Application of AD for biogas production could be a sustainable and cost-effective management route. Since septage is rich in phosphorus and nitrogen, it can be used as a chemical-free biofertilizer for land application when managed properly.

REFERENCES

Afroze, N., Nakhla, G., Kim, M., Yazdanpanah, A., 2022. Effects of trace elements on digester performance and microbial community response in anaerobic digestion systems. Environ. Technol. 44(27), 4157–4172.

Ajmal, Z., Muhmood, A., Usman, M., Kizito, S., Lu, J., Dong, R., Wu, S., 2018. Phosphate removal from aqueous solution using iron oxides: adsorption, desorption and regeneration characteristics. J. Colloid Interface Sci. 528, 145–155.

Al-Amshawee, S., Yunus, M.Y.B.M., Azoddein, A.A.M., Hassell, D.G., Dakhil, I.H., Hasan, H.A., 2020. Electrodialysis desalination for water and wastewater: a review. Chem. Eng. J. 380, 122231.

Alberta Environment, 2000. Guidelines for Municipal Wastewater Irrigation. Municipal Program Development Branch, Environmental Sciences Division, Environmental Service.

Alsawy, T., Rashad, E., El-Qelish, M., Mohammed, R.H., 2022. A comprehensive review on the chemical regeneration of biochar adsorbent for sustainable wastewater treatment. npj Clean Water 5, 29. https://doi.org/10.1038/s41545-022-00172-3

Andrade, N.C., 2004. Anaerobic filter applied for the treatment of domestic sewage. Ph.D. thesis, Federal University of Campina Grande, Brazil.

Azam, H.M., Finneran, K.T., 2014. Fe(III) reduction-mediated phosphate removal as vivianite ($Fe_3(PO_4)_2{\cdot}8H_2O$) in septic system wastewater. Chemosphere 97, 1–9.

Barber, W.P., Stuckey, D.C., 1999. The use of the anaerobic baffled reactor (ABR) for wastewater treatment: a review. Water Res. 33(7), 1559–1578.

Beckinghausen, A., Odlare, M., Thorin, E., Schwede, S., 2020. From removal to recovery: an evaluation of nitrogen recovery techniques from wastewater. Appl. Energy 263, 114616.

Bella, K., Venkateswara Rao, P., 2022. Anaerobic co-digestion of cheese whey and septage: effect of substrate and inoculum on biogas production. J. Environ. Manage. 308, 114581. https://doi.org/10.1016/j.jenvman.2022.114581

Bilstad, T., 1995. Nitrogen separation from domestic wastewater by reverse osmosis. J. Memb. Sci. 102, 93–102.

Bradley, R.B., Daigger, G.T., Rubin, R., Tchobanoglous, G., 2002. Evaluation of onsite wastewater treatment technologies using sustainable development criteria. Clean Technol. Environ. Policy 4, 87–99.

Carroll, S., Goonetilleke, A., Thomas, E., Hargreaves, M., Frost, R., Dawes, L., 2006. Integrated risk framework for onsite wastewater treatment systems. Environ. Manag. 38(2), 286–303.

Cerda, A., Artola, A., Font, X., Barrena, R., Gea, T., Sánchez, A., 2018. Composting of food wastes: status and challenges. Bioresour. Technol. 248, 57–67. https://doi.org/10.1016/j.biortech.2017.06.133

Chen, T.-L., Chen, L.-H., Lin, Y.J., Yu, C.-P., Ma, H., Chiang, P.-C., 2021. Advanced ammonia nitrogen removal and recovery technology using electrokinetic and stripping process towards a sustainable nitrogen cycle: a review. J. Clean. Prod. 309, 127369.

Chen, M., Graedel, T.E., 2016. A half-century of global phosphorus flows, stocks, production, consumption, recycling, and environmental impacts. Glob. Environ. Chang. 36, 139–152.

Chrispim, M.C., Scholz, M., Nolasco, M.A., 2019. Phosphorus recovery from municipal wastewater treatment: critical review of challenges and opportunities for developing countries. J. Environ. Manage. 248, 109268.

Coelho, B.B., Murray, R., Lapen, D., Topp, E., Bruin, A., 2012. Phosphorus and sediment loading to surface waters from liquid swine manure application under different drainage and tillage practices. Agric. Water Manag. 104, 51–61.

Conley, D. J., Paerl, H. W., Howarth, R. W., Boesch, D. F., Seitzinger, S. P., Havens, K. E., ... & Likens, G. E. (2009). Controlling eutrophication: nitrogen and phosphorus. Science, 323(5917), 1014–1015.

Cooper, J., Lombardi, R., Boardman, D., Carliell-Marquet, C., 2011. The future distribution and production of global phosphate rock reserves. Resour. Conserv. Recycl. 57, 78–86.

Cruz, H., Law, Y.Y., Guest, J.S., Rabaey, K., Batstone, D., Laycock, B., Verstraete, W., Pikaar, I., 2019. Mainstream ammonium recovery to advance sustainable urban wastewater management. Environ. Sci. Technol. 53, 11066–11079.

de Maeseneer, J.L., 1997. Constructed wetlands for sludge dewatering. Water Sci. Technol. 35(5), 279–285. https://doi.org/10.1016/S0273-1223(97)00080-2

Desmidt, E., Ghyselbrecht, K., Zhang, Y., Pinoy, L., Van der Bruggen, B., Verstraete, W., Rabaey, K., Meesschaert, B., 2015. Global phosphorus scarcity and full-scale P-recovery techniques: a review. Crit. Rev. Environ. Sci. Technol. 45, 336–384.

Di Capua, F., de Sario, S., Ferraro, A., Petrella, A., Race, M., Pirozzi, F., Fratino, U., Spasiano, D., 2022. Phosphorous removal and recovery from urban wastewater: current practices and new directions. Sci. Total Environ. 823, 153750. https://doi.org/10.1016/j.scitotenv.2022.153750

Di Capua, F., Spasiano, D., Giordano, A., Adani, F., Fratino, U., Pirozzi, F., Esposito, G., 2020. High-solid anaerobic digestion of sewage sludge: challenges and opportunities. Appl. Energy 278, 115608.

Dittmann, E., Wiegand, C., 2006. Cyanobacterial toxins–occurrence, biosynthesis and impact on human affairs. Mol. Nutr. Food Res. 50, 7–17.

Doma, H.S., El-Kamah, H.M., El-Qelish, M., 2016. Slaughterhouse wastewater treatment using UASB reactor followed by down flow hanging sponge unit. Res. J. Pharm. Biol. Chem. Sci. 7, 568–576.

dos Reis, G.S., Thue, P.S., Cazacliu, B.G., Lima, E.C., Sampaio, C.H., Quattrone, M., Ovsyannikova, E., Kruse, A., Dotto, G.L., 2020. Effect of concrete carbonation on phosphate removal through adsorption process and its potential application as fertilizer. J. Clean. Prod. 256, 120416.

Edwards, P., 1980. A review of recycling organic wastes into fish, with emphasis on the Tropics. Aquaculture 21(3), 261–279. https://doi.org/10.1016/0044-8486(80)90136-2

Egle, L., Rechberger, H., Krampe, J., Zessner, M., 2016. Phosphorus recovery from municipal wastewater: an integrated comparative technological, environmental and economic assessment of P recovery technologies. Sci. Total Environ. 571, 522–542.

Elgarahy, A.M., Maged, A., Elwakeel, K.Z., El-Gohary, F., El-Qelish, M., 2022. Tuning cationic/anionic dyes removal from aqueous solution onto green algal biomass for biohydrogen production. Environ. Res., 216(Pt 1):114522. doi:10.1016/j.envres.2022.114522.

El-Qelish, M., Chatterjee, P., Dessì, P., Kokko, M., El-Gohary, F., Abo-Aly, M., Rintala, J., 2020. Bio-hydrogen production from sewage sludge: screening for pretreatments and semi-continuous reactor operation. Waste Biomass Valorization 11, 4225–4234. https://doi.org/10.1007/s12649-019-00743-5

El-Qelish, M., Chatterjee, P., Kokko, M., El-Gohary, F., Abo-Aly, M., Rintala, J., 2022a. Conversion of boreal lake sedimented pulp mill fibre into biogas: a two-stage hydrogen and methane production. Biomass Convers. Biorefinery 1–10.

El-Qelish, M., Hassan, G.K., Leaper, S., Dessì, P., Abdel-Karim, A., 2022b. Membrane-based technologies for biohydrogen production: a review. J. Environ. Manage. 316, 115239. https://doi.org/10.1016/j.jenvman.2022.115239

El-Qelish, M., Mahmoud, M., 2022. Overcoming organic matter limitation enables high nutrient recovery from sewage sludge reject water in a self-powered microbial nutrient recovery cell. Sci. Total Environ. 802, 149851. https://doi.org/10.1016/j.scitotenv.2021.149851

El-Shafai, S.A., El-Qelish, M., Zahid, W.M., 2022. Significance of ferric chloride addition on the treatment performance of cloth-media MBR treating municipal wastewater. Arab. J. Sci. Eng. https://doi.org/10.1007/s13369-022-07373-5

El-Shafai, S.A., Gijzen, H.J., Nasr, F.A., El-Gohary, F.A., 2004. Microbial quality of tilapia reared in fecal-contaminated ponds. Environ. Res. 95(2), 231–238. https://doi.org/10.1016/j.envres.2004.01.002

Errico, M., Fjerbaek Sotoft, L., Kjærhuus Nielsen, A., Norddahl, B., 2018. Treatment costs of ammonia recovery from biogas digestate by air stripping analyzed by process simulation. Clean Technol. Environ. Policy 20, 1479–1489.

Eves, A., Turner, C., Yakupitiyage, A., Tongdee, N., Ponza, S., 1995. The microbiological and sensory quality of septage-raised Nile tilapia (*Oreochromis niloticus*). Aquaculture 132(3), 261–272. https://doi.org/10.1016/0044-8486(94)00328-L

Field, H.R., Whitaker, A.H., Henson, J.A., Duckworth, O.W., 2019. Sorption of copper and phosphate to diverse biogenic iron (oxyhydr) oxide deposits. Sci. Total Environ. 697, 134111.

Gao, Y., Fang, Z., Liang, P., Huang, X., 2018. Direct concentration of municipal sewage by forward osmosis and membrane fouling behavior. Bioresour. Technol. 247, 730–735.

Goel, S., Kansal, A., 2020. Phosphorous recovery from septic tank liquor: optimal conditions and effect of tapered velocity gradient. J. Clean. Prod. 275, 124056. https://doi.org/10.1016/j.jclepro.2020.124056

Gong, H., Wang, Z., Zhang, X., Jin, Z., Wang, C., Zhang, L., Wang, K., 2017. Organics and nitrogen recovery from sewage via membrane-based pre-concentration combined with ion exchange process. Chem. Eng. J. 311, 13–19.

Grau, F., Drechsel, N., Haering, V., Trautz, D., Weerakkody, W.J.S.K., Drechsel, P., Marschner, B., Dissanayake, D.M.P.S., Sinnathamby, V., 2017. Impact of fecal sludge and municipal solid waste co-compost on crop growth of *Raphanus sativus* L. and *Capsicum anuum* L. under stress conditions. Resources 6. https://doi.org/10.3390/resources6030026

Gray, N.F., 1995. The influence of sludge accumulation rate on septic tank design. Environ. Technol. 16(8), 795–800.

Green, B. W. "Fertilizers in aquaculture. Feed and Feeding Practices in Aquaculture, Vol." (2015): 27–52.

Guo, Y.-X., Chen, Q.-J., Qin, Y., Yang, Y.-R., Yang, Q.-Z., Wang, Y.-X., Cheng, Z., Cao, N., Zhang, G.-Q., 2021. Succession of the microbial communities and function prediction during short-term peach sawdust-based composting. Bioresour. Technol. 332, 125079 (1–9).

Gutterer, B., Sasse, L., Panzerbieter, T., Reckerzügel, T., 2009. Decentralised Wastewater Treatment Systems (DEWATS) and Sanitation in Developing Countries. BORDA, Bremen.

Guven, H., Dereli, R.K., Ozgun, H., Ersahin, M.E., Ozturk, I., 2019. Towards sustainable and energy efficient municipal wastewater treatment by up-concentration of organics. Prog. Energy Combust. Sci. 70, 145–168.

Huang, J., Zhou, Z., Zheng, Y., Sun, X., Yu, S., Zhao, X., Yang, A., Wu, C., Wang, Z., 2020. Biological nutrient removal in the anaerobic side-stream reactor coupled membrane bioreactors for sludge reduction. Bioresour. Technol. 295, 122241.

Jasinski, S.M., 2021. Mineral commodity summaries: phosphate rock. US Geol. Surv.

Khin, T., Annachhatre, A.P., 2004. Novel microbial nitrogen removal processes. Biotechnol. Adv. 22, 519–532.

Kim, M., Abdulazeez, M., Haroun, B.M., Nakhla, G., Keleman, M., 2019. Microbial communities in co-digestion of food wastes and wastewater biosolids. Bioresour. Technol. 289, 121580. https://doi.org/10.1016/j.biortech.2019.121580

Kim, B., Bel, T., Bourdoncle, P., Dimare, J., Troesch, S., Molle, P., 2018. Septage unit treatment by sludge treatment reed beds for easy management and reuse: performance and design considerations. Water Sci. Technol. 77, 279–285. https://doi.org/10.2166/wst.2017.461

Kinsley, C., Kennedy, K., Crolla, A., 2017. A combined reed bed/freezing bed technology for septage treatment and reuse in cold climate regions. Water Sci. Technol. 76, 286–293. https://doi.org/10.2166/wst.2017.189

Kumar, R., Pal, P., 2015. Assessing the feasibility of N and P recovery by struvite precipitation from nutrient-rich wastewater: a review. Environ. Sci. Pollut. Res. 22, 17453–17464.

Kumar, I.A., Viswanathan, N., 2018. A facile synthesis of magnetic particles sprayed gelatin embedded hydrotalcite composite for effective phosphate sorption. J. Environ. Chem. Eng. 6, 208–217.

Lahav, O., Green, M., 1998. Ammonium removal using ion exchange and biological regeneration. Water Res. 32, 2019–2028. https://doi.org/10.1016/S0043-1354(97)00453-3

Landsman, M.R., Lawler, D.F., Katz, L.E., 2020. Application of electrodialysis pretreatment to enhance boron removal and reduce fouling during desalination by nanofiltration/reverse osmosis. Desalination 491, 114563.

Lei, Y., Remmers, J.C., Saakes, M., van der Weijden, R.D., Buisman, C.J.N., 2018. Is there a precipitation sequence in municipal wastewater induced by electrolysis? Environ. Sci. Technol. 52, 8399–8407.

Li, N., Wan, Y., Wang, X., 2020. Nutrient conversion and recovery from wastewater using electroactive bacteria. Sci. Total Environ. 706, 135690.

Liu, J., Cheng, X., Qi, X., Li, N., Tian, J., Qiu, B., Xu, K., Qu, D., 2018. Recovery of phosphate from aqueous solutions via vivianite crystallization: thermodynamics and influence of pH. Chem. Eng. J. 349, 37–46.

Liu, Y., Kumar, S., Kwag, J., Ra, C., 2013. Magnesium ammonium phosphate formation, recovery and its application as valuable resources: a review. J. Chem. Technol. Biotechnol. 88, 181–189.

Lu, D., Liu, X., Apul, O.G., Zhang, L., Ryan, D.K., Zhang, X., 2019. Optimization of biomethane production from anaerobic co-digestion of microalgae and septic tank sludge. Biomass Bioenergy 127, 105266.

Lundin, M., Bengtsson, M., Molander, S., 2000. Life cycle assessment of wastewater systems: influence of system boundaries and scale on calculated environmental loads. Environ. Sci. Technol. 34, 180–186.

Mahmoud, M., El-Qelish, M. (2022). Efficient Biogas Production Through Syntrophic Microbial Partnerships in the Presence of Conductive Materials in Anaerobic Digesters Treating Organic Waste Streams: A Critical Assessment. In: Meghvansi, M.K., Goel, A.K. (eds) Anaerobic Biodigesters for Human Waste Treatment. Environmental and Microbial Biotechnology. Springer, Singapore 249–261.

Majumder, A., Gupta, A.K., Ghosal, P.S., Varma, M., 2021. A review on hospital wastewater treatment: a special emphasis on occurrence and removal of pharmaceutically active compounds, resistant microorganisms, and SARS-CoV-2. J. Environ. Chem. Eng. 9(2), 104812.

Malekian, R., Abedi-Koupai, J., Eslamian, S.S., Mousavi, S.F., Abbaspour, K.C., Afyuni, M., 2011. Ion-exchange process for ammonium removal and release using natural Iranian zeolite. Appl. Clay Sci. 51, 323–329.

Manto, M.J., Xie, P., Keller, M.A., Liano, W.E., Pu, T., Wang, C., 2018. Recovery of ammonium from aqueous solutions using ZSM-5. Chemosphere 198, 501–509.

Manyuchi, M.M., Mbohwa, C., Muzenda, E., 2018. Potential to use municipal waste bio char in wastewater treatment for nutrients recovery. Phys. Chem. Earth, Parts A/B/C 107, 92–95.

Mara, D., 2013. Domestic Wastewater Treatment in Developing Countries. Routledge, Earthscan, London, UK.

Mara, D.D., Cairncross, S., World Health OrganizationOrganization. 1989. Guidelines for the Safe Use of Wastewater and Excreta in Agriculture and Aquaculture: Measures for Public Health Protection. World Health Organization.

Maurer, M., Schwegler, P., Larsen, T.A., 2003. Nutrients in urine: energetic aspects of removal and recovery. Water Sci. Technol. 48, 37–46.

McConville, J.R., Kvarnström, E., Maiteki, J.M., Niwagaba, C.B., 2019. Infrastructure investments and operating costs for fecal sludge and sewage treatment systems in Kampala, Uganda. Urban Water J. 16, 584–593.

McGaughy, K., Reza, M.T., 2018. Recovery of macro and micro-nutrients by hydrothermal carbonization of septage. J. Agric. Food Chem. 66 (8), 1854–1862. https://doi.org/10.1021/acs.jafc.7b05667

Means, E.G., 2004. Water and Wastewater Industry Energy Efficiency: A Research Roadmap.Awwa Research Foundation. CEC-500-2001-901

Mo, W., Zhang, Q., 2013. Energy-nutrients-water nexus: integrated resource recovery in municipal wastewater treatment plants. J. Environ. Manage. 127, 255–267. https://doi.org/10.1016/j.jenvman.2013.05.007

Mohammadi, R., Ramasamy, D.L., Sillanpää, M., 2021. Enhancement of nitrate removal and recovery from municipal wastewater through single-and multi-batch electrodialysis: process optimisation and energy consumption. Desalination 498, 114726.

Nasr, F.A., Mikhaeil, B., 2013. Treatment of domestic wastewater using conventional and baffled septic tanks. Environ. Technol. 34(16), 2337–2343.

Nasr, F.A., Mikhaeil, B., 2015. Treatment of domestic wastewater using modified septic tank. Desalination Water Treatment 56(8), 2073–2081.

Nasr, M., Tawfik, A., Awad, H.M., Galal, A., El-qelish, M., Abdul, M., Mumtaz, M., Khan, A., Rehan, M., Nizami, A., Lee, M., 2021. Dual production of hydrogen and biochar from industrial effluent containing phenolic compounds. Fuel 301, 121087. https://doi.org/10.1016/j.fuel.2021.121087

Ott, C., Rechberger, H., 2012. The European phosphorus balance. Resour. Conserv. Recycl. 60, 159–172.

Paraskevas, P.A., Giokas, D.L., Lekkas, T.D., 2002. Wastewater management in coastal urban areas: the case of Greece. Water Sci. Technol. 46(8), 177–186.

Park, S., Li, Y., 2012. Evaluation of methane production and macronutrient degradation in the anaerobic co-digestion of algae biomass residue and lipid waste. Bioresour. Technol. 111, 42–48.

Qin, Y., Wang, K., Xia, Q., Yu, S., Zhang, M., An, Y., Zhao, X., Zhou, Z., 2023. Up-concentration of nitrogen from domestic wastewater: a sustainable strategy from removal to recovery. Chem. Eng. J. 451, 138789. https://doi.org/10.1016/j.cej.2022.138789

Ravishankara, A.R., Daniel, J.S., Portmann, R.W., 2009. Nitrous oxide (N_2O): the dominant ozone-depleting substance emitted in the 21st century. Science (80-.) 326, 123–125.

Reinecke, D., Bischoff, L.-S., Klassen, V., Blifernez-Klassen, O., Grimm, P., Kruse, O., Klose, H., Schurr, U., 2023. Nutrient recovery from wastewaters by algal biofilm for fertilizer production part 1: case study on the techno-economical aspects at pilot-scale. Sep. Purif. Technol. 305, 122471. https://doi.org/10.1016/j.seppur.2022.122471

Rittmann, B.E., Mayer, B., Westerhoff, P., Edwards, M., 2011. Capturing the lost phosphorus. Chemosphere 84, 846–853.

Rodrigues, M., Paradkar, A., Sleutels, T., Ter Heijne, A., Buisman, C.J.N., Hamelers, H.V.M., Kuntke, P., 2021. Donnan dialysis for scaling mitigation during electrochemical ammonium recovery from complex wastewater. Water Res. 201, 117260.

Rohilla, S.K., Luthra, B., Bhatnagar, A., Matto, M., Bhonde, U., 2017. Septage management: a practitioner's guide. Cent. Sci. Environ. New Delhi.

Saeed, T., Afrin, R., Al-Muyeed, A., Miah, M.J., Jahan, H., 2021. Bioreactor septic tank for on-site wastewater treatment: floating constructed wetland integration. J. Environ. Chem. Eng. 9(4), 105606.

Sasse L (1998) DEWATS - decentralized wastewater treatment in developing countries. Bremen Overseas Research and Development Association, Delhi, India, p 161.

Schott, A.B.S., Wenzel, H., la Cour Jansen, J., 2016. Identification of decisive factors for greenhouse gas emissions in comparative life cycle assessments of food waste management—an analytical review. J. Clean. Prod. 119, 13–24.

Sharma, M.K., Khursheed, A., Kazmi, A.A., 2014. Modified septic tank-anaerobic filter unit as a two-stage on-site domestic wastewater treatment system. Environ. Technol. 35(17) 2183–2193. https://doi.org/10.1080/09593330.2014.896950

Shou, Z., Zhu, N., Yuan, H., Dai, X., Shen, Y., 2019. Buffering phosphate mitigates ammonia emission in sewage sludge composting: enhanced organics removal coupled with microbial ammonium assimilation. J. Clean. Prod. 227, 189–198.

Shukla, A., Patwa, A., Parde, D., & Vijay, R. (2022). A review on generation, characterization, containment, transport and treatment of fecal sludge and septage with resource recovery-oriented sanitation. Environ. Res. 216, 114389. 114389. https://doi.org/10.1016/j.envres.2022.114389

Smith, C., Hill, A.K., Torrente-Murciano, L., 2020. Current and future role of Haber-Bosch ammonia in a carbon-free energy landscape. Energy Environ. Sci. 13, 331–344. https://doi.org/10.1039/c9ee02873k

Stumpf, L., Schöggl, J.-P., Baumgartner, R.J., 2021. Climbing up the circularity ladder?–A mixed-methods analysis of circular economy in business practice. J. Clean. Prod. 316, 128158.

Swan, J.R.M., Crook, B., Gilbert, E.J., 2002. Microbial emission from composting sites Issues Environ. Sci. Technol., 18, 23–85.

Tang, H.L., Chen, H., 2015. Nitrification at full-scale municipal wastewater treatment plants: evaluation of inhibition and bioaugmentation of nitrifiers. Bioresour. Technol. 190, 76–81.

Tarpeh, W.A., Barazesh, J.M., Cath, T.Y., Nelson, K.L., 2018. Electrochemical stripping to recover nitrogen from source-separated urine. Environ. Sci. Technol. 52, 1453–1460.

Toreti, A., Bavera, D., Acosta Navarro, J., Cammalleri, C., de Jager, A., Di Ciollo, C., Hrast Essenfelder, A., Maetens, W., Magni, D., Masante, D., Mazzeschi, M., Niemeyer, S., Spinoni, J., 2022. Drought in Europe August 2022. Publications Office of the European Union, Luxembourg. https://doi.org/10.2760/264241, JRC130493.

Tsagarakis, K.P., Mara, D.D., Angelakis, A.N., 2001. Wastewater management in Greece: experience and lessons for developing countries. Water Sci. Technol. 44(6), 163–172.

U.S. EPA, 1994a. A Plain English Guide to the EPA Part 503 Biosolids Rule. US Environmental Protection Agency: Washington, DC, USA, 1994; pp. 1–55.

U.S. EPA, 1994b. Land application of sewage sludge: A guide for land appliers on the requirements of the federal standards for the use or disposal of sewage sludge, 40 CFR Part 503. Washington DC: Office of Enforcement and Compliance Assurance, United States Environmental Protection Agency; Report No.: EPA/831-B-93-002b.

U.S. EPA, 2000. Decentralized systems technology fact sheet: septage treatment and disposal. Environ. Prot.., EPA 832-F-00-079, 1-10, September 2000.

U.S. EPA, 2005. Handbook for Managing Onsite and Clustered (Decentralized) Wastewater Treatment Systems, EPA/832-B-05-001. Office of Water, Washington, DC. 66 pp.

van Linden, N., Spanjers, H., van Lier, J.B., 2019. Application of dynamic current density for increased concentration factors and reduced energy consumption for concentrating ammonium by electrodialysis. Water Res. 163, 114856.

Walker, W.S., Kim, Y., Lawler, D.F., 2014. Treatment of model inland brackish groundwater reverse osmosis concentrate with electrodialysis—Part III: Sensitivity to composition and hydraulic recovery. Desalination 347, 158–164.

Wankhade, K., 2015. Urban sanitation in India: key shifts in the national policy frame. Environ. Urban. 27, 555–572.

World Health Organization (WHO), 2002. Environmental Health. Eastern Mediterranean Regional Center for Environmental Health Activities (CEHA).

Withers, P.J.A., Jarvie, H.P., Stoate, C., 2011. Quantifying the impact of septic tank systems on eutrophication risk in rural headwaters. Environ. Int. 37, 644–653.

Wood, S.W., Cowie, A., 2004. A Review of Greenhouse Gas Emission Factors for Fertiliser Production. (IEA Bioenergy Task 38, Armidale, Australia, 2004).

World Bank, 2015. WSP. Improving On-site Sanitation and Connections to Sewers in Southeast Asia: Insights from Indonesia and Vietnam; World Bank: Washington, DC, USA.

World Bank, 2017. Sanitation [Topics]. http://www.worldbank.org/en/topic/sanitation

Xiao, R., Ni, B.-J., Liu, S., Lu, H., 2021. Impacts of organics on the microbial ecology of wastewater anammox processes: recent advances and meta-analysis. Water Res. 191, 116817.

Xie, M., Shon, H.K., Gray, S.R., Elimelech, M., 2016. Membrane-based processes for wastewater nutrient recovery: technology, challenges, and future direction. Water Res. https://doi.org/10.1016/j.watres.2015.11.045

Zahid, W.M., El-Qelish, M. & El-Shafai, S.A. 2023. Effect of operating regime on the performance of biofilm reactor treating municipal wastewater. Int. J. Environ. Sci. Technol. 20, 8799-8812. https://doi.org/10.1007/s13762-022-04589-8.

Zamora, P., Georgieva, T., Ter Heijne, A., Sleutels, T.H., Jeremiasse, J.A., Saakes, A.W., Buisman, M., Kuntke, C.J.N., 2017. Ammonia recovery from urine in a scaled-up microbial electrolysis cell. J. Power Sources 356, 491–499.

Zhang, X., Liu, Y., 2021. Reverse osmosis concentrate: an essential link for closing loop of municipal wastewater reclamation towards urban sustainability. Chem. Eng. J. 421, 127773.

Zhang, C., Ma, J., Waite, T.D., 2019. Ammonia-rich solution production from wastewaters using chemical-free flow-electrode capacitive deionization. ACS Sustain. Chem. Eng. 7, 6480–6485.

Zheng, Y., Zimmerman, A.R., Gao, B., 2020. Comparative investigation of characteristics and phosphate removal by engineered biochars with different loadings of magnesium, aluminum, or iron. Sci. Total Environ. 747, 141277.

Zhou, K., Remy, C., Kabbe, C., Barjenbruch, M., 2019. Comparative environmental life cycle assessment of phosphorus recovery with different generations of the AirPrex® systems. Int. J. Environ. Sci. Technol. 16, 2427–2440.

Zhou, Z., Wang, K., Qiang, J., Pang, H., Yuan, Y., An, Y., Zhou, C., Ye, J., Wu, Z., 2021. Mainstream nitrogen separation and side-stream removal to reduce discharge and footprint of wastewater treatment plants. Water Res. 188, 116527.

Zou, H., Wang, Y., 2016. Phosphorus removal and recovery from domestic wastewater in a novel process of enhanced biological phosphorus removal coupled with crystallization. Bioresour. Technol. 211, 87–92.

11 Sustainable Energy Source for Water and Wastewater Decontamination

Innovative and Green Technologies Adapted to Decentralized Sanitation

Song Yan[1], Patrick Drogui[1,3], Rajeshwar D. Tyagi[2,3], and Jonathan W.C. Wong[2]

[1] Université du Québec, Québec, Canada

[2] Dongguan University of Technology, Dongguan, China

[3] BOSK Bioproducts, Québec, Canada

11.1 INTRODUCTION

Energy costs make up a significant portion of the total operating costs, ranging from 5 to 30%, for water and wastewater utilities across the globe. As climate change and carbon footprints become increasingly important issues, it is crucial to optimize energy use and reduce greenhouse gas (GHG) emissions from operations by adopting sustainable energy sources (Liu et al. 2012).

Sustainable energy resources, which are applicable to water and wastewater treatment (WWT) plants (WWTPs), include solar thermal energy, solar photovoltaic (PV) cells, wind power, small hydropower (SHP), and thermal energy recovery from wastewater using heat pumps (Chae and Kang 2013; Fernández-Marchante et al. 2021; Frijns, Hofman and Nederlof 2013; Ganiyu, Martínez-Huitle and Rodrigo 2020; Karaca, Dincer and Nitefor 2022; Mook, Aroua and Issabayeva 2014; Musabandesu and Loge 2021). Among these, PV-based power and wind power are relatively universal energy sources, and they have been employed as energy sources for energy-intensive water desalinization systems. Due to technological improvements, many renewable technologies, including PVs, wind power, SHP, and sewage heat recovery, can be used individually or in combination with other technologies in WWTPs.

Renewable energy sources have been used and will continue to be used, either directly or indirectly, in water and WWT. Renewable energy sources were proposed as an alternative to the depletion of the fossil fuels, and it presently represents an optimal solution for achieving sustainable energy system. The global energy demand is anticipated to grow by 37% by 2040 (Climate Council 2019).

However, renewable energy sources, unlike conventional power sources (petroleum-based gensets and power from the grid), are mostly used for small-to-medium applications because of their high initial investment costs. The power needed to treat a rural water supply is relatively small, and for this reason renewable energy power sources are widely used in many decentralized water and WWT systems or developing countries (Argaw 2003).

Thus, there are also increasing interests in the development of self-powered electrochemical system, designed by exploring the possibility of applying renewable energy source for powering the

DOI: 10.1201/9781003381419-13

electrochemical environmental applications either directly or regulated by energy storage systems. Solar PV, wind, and biomass are some of the renewable energy sources that have been utilized to provide electricity for different electrochemical technologies (Ganiyu, Martínez-Huitle and Rodrigo 2020).

This chapter presents a comprehensive analysis of the application of sustainable energy sources in decentralized water treatment, specifically solar, wind, geothermal, and solar electro-technology. Each energy source is discussed in detail, including power generation efficiency, system energy efficiency, and system operation state under various weather conditions. Additionally, we consider the challenges and perspectives of these technologies.

The use of environmentally friendly energy sources for water and WWT systems is proven feasible through our analysis of environmental conditions, operational efficiency, treatment effects, and pollutant removal mechanisms. The examples presented in this chapter demonstrate the potential of sustainable energy sources for decentralized water treatment, highlighting the advantages of each technology and its specific applications.

Furthermore, future research should focus on addressing the challenges that arise with these technologies, such as the need for energy storage systems and the lack of standardized evaluation methods. Overall, sustainable energy sources hold great potential for decentralized water treatment systems, with the ability to improve operational efficiency, reduce energy costs, and provide environmentally friendly solutions for water and WWT.

11.2 SOLAR ENERGY SOURCE

Solar energy, which is typically stabilization ponds and solar detoxification, is often used for WWT and is still used in many countries. Among all the alternative energy resources, solar energy is at the top of all the sources for its potential to provide for the future energy needs (Li, Goswami and Stefanakos 2013).

Solar energy is the simplest technology for desalination of salty waters, and for water disinfection. It can be converted into electricity, which can be used to provide electricity for pumps, ultraviolet (UV) systems, photocatalysis, reverse osmosis (RO), and conventional surface water treatment systems (Argaw 2003).

Renewable energy sources can provide electricity for traditional or conventional water treatment technologies as well as new emerging technologies (UV disinfection, desalination plants, and distillation, direct heat, or photocatalytic oxidation to destroy pathogens).

Solar energy can be used either directly or indirectly (thermally or electrically) to pump or treat water and wastewater. Solar thermal energy can be best utilized for desalination of salty or brackish water, pasteurization, various methods of distillation, or indirectly even for water pumping applications (Argaw 2003; Sharon and Reddy 2015).

11.2.1 Solar Energy for Drinking Water Treatment

Solar energy is a promising energy source that can effectively remove pollutants from water. To achieve this, a new system was developed that utilizes peroxydisulfate (PDS) as an oxidant and a xenon lamp as a solar irradiation light source. This system has demonstrated the ability to break down organic matter and reduce fouling in ultrafiltration (UF) membrane when treating surface water (Guo et al. 2021).

The researchers also compared heat-activated PDS pre-treatment to determine the proportion of solar light and heating effects. The study found that solar light was more effective than heating for activating PDS to break down natural organic matter (NOM) at temperatures below 50°C. However, both solar light and heating contributed to the production of free radicals at temperatures above 50°C. When solar light was present, dissolved organic matter (DOC), UV254, and the maximum fluorescent intensity, as well as micropollutant atrazine in the feed water, were degraded more efficiently. Total fouling index (TFI) and hydraulic irreversible fouling index (HIFI) were moderately correlated with DOC and UV254 but highly correlated with the maximum fluorescent intensity (Fmax) of component 2 (C2) and component 3 (C3) (Guo et al. 2021).

Solar energy, regarded as a green and sustainable resource, could stimulate the production of free radicals in advanced oxidation. Miralles and co-workers compared UV and solar-combined advanced oxidation technologies to treat five kinds of micro-contaminants and reported that the final total costs of solar-combined advanced oxidation technology were the lowest when the target pollutant degradation rate reached 90%. Thus, solar energy can be an alternative to UV light sources in some specific cases, especially in some decentralized systems that treat the potable water for a single household or a community. Recently, the utilization of full solar spectrum for the degradation of pollutants in water and WWTs has attracted the attention due to its unique advantages. Currently, the main means of using solar energy to treat water or wastewater include direct use of solar energy (using some UV-absorbing or visible light-absorbing materials and conversion of solar energy into electric (using PV panels to generate electricity or thermal (using infrared-absorbing materials or light-concentrating elements energy. For instance, Wang and co-workers constructed a new hybrid system that integrated the photoelectrochemical reactor with PV, which activated persulfate in UV bands, generated power in the visible and near-infrared bands and heated solution temperature in the far infrared bands, respectively. Lin and co-workers established a system combining sunlight-focusing with solar tracking for the treatment of pharmaceutical pollutants and gained significant effect in the presence of persulfate.

Saline water is converted to potable water by desalination. Conventional desalination plants operated by fossil fuel have also contributed to GHG emissions. This has forced researchers to look for alternate way of powering desalination units by renewable energy (Sharon and Reddy 2015).

Seawater desalination is an important process to meet the increasing demand for freshwater demand; solar, wind, and geothermal are commonly considered renewable energy for desalination, and among these, solar energy occupies nearly 57% of renewable energy-based desalination market (Sharon and Reddy 2015). Renewable energy-based desalination systems/units are highly suitable for arid, semi-arid, remote, and rural areas where other modes of power supply and freshwater pipe lines are not possible. The solar thermal collectors are very well developed, and now attention has been paid toward solar PV cells to reduce their cost and increase their efficiency; thereby considerable reduction in water production cost could be achieved (Li, Goswami and Stefanakos 2013; Sharon and Reddy 2015).

11.2.2 Solar Energy for Wastewater Treatment

There are many ways to clean wastewater, like using chemicals, filters, and biological processes. Solar power has a gross potential for about 600 TW (terawatt) with technical feasibility for 60 TW; the current total installed capacity of solar power is only 0.005 TW. Though the present technology contributes to very less fraction of overall energy consumption, developments in the field of solar thermal system are continuously improving over the decades. Solar photocatalysis, solar desalination, solar disinfection, solar detoxification, and solar pasteurization are the common technologies employed for treating wastewater.

The major focus is to systematically analyze the chemical and mechanical solar WWT technique like solar photocatalysis (heterogeneous and homogeneous), solar thermal electrochemical technique, solar desalination process, and solar advance oxidation process.

Photocatalysis is catalysis technology that is used to speed up light-relevant chemical reactions. In the concept of photocatalysis, TiO_2 semi-conductor materials absorb light and create electron-holes that enable chemical transformation to be carried out by reaction participants. This is due to the UV irradiation of TiO_2 with an energy equal or superior to the band gap (>3.2 eV). These charge carriers (e−/h+) can migrate to the surface of the catalyst, where they are then available to undergo redox reactions with pollutants (Daghrir, Drogui and Robert 2012).

Solar (PV cell) energy is one of the most promising technologies. It has been projected as being a strategic electricity source that has becoming one of the major suppliers of electricity both for industrial and household use. PV cells convert sunlight into electric energy directly, via PV effect, and have emerged as a valuable and sustainable approach to overcome global energy and environmental crisis.

Solar energy is particularly important in treating wastewater. Direct solar radiation is used for WWT. The three most common WWT methods are stabilization ponds, aerated lagoons, and oxidation ditches.

Solar detoxification can be an effective method when chemicals and biological treatments are used together. By adding certain chemicals, the performance of the treatment plant can be improved. TiO_2, also known as hydroxyl radical, is a strong oxidizing agent that can break down almost any organic compound. This substance is commonly used in catalytic treatments for solar wastewater detoxification (Argaw 2003).

11.3 WIND ENERGY SOURCE

People have been using wind energy for a really long time since 1200 BC. Nowadays, there are over a million windmills in countries like the United States, Argentina, and Australia. Just like solar panels, wind turbines can turn wind energy into electricity, which can be used to power water treatment systems. However, wind turbines are not usually used for WWT because these systems require a lot of power or need direct sunlight (like stabilization ponds) (Argaw 2003).

Wind is non-exhaustible, clean, and environmental friendly source of energy found abundantly in most part of the world. Wind energy is one of the most rapidly advancing technologies in renewable and sustainable energy and has gained more popularity due to its low and stable cost compared to conventional fossil fuel. It also has advantage of being available in remote and rural areas with limited access to the main electricity grid. Wind energy is suitable for location with wind speed greater than the cut-in speed. Indeed, the power generation rate in wind turbine depends mainly on the wind speed in a similar manner with the dependence of solar energy on sunlight irradiation intensity. The global wind energy capacity stands at 433 GW as at the end of year 2017, with over 52 GW installations in 2017 representing 10% capacity addition (Qin et al. 2005).

11.3.1 Wind Energy for Water Treatment

Similarly, wind energy can be used to either pump water mechanically (using windmills), or the electricity produced from the wind turbine can be used to pump, treat, or disinfect water. Mechanical wind pumps (windmills) operate at lower wind speeds compared to electric wind turbines. Windmills start pumping at speeds between 2.5 and 3.5 m/s, while electrical wind turbines need an average wind speed of 5–6 m/s to become competitive with windmills for water pumping applications.

Hybrid systems are also becoming more attractive; a hybrid system can be a combination of PV, a wind turbine with or without a backup generator, and battery storage.

In rural and remote communities, especially in developing countries, there is often not enough money, expertise, or electricity to build and operate water treatment facilities. As a result, these communities often struggle to access clean water. Jeon et al. (2016) reported a fully self-powered electro-coagulation (SPEC) system for decentralized water treatment. For this, a triboelectric nanogenerator harvests ambient wind energy to provide the needed power. A simple electro-coagulation unit, composed of a simple pair of electrodes, successfully removes dissolved colloidal pollutants from water, while coagulating and precipitating pollutants, without the use of chemicals or an external power supply. The SPEC system removes 90% of algae and 97% of organic dye with self-powered treatment for 72 h. Considering that the SPEC system does not require a large facility, additive chemicals, or an external power supply, the SPEC system could become a powerful candidate for decentralized treatment of water (Jeon et al. 2016).

11.3.2 Wind Energy for Wastewater Treatment

RO is a highly advanced method used for treating wastewater and recycling it. It is commonly used for purifying drinking water from brackish water and seawater, making polluted water sources reusable, and lowering the salt content in water for industrial uses. But the application of RO membrane for aquaculture WWT has been largely limited. One of the major problems is the energy cost during the membrane filtration process. In order to reduce the energy cost for RO membrane operation, researchers have turned to renewable energy sources for solutions.

A regular windmill with multiple blades can be used to convert wind energy into hydraulic pressure, which can then be used to operate RO membranes (Qin et al. 2005). Aquaculture wastewater passing through the RO membrane is separated into permeate (freshwater) and brine (concentrated wastewater). The permeate is recirculated to the fish tanks, while the brine is collected for possible treatment or reuse. As a result, no wastewater discharge is made to the ambient coastal water. It showed that the wind-driven RO system for nitrogen removal was technically feasible and environmentally friendly. The most unique and important feature of this system is that it can treat and recycle aquaculture wastewater using renewable energy, making it suitable for use in remote areas where electricity is hard to obtain.

The system is able to work at an average wind speed as low as 3.0 m/s. Depending on the wind speed, it can generate and recycle freshwater at a flow rate of 228–366 L/h. The permeate production increases linearly with the wind speed. About 70–84% of aquaculture wastewater can be recycled using this system, which is capable of removing 90–97% of nitrogenous waste present in tilapia culture effluent. The average recovery rate of the membrane used in this system is about 39.2–57.5% (Qin et al. 2005).

The estimated cost of producing 1 m^3 of freshwater from the aquaculture effluent is US$ 4.00. This cost will decrease substantially, and the technology will become economically feasible if the system capacity is increased (Qin et al. 2005).

The cost of electricity for the system driven by wind energy would only be about 0.3% of the total unit water production cost. Although producing water using the wind-driven RO system seems too expensive in the current situation, this technology has great promise once the system scale can be upgraded (Qin et al. 2005).

One alternative solution that could be applied is an integrated system with both hydropower and wind energy. A system (Kose and Kaya 2013) consisting of a hydroelectric power plant (HEPP) and/or wind power plant(s) (WPP) was designed to resolve the energy demand for the Konya water treatment plant. The wind power calculator program and local wind energy measurement data were used to design the WPP. There are plans to install a 250-kW HEPP and a 250-kW WPP to meet the electrical energy demand of the Konya water treatment plant. Despite the capacity increasing by 20%, the electricity demand can still be met. If two 250-kW or one 500-kW wind turbine were additionally installed with the HEPP, the energy demand would still be met, even if there was up to a 50% capacity increase. It is remarkable that almost all the energy demanded of the water treatment plant can be provided from the HEPP. The payback period of investment cost will be 4.54 years if a 250-kW HEPP and a WPP are installed, 5.1 years if a 250-kW HEPP and a 500-kW WPP are installed, which are quite reasonable payback periods. The HEPP and WPP have short payback periods because the HEPP is constructed on an existing pipeline, the HEPP is domestic production, and the distance between the WPP and the WTP is short (Kose and Kaya 2013).

Most rural WWT facilities require aeration equipment to ensure sufficient dissolved oxygen (DO) during processing. Operation and maintenance are costly and cannot be met in many areas with poor economic levels. This has led to further deterioration of the rural water environment and aroused much attention. A work (Li et al. 2020) reported a plug-flow step feed system utilizing wind and solar hybrid energy for rural WWT. Under certain climatic conditions, the wind energy and solar energy provided complimentary power generation, and an automatic control system (without batteries) was constructed. The corresponding control logic for multi-energy level operation was developed. Furthermore, the power generation efficiency of the system, the pollutant removal, and its mechanism on the bioreactor were also analyzed. According to the monitoring of meteorological conditions, wind and solar resources at the test site were abundant, and the electricity generated by the power generation was sufficient to meet the operational needs of the equipment. Energy efficiency can reach 80.0%. The characteristics of pollutant removal in each process section were studied on spatial and temporal dimensions. Results showed that the WWT process reached mean removal efficiencies of chemical oxygen demand (COD_{cr}), NH_4^+-N, total nitrogen (TN), and total phosphorus (TP) were 90.2, 94.3, 61.4, and 63.1%, respectively. Analyses of microbial community richness and group changes in each anoxic/aerobic reaction chamber in the biofilm reactor showed that the population structure was relatively stable and that there were abundant functional bacteria capable of degrading pollutants in each aerobic and anoxic unit. This system can thus be a more sustainable treatment process than traditional techniques used for rural WWT, providing a new design approach for low-energy consumption and unattended rural WWT (Li et al. 2020).

A decentralized WWT device was designed (Li et al. 2020) that was powered by wind and solar energy (without lead-acid batteries) and was capable of automatic stable operation according to the power supply strength. The results indicated that the power generation of the equipment was stable with a peak value at 12:00, but it varied with the season and weather conditions. The removal efficiencies of COD_{cr}, NH_4^+-N, TN, and TP were 90.2, 94.3, 61.4, and 63.1%, respectively, during the steady operational periods. This new environmentally friendly WWT system was proved to be feasible (Li et al. 2020).

11.4 GEOTHERMAL ENERGY SOURCE

Geothermal energy can be directly used when exploited for heating, whereas it is indirectly employed in the case of electricity production. When geothermal energy is used for power production, conventional systems or binary cycles power plants can be used, depending on the source temperature.

11.4.1 Geothermal Energy for Water Treatment

The integration of renewable energy generation with water treatment techniques, based on geothermal water and energy resources, represents a new and interesting research direction. Kaczmarczyk, Tomaszewska and Bujakowski (2022) presented a model study aimed at evaluating the two-fold utilization of geothermal water potential: as an energy resource necessary for the desalination process and as a source of wastewater used in the desalination process. With respect to the feasibility of implementing the organic Rankine cycle or the Kalina cycle, an analysis was carried out that focused on using geothermal water extracted in the temperature range of 80–95°C. The mass flow rate of geothermal water directed to the power plant was assumed to be in the range 10–150 kg/s. The mineral content of the geothermal water used in the model was 3 g/L (Kaczmarczyk, Tomaszewska and Bujakowski 2022).

The thermal energy reserves in wastewater are dependent on the temperature, flow rate, heat transfer efficiency, and specific heat capacity of the water.

Electricity generation from sewage is a direction of regional integrated energy system development. The sewage plant uses sludge anaerobic digestion and biogas trigeneration, which can convert waste into clean renewable energy and reduce the operating cost of the wastewater plant. Japan's sewage plant biogas power generation accounts for 55–70% of its own electricity consumption; in Germany, the anaerobic treatment of sewage sludge is used for its own use, replacing about 80% of electricity. The Heboygan Wastewater Treatment Plant in the United States uses methane gas from waste and sewage to trigeneration.

11.4.2 Geothermal Energy for Wastewater Treatment

WWT is one of the most energy-consuming processes in the water sector, requiring both electric and thermal energy. In order to improve their sustainability, renewable energy sources can supply such energy demands in WWTPs, especially in decentralized small treatment systems.

Di Fraia et al. (2019) carried out a study for a district WWTP on the island of Ischia, in southern Italy, which presents diffused low-medium enthalpy geothermal sources, considered to power an organic Rankine cycle system for electric energy production and to heat the desiccant flow for sludge drying. The proposed system reduced sludge to be disposed by around 70%, and it covered 100% of the electrical energy demand of the WWTP, with a production of 1.68 kWh/kg of processed sludge. Despite the high investment, the revenues deriving from the avoided sludge disposal and the electricity production make the system convenient, with a simple payback of about 5 years and a saving of CO_2 equivalent emissions equal to 628 t/year. The results showed that the geothermal source is a viable and environmentally sustainable solution to produce electrical and thermal energy for wastewater and sludge treatment plants, especially in small islands or energy-isolated systems, where electric energy supply and sewage sludge treatment are commonly critical challenges (Di Fraia et al. 2019).

An electrochemical WWT system, powered through the conversion of temperature difference into energy, was designed (Oh et al. 2018). The WWT system was applied to drainage flowing from two different mines: one contaminated by arsenic (As) and the other ferrous iron (Fe^{2+}). Arsenic was adsorbed on, or co-precipitated with, iron hydroxide generated from iron electrodes. A pseudo second-order model well described the tendency for As removal. Ferrous iron oxidation occurred directly on graphite electrodes and followed a first-order reaction model. The efficiency of As and Fe^{2+} removal was up to 99.7 and 97.9%, respectively. The rate constants for each model were proportional to given temperature differences, as the relationship between current generation and temperature difference was linear. The applicability of the system to low-temperature wastewaters will expand as energy production capacity per unit cost of thermocouples increases, as occurred with the PV and heat pump systems currently in use (Oh et al. 2018).

A novel system design (Almtairi et al. 2021) of using geothermal well energy acting as a brine heater for desalination purposes has been numerically developed and investigated. The system contains a pumping unit, nanofiltration, flash evaporation unit, and end condenser for condensation and freshwater production. It is anticipated by the geothermal well to increase the brine temperature up to an optimized value (40–50°C). A flash evaporation tank has been used as a steam generator. The brine blowdown will be dumped into the sea. To prevent corrosion and tubes deterioration, the nanofiltration system has been used as a pre-stage before pumping the saline flow into the geothermal well. The salt-free steam will be directly flowing toward the condenser unit for condensation and freshwater production. It is expected to produce an amount of freshwater in the range of 500–1500 m^3/day. Results reveal that the total hourly costs are 1.185$/h, and total water price was in the range of 0.12–1.2$/$m^3$ depending on the performance and salinity concentration of the nano filtration system (Almtairi et al. 2021).

11.5 SOLAR ELECTRO-TECHNOLOGY FOR DECENTRALIZED WATER TREATMENT

11.5.1 Water

The feasibility of using electrocoagulation (EC) as a primary surface water treatment process that could be used with solar panels in decentralized locations was investigated (St-Onge et al. 2020). In this study, the removal of turbidity, UV light at 254 nm (UV254), and dissolved organic carbon (DOC) were studied. Operating parameters such as current density (0.91–10 mA/cm^2) and charge loading (0.083–1.250 A h/L) in bipolar and monopolar configuration were studied using a direct current (DC) power supply. Experimental results showed that a current density of 4.55 mA/cm^2 and a treatment time of 60 min in the bipolar configuration allowed the best removal of turbidity, UV254, and DOC (97.0, 93.0, and 95.2%, respectively). For the second part of the study, the DC power supply was replaced by two PV panels without the use of a set of batteries and a charge controller. According to the results, the turbidity removal is directly proportional to the solar radiation intensity. This suggests that it is possible to control the charge loading by adjusting the treatment time as a function of solar intensity. Such outcomes suggest that solar-powered EC would be a suitable technology for areas not connected to centralized water treatment networks (St-Onge et al. 2020).

Rural water treatment is generally more challenging than urban water treatment. Du et al. (2021) proposed a novel rural water treatment system effectively harnessing solar energy and gravitational hydropower. Influent was initially fed to a solar-driven EC unit, in which absorbance UV254 level was reduced (by almost 60%) and large flocs (averaging up to 66.2 μm in diameter) were formed. The effluent was then introduced to a gravity-driven ceramic membrane bioreactor (GDCMBR) at different water head differences ($\Delta H = 0.5$ or 1.0 m). Generally, applying the greater water head difference ($\Delta H = 1.0$ m) did not enhance removal of turbidity and organic matter, but it led to a reasonably high flux (25.9 $L/m^2/h$). Overall, the findings of this study suggest the possibility of sustainable operation during rural water treatment (Du et al. 2021).

Solar-assisted electro-dialysis: electro-dialysis (ED), driven by electricity only, is a type of technology that arranges cationic and anionic ion-exchange membranes alternately in a DC field, where the salt ions migrate from the dilute solution side to the concentrated solution side through ion-exchange membranes under the influence of an applied electric potential difference. So far, only a few studies have been carried out on solar-assisted ED seawater desalination (Li, Goswami and Stefanakos 2013).

PV system is a promising alternative method to provide electrical power in hydrogen production. Navarro-Solís and co-workers in an earlier study demonstrated H_2 production as a byproduct of synthetic textile effluent treatment. A hybrid PV electrochemical reactor was developed by Park and colleagues to generate molecular hydrogen at the stainless steel cathode and to oxidize organic substrate at the titanium anode coated with BiO_x-TiO_2. A commercial amorphous silicon PV panel with power rating of 6.4 W and active surface area of 1280 cm^2 was connected to the electrodes to supply the voltage and current to the system. Fikret Kargi (2011) also investigated the hydrogen gas generation from electro-hydrolysis of organic compounds present in industrial wastewater by using electrical power produced by PV system. Three types of electrodes (graphite, stainless steel, and aluminum) having diameter of 0.9 cm and length of 49.5 cm were used, and the total hydrogen production was compared. The system is fast and energy efficient to produce hydrogen gas and to remove organic compounds from the wastewater simultaneously.

Combination of electrochemical methods with PV solar cell is a novel technological approach to enhance the sustainability of WWT processes. Installation of PV technology is especially beneficial for tropical and subtropical countries. The significant operating variables are the light absorbance capacity of the material, PV array configuration, tilt angle, electrode materials, their shape, number, and arrangement (Mook, Aroua and Issabayeva 2014).

The results showed that annual energy production equal to 107 t of oil equivalent could be expected when the proposed green energy resources are implemented in the WWTP. The energy independence, which was defined as the percent ratio of green energy production to energy consumption, was estimated to be a maximum of 6.5% and to vary with on-site energy consumption in the WWTP (Chae and Kang 2013).

11.5.2 Saline Water

Fan et al. (2022) developed a unique water desalination process by integrating solar steam generation with electrochemical degradation to treat saline water-containing organics, and strong synergistic effects. The process used a dual-functional solar absorber that simultaneously served as a cathode of the electrochemical reactor, whose structural design was optimized by numerical simulation to balance heat transfer and mass transport. The concentration of VOCs in the condensate was reduced by 20 folds when electrochemical degradation of feed water was applied. In addition, the electrochemical degradation effectively mitigated humic acid fouling on the solar absorber, improving the steam generation rate by 20% after 12 h treatment, compared to the conventional solar evaporation process. Finally, the integrated solar desalination system achieved a thermal efficiency of 92.6% under real sunlight testing (Fan et al. 2022).

Karaca, Dincer and Nitefor (2022) created a new energy system that uses renewable sources to generate power and produce drinking water. The system was specifically designed for Antigua and Barbuda, a Caribbean Island nation, to reduce their reliance on polluting and unpredictable fossil fuels. The system uses solar and wind energy, which are abundant in the region, to create electricity and clean drinking water. The extra energy generated by the solar panels and wind turbines is stored in underwater balloons offshore by compressing air. This creates a reliable and sustainable energy source for Antigua and Barbuda. The proposed system was capable of operating with energy and exergy efficiencies of 45.4 and 37.6%, respectively. The system's annual electricity production was evaluated as 316 GWh corresponding to 102.6% of annual electricity requirement of Antigua and Barbuda. In addition, the system supplies 124 t of freshwater on a daily basis and compressed air to fuel 140 short-range (80 km) pneumatic vehicles. It is, furthermore, found that potential system implementation can reduce GHG emissions in the region by 88%. Wind and solar energy capacities are adequate for the development of effective and sufficient wind and solar PV-based power systems (Karaca, Dincer and Nitefor 2022).

11.5.3 Wastewater Post Treatment

Vivar et al. (2021) evaluated the SolWat hybrid system for solar water disinfection and PV energy generation, for its implementation in tertiary treatment plants, using real wastewater directly from the effluent after its secondary treatment. Results showed that *Escherichia coli*, *Enterococcus faecalis*, and *Clostridium perfringens* did not complete total bacterial inactivation after 4 h of treatment in the SolWat system, but that the inactivation levels achieved were sufficient as to allow for the reuse of water for various uses (urban, agricultural, industrial, etc.). *C. perfringens* continued to be the most resistant bacteria vs. *E. coli* and *E. faecalis*. The total PV energy production in the hybrid system compared to the reference system was the same, generating both the SolWat module and the reference module identical electrical power due to the compensating effect of module water cooling vs. radiation losses (Vivar et al. 2021).

11.6 CHALLENGES

Sustainable energy sources for water and WWT have been gaining popularity due to their numerous advantages over conventional energy sources, such as reduced energy costs, increased efficiency, and decreased carbon emissions. However, there are still significant challenges that must be addressed to ensure that these technologies can be implemented on a larger scale (Guven, Ersahin and Ozgun 2022; Hafeez et al. 2021; Pandey et al. 2021).

One of the main challenges associated with sustainable energy sources for water and WWT is their variability. Solar and wind energy, for example, are dependent on weather conditions and are not always available at the same time or in the same quantity (Karaca, Dincer and Nitefor 2022; Vivar et al. 2021). This variability can make it difficult to maintain a consistent level of energy production, which can impact the overall efficiency of the system. Therefore, it is necessary to develop energy storage systems to ensure that energy can be stored when it is available and used when it is needed.

Another challenge associated with sustainable energy sources for water and WWT is their scalability. Many of these technologies are still in the research and development phase and have not yet been scaled up to the level required for large-scale implementation. The development of appropriate infrastructure and the establishment of regulatory frameworks is also necessary for successful deployment. This requires significant investment, which may be a barrier for many regions with limited resources (Karaca, Dincer and Nitefor 2022; Vivar et al. 2021).

In addition, there is a lack of standardized evaluation methods for sustainable energy sources for water and WWT. This makes it difficult to compare the effectiveness and efficiency of different systems and technologies and can make it difficult to assess their overall impact. The development of standardized evaluation methods is therefore necessary to ensure that these technologies are properly evaluated and compared.

There are also challenges associated with the maintenance and operation of sustainable energy sources for water and WWT. Many of these technologies require specialized knowledge and expertise to maintain and operate, which may not be available in all regions. Therefore, training and capacity-building programs are necessary to ensure that local communities can maintain and operate these systems effectively (Karaca, Dincer and Nitefor 2022; Vivar et al. 2021).

Moreover, the cost of sustainable energy sources for water and WWT can be a barrier to their implementation. While these technologies can be cost-effective in the long term, the upfront costs can be significant. Therefore, it is necessary to develop financing mechanisms to support the implementation of these technologies, especially in low-income and developing countries where the need for sustainable water and WWT is greatest.

The regulatory framework also presents a challenge to the deployment of sustainable energy sources for water and WWT. The regulatory framework needs to be flexible and adaptable to accommodate the changing landscape of energy production and consumption. Additionally, the regulatory framework needs to be tailored to the unique needs of each region, taking into account factors such as weather patterns, energy consumption patterns, and available resources (Karaca, Dincer and Nitefor 2022; Vivar et al. 2021).

Furthermore, the integration of sustainable energy sources into existing water and WWT infrastructure can present challenges. Many existing infrastructure systems were not designed to accommodate sustainable energy sources and may require significant modifications. This can be costly and time-consuming and may require the development of new technologies and solutions.

Finally, the social and cultural acceptance of sustainable energy sources for water and WWT is a significant challenge. Many communities may be resistant to the implementation of new technologies and may require education and awareness programs to build acceptance and understanding of the benefits of these technologies (Karaca, Dincer and Nitefor 2022; Vivar et al. 2021).

Therefore, while sustainable energy sources hold great promise for water and WWT, there are significant challenges that must be addressed to ensure that these technologies can be implemented effectively on a larger scale. The variability of these energy sources, their scalability, the lack of standardized evaluation methods, the need for specialized knowledge and expertise, the upfront costs, the regulatory framework, the integration with existing infrastructure, and the social and cultural acceptance are all significant challenges that need to be addressed. Overcoming these challenges will require a collaborative effort from governments, academia, industry, and civil society to ensure that sustainable energy sources for water and WWT can be effectively deployed to meet the growing demand for clean water and sanitation.

11.7 CONCLUSIONS AND PERSPECTIVES

The integration of sustainable energy sources, such as solar, wind, and thermal energy, holds tremendous promise for revolutionizing water and WWT in decentralized sanitation systems. While these technologies offer significant advantages, transitioning from laboratory research to full-scale industrial implementation requires further development and overcoming certain challenges. However, with dedication and innovation, these obstacles can be addressed, paving the way for a greener and more efficient future in water and WWT.

Solar energy, in particular, has proven to be a reliable and renewable energy source for powering water and WWT processes. Its widespread availability and low environmental impact make it an attractive option for sustainable sanitation systems. Nonetheless, the intermittency of solar power, especially during night-time, poses a challenge for continuous operation. To overcome this

TABLE 11.1
Examples of Various Water and Wastewater Treatment Using Renewable Energy

SN	Water/ Wastewater Treatment	Energy Type (Solar Thermal Energy, Solar Photovoltaic (PV) Cells, Wind Power, Small Hydropower (SHP), or Combinations)	Condition	Results and/or Efficiency	Reference
1	Drinking water	Solar energy A new system using peroxydisulfate (PDS) as an oxidant and a xenon lamp as a solar irradiation light	Solar light was present, UV254	Solar light was more effective than heating for activating PDS to break down natural organic matter (NOM) at temperatures below 50°C Total fouling index (TFI) and hydraulic irreversible fouling index (HIFI) were moderately correlated with DOC and UV254 but highly correlated with the maximum fluorescent intensity (Fmax) of component 2 (C2) and component 3 (C3)	Guo et al. (2021)
2	Potable water	Solar energy	Compared ultraviolet and solar-combined advanced oxidation technologies to treat five kinds of micro-contaminants	The final total costs of solar-combined advanced oxidation technology were the lowest when the target pollutant degradation rate reached 90%	Miralles and colleagues
3	Decentralized water treatment	Wind energy Fully self-powered electro-coagulation (SPEC) system	A simple electro-coagulation unit, composed of a simple pair of electrodes, successfully removes dissolved colloidal pollutants from water, while coagulating and precipitating pollutants, without any use of chemicals or an external power supply	The SPEC system removes 90% of algae and 97% of organic dye with self-powered treatment for 72 h	Jeon et al. (2016)
4	Wastewater	Wind energy A conventional multi-blade windmill is used to convert wind energy directly to hydraulic pressure for reverse osmosis (RO) membrane operation	The system is able to work at an average wind speed as low as 3.0 m/s It can generate and recycle freshwater at a flow rate of 228–366 L/h The permeate production increases linearly with the wind speed	The estimated cost of producing 1 m^3 of freshwater from the aquaculture effluent is US$ 4.00 About 70–84% of aquaculture wastewater can be recycled using this system, which is capable of removing 90–97% of nitrogenous waste present in tilapia culture effluent. The average recovery rate of the membrane used in this system is about 39.2–57.5%	Qin et al. (2005)

(Continued)

TABLE 11.1 (*Continued*)
Examples of Various Water and Wastewater Treatment Using Renewable Energy

SN	Water/Wastewater Treatment	Energy Type (Solar Thermal Energy, Solar Photovoltaic (PV) Cells, Wind Power, Small Hydropower (SHP), or Combinations)	Condition	Results and/or Efficiency	Reference
5	Rural wastewater	Both hydropower and wind energy	A plug-flow step feed system utilizing wind and solar hybrid energy for rural wastewater treatment	Energy efficiency can reach 80.0% Reached mean removal efficiencies of chemical oxygen demand (COD_{cr}), NH_4^+-N, total nitrogen (TN), and total phosphorus (TP) were 90.2, 94.3, 61.4, and 63.1%, respectively The population structure was relatively stable and that there were abundant functional bacteria capable of degrading pollutants in each aerobic and anoxic unit	Li et al. (2020)
6	Water treatment	Geothermal energy	The integration of renewable energy generation with water treatment techniques, based on geothermal water and energy resources, represents a new and interesting research direction	Using geothermal water extracted in the temperature range of 80–95°C. The mass flow rate of geothermal water directed to the power plant was assumed to be in the range 10–150 kg/s. The mineral content of the geothermal water used in the model was 3 g/L	Kaczmarczyk et al. (2022)
7	Wastewater treatment	Geothermal energy for Diffused low-medium enthalpy geothermal sources	Despite the high investment, the revenues deriving from the avoided sludge disposal and the electricity production make the system convenient, with a simple payback of about five years and a saving of CO_2 equivalent emissions equal to 628 t/year	Reduced sludge to be disposed by around 70% and it covered 100% of the electrical energy demand of the wastewater treatment plant, with a production of 1.68 kWh/kg of processed sludge The geothermal source is a viable and environmentally sustainable solution	Di Fraia et al. (2019)

(*Continued*)

TABLE 11.1 *(Continued)*
Examples of Various Water and Wastewater Treatment Using Renewable Energy

SN	Water/ Wastewater Treatment	Energy Type (Solar Thermal Energy, Solar Photovoltaic (PV) Cells, Wind Power, Small Hydropower (SHP), or Combinations)	Condition	Results and/or Efficiency	Reference
8	Electrochemical wastewater treatment system	Power through the conversion of temperature difference into energy was designed	Drainage flowing from two different mines: one contaminated by arsenic (As) and the other ferrous iron (Fe^{2+}). Arsenic was adsorbed on, or co-precipitated with, iron hydroxide generated from iron electrodes. A pseudo second-order model well described the tendency for As removal. Ferrous iron oxidation occurred directly on graphite electrodes and followed a first-order reaction model.	The efficiency of As and Fe^{2+} removal was up to 99.7 and 97.9%, respectively. The rate constants for each model were proportional to given temperature differences, as the relationship between current generation and temperature difference was linear. The applicability of the system to low-temperature wastewaters will expand as energy production capacity per unit cost of thermocouples increases, as occurred with the photovoltaic and heat pump systems currently in use.	Oh et al. (2018)
9	Desalination	Geothermal well energy acting as a brine heater The system contains a pumping unit, nanofiltration, flash evaporation unit, and end condenser for condensation and freshwater production	It is anticipated by the geothermal well to increase the brine temperature up to an optimized value (40–50°C) A flash evaporation tank has been used as a steam generator. The brine blowdown will be dumped into the sea.	Results reveal that the total hourly costs are 1.185$/h and total water price was in the range of 0.12–1.2$/m^3 depending on the performance and salinity concentration of the nanofiltration system	Almtairi et al. (2021)
10	Solar electro-technology for decentralized water treatment	Using electrocoagulation (EC) as a primary surface water treatment process that could be used with solar panels in decentralized locations was investigated The removal of turbidity, ultraviolet light at 254 nm (UV254), and dissolved organic carbon (DOC) was studied	Operating parameters such as current density (0.91–10 mA/cm^2) and charge loading (0.083–1.250 A h/L) in bipolar and monopolar configuration	A current density of 4.55 mA/cm^2 and a treatment time of 60 min in the bipolar configuration allowed the best removal of turbidity, UV254, and DOC (97.0, 93.0, and 95.2%, respectively) Solar-powered EC would be a suitable technology for areas not connected to centralized water treatment networks	St-Onge et al. (2020)

(Continued)

TABLE 11.1 (*Continued*)
Examples of Various Water and Wastewater Treatment Using Renewable Energy

SN	Water/ Wastewater Treatment	Energy Type (Solar Thermal Energy, Solar Photovoltaic (PV) Cells, Wind Power, Small Hydropower (SHP), or Combinations)	Condition	Results and/or Efficiency	Reference
11	Rural water	Solar energy and gravitational hydropower	Absorbance UV254 level was reduced (by almost 60%) and large flocs (averaging up to 66.2 μm in diameter) were formed The effluent was then introduced to a gravity-driven ceramic membrane bioreactor (GDCMBR) at different water head differences (ΔH = 0.5 or 1.0 m)	Generally, applying the greater water head difference (ΔH = 1.0 m) did not enhance removal of turbidity and organic matter, but it led to a reasonably high flux (25.9 $L/m^2/h$). Overall, the findings of this study suggest the possibility of sustainable operation during rural water treatment	Du et al. (2021)
12	Industrial wastewater	Solar photovoltaic (PV) system	Using electrical power produced by three types of electrodes (graphite, stainless steel, and aluminum) having diameter of 0.9 cm and length of 49.5 cm were used	The total hydrogen production was compared. The system is fast and energy efficient to produce hydrogen gas and to remove organic compounds from the wastewater simultaneously.	Fikret Kargi
13	Wastewater post-treatment	Solar water disinfection and photovoltaic energy generation	The total photovoltaic energy production in the hybrid system compared to the reference system was the same	*Escherichia coli, Enterococcus faecalis,* and *Clostridium perfringens* did not complete total bacterial inactivation after 4 h of treatment in the SolWat system, but that the inactivation levels achieved were sufficient as to allow for the reuse of water for various uses (urban, agricultural, industrial, etc.)	Vivar et al. (2021)

limitation, the development of energy storage systems becomes imperative. Implementing efficient and cost-effective energy storage solutions will ensure a stable power supply, enabling the seamless functioning of treatment facilities round-the-clock.

Furthermore, standardizing evaluation methods for sustainable energy-driven water and WWT systems is essential. The lack of established evaluation criteria hinders a comprehensive assessment of these technologies' performance and efficiency. By establishing industry-wide standards, researchers and practitioners can reliably compare different solutions and identify areas for improvement. Standardization will instill confidence among stakeholders and encourage broader adoption of sustainable energy sources in the water and WWT sector.

Additionally, enhancing the stability and resilience of these systems is crucial for their successful application. Rigorous testing and ongoing research are necessary to identify and address vulnerabilities, ensuring that these technologies can withstand varying environmental conditions and maintain consistent performance over extended periods. Strengthening the reliability of sustainable energy-driven systems will foster trust among users, investors, and policymakers, driving further investment and expansion of these technologies.

Despite these challenges, the opportunities for advancing innovative and eco-friendly technologies in decentralized water and WWT are significant. Sustainable energy sources present a transformative path toward achieving efficient, cost-effective, and environmentally responsible sanitation systems. Their integration not only helps in reducing carbon emissions and minimizing environmental impact but also promotes energy self-sufficiency and resilience in water treatment facilities.

As move forward, continued research and development in this field will be vital in fully realizing the potential of sustainable energy sources for water and WWT. Collaborations between academic institutions, technology developers, and industry stakeholders will be key to driving progress and creating cutting-edge solutions. Government support and investment in research initiatives can accelerate the transformation from concept to practical application, expediting the adoption of sustainable energy-driven water and WWT technologies on a larger scale.

By capitalizing on the benefits of solar, wind, and thermal energy, it can revolutionize the way to treat water and wastewater, promoting environmental conservation and meeting the global challenge of sustainable water management. Embracing these innovative solutions will not only safeguard our natural resources but also foster a greener, healthier, and more resilient world for generations to come.

REFERENCES

Almtairi, Abdullah, Mohamed A. Sharaf Eldean, A. M. Soliman, Abdelnasser Mabrouk, and Hassan E. S. Fath. 2021. "A new preliminary system design of using geothermal well brine heater for desalination/nanofiltration process." *Cleaner Engineering and Technology* 4:100213. doi: 10.1016/j.clet.2021.100213

Argaw, Neway. 2003. *Renewable energy in water and wastewater treatment applications; period of performance: April 1, 2001–September 1, 2001.* National Renewable Energy Lab. (NREL), Golden, CO (United States).

Chae, Kyu-Jung, and Jihoon Kang. 2013. "Estimating the energy independence of a municipal wastewater treatment plant incorporating green energy resources." *Energy Conversion and Management* 75:664–672. doi: 10.1016/j.enconman.2013.08.028

Climate Council. 2019. "11 Countries Leading the Charge on Renewable Energy." https://www.climatecouncil.org.au/11-countries-leading-the-charge-on-renewable-energy/

Daghrir, Rimeh, Patrick Drogui, and Didier Robert. 2012. "Photoelectrocatalytic technologies for environmental applications." *Journal of Photochemistry and Photobiology A: Chemistry* 238:41–52.

Di Fraia, Simona, Adriano Macaluso, Nicola Massarotti, and Laura Vanoli. 2019. "Energy, exergy and economic analysis of a novel geothermal energy system for wastewater and sludge treatment." *Energy Conversion and Management* 195:533–547. doi: 10.1016/j.enconman.2019.05.035

Du, Xing, Wentao Zhao, Zhen Wang, Rong Ma, Yunlong Luo, Zhihong Wang, Qiong Sun, and Heng Liang. 2021. "Rural drinking water treatment system combining solar-powered electrocoagulation and a gravity-driven ceramic membrane bioreactor." *Separation and Purification Technology* 276:119383. doi: 10.1016/j.seppur.2021.119383

Fan, Tianzhu, Wei Deng, Xuhui Feng, Shoufeng Lan, John Pellessier, and Ying Li. 2022. "Integrating solar steam generation with electrocatalysis to achieve simultaneous fouling–resistant desalination and accelerated organics degradation." *Desalination* 532:115763. doi: 10.1016/j.desal.2022.115763

Fernández-Marchante, C. M., F. L. Souza, M. Millán, J. Lobato, and M. A. Rodrigo. 2021. "Improving sustainability of electrolytic wastewater treatment processes by green powering." *Science of The Total Environment* 754:142230. doi: 10.1016/j.scitotenv.2020.142230

Frijns, Jos, Jan Hofman, and Maarten Nederlof. 2013. "The potential of (waste) water as energy carrier." *Energy Conversion and Management* 65:357–363.

Ganiyu, Soliu O., Carlos A. Martínez-Huitle, and Manuel A. Rodrigo. 2020. "Renewable energies driven electrochemical wastewater/soil decontamination technologies: A critical review of fundamental concepts and applications." *Applied Catalysis B: Environmental* 270:118857. doi: 10.1016/j.apcatb.2020.118857

Guo, Yuanqing, Heng Liang, Guibai Li, Daliang Xu, Zhongsen Yan, Rui Chen, Jing Zhao, and Xiaobin Tang. 2021. "A solar photo-thermochemical hybrid system using peroxydisulfate for organic matters removal and improving ultrafiltration membrane performance in surface water treatment." *Water Research* 188:116482. doi: 10.1016/j.watres.2020.116482

Guven, Huseyin, Mustafa Evren Ersahin, and Hale Ozgun. 2022. "Chapter 7 – Energy self-sufficiency in wastewater treatment plants: perspectives, challenges, and opportunities." In *Clean energy and resource recovery*, edited by Alicia An, Vinay Tyagi, Manish Kumar and Zeynep Cetecioglu, 105–122. Elsevier.

Hafeez, Ainy, Zufishan Shamair, Nasir Shezad, Fahed Javed, Tahir Fazal, Saif ur Rehman, Aqeel Ahmed Bazmi, and Fahad Rehman. 2021. "Solar powered decentralized water systems: A cleaner solution of the industrial wastewater treatment and clean drinking water supply challenges." *Journal of Cleaner Production* 289:125717. doi: 10.1016/j.jclepro.2020.125717

Jeon, Seung-Bae, Sunmin Kim, Sang-Jae Park, Myeong-Lok Seol, Daewon Kim, Yong Keun Chang, and Yang-Kyu Choi. 2016. "Self-powered electro-coagulation system driven by a wind energy harvesting triboelectric nanogenerator for decentralized water treatment." *Nano Energy* 28:288–295. doi: 10.1016/j.nanoen.2016.08.051

Kaczmarczyk, Michał, Barbara Tomaszewska, and Wiesław Bujakowski. 2022. "Innovative desalination of geothermal wastewater supported by electricity generated from low-enthalpy geothermal resources." *Desalination* 524:115450. doi: 10.1016/j.desal.2021.115450

Karaca, Ali Erdogan, Ibrahim Dincer, and Michael Nitefor. 2022. "Development of an integrated solar and wind driven energy system for desalination and power generation." *Sustainable Energy Technologies and Assessments* 52:102249. doi: 10.1016/j.seta.2022.102249

Kose, Faruk, and Mehmet Numan Kaya. 2013. "Analysis on meeting the electric energy demand of an active plant with a wind-hydro hybrid power station in Konya, Turkey: Konya water treatment plant." *Renewable Energy* 55:196–201.

Li, Chennan, Yogi Goswami, and Elias Stefanakos. 2013. "Solar assisted sea water desalination: A review." *Renewable and Sustainable Energy Reviews* 19:136–163. doi: 10.1016/j.rser.2012.04.059

Li, Pengyu, Tianlong Zheng, Lin Li, Yingqun Ma, Xu Sun, and Junxin Liu. 2020. "An appropriate technique for treating rural wastewater by a flow step feed system driven by wind-solar hybrid power." *Environmental Research* 187:109651. doi: 10.1016/j.envres.2020.109651

Liu, Feng, Alain Ouedraogo, Seema Manghee, and Alexander Danilenko. 2012. "A Primer on Energy Efficiency for Municipal Water and Wastewater Utilities." Energy Sector Management Assistance Program; Technical Report 001/12. ©World Bank, Washington, DC. License: CC BY 3.0 IGO. http://hdl.handle.net/10986/18060

Mook, W. T., M. K. Aroua, and G. Issabayeva. 2014. "Prospective applications of renewable energy based electrochemical systems in wastewater treatment: A review." *Renewable and Sustainable Energy Reviews* 38:36–46. doi: 10.1016/j.rser.2014.05.042

Musabandesu, Erin, and Frank Loge. 2021. "Load shifting at wastewater treatment plants: A case study for participating as an energy demand resource." *Journal of Cleaner Production* 282:124454. doi: 10.1016/j.jclepro.2020.124454

Oh, Chamteut, Sangwoo Ji, Youngwook Cheong, and Giljae Yim. 2018. "Applicability of electrochemical wastewater treatment system powered by temperature difference energy." *Journal of Hazardous Materials* 351:108–116. doi: 10.1016/j.jhazmat.2018.02.026

Pandey, A. K., R. Reji Kumar, B. Kalidasan, Imtiaz Ali Laghari, M. Samykano, Richa Kothari, Abdullah M. Abusorrah, Kamal Sharma, and V. V. Tyagi. 2021. "Utilization of solar energy for wastewater treatment: Challenges and progressive research trends." *Journal of Environmental Management* 297:113300. doi: 10.1016/j.jenvman.2021.113300

Qin, Gang, Clark C. K. Liu, N. Harold Richman, and James E. T. Moncur. 2005. "Aquaculture wastewater treatment and reuse by wind-driven reverse osmosis membrane technology: A pilot study on Coconut Island, Hawaii." *Aquacultural Engineering* 32 (3):365–378. doi: 10.1016/j.aquaeng.2004.09.002

Sharon, H., and K. S. Reddy. 2015. "A review of solar energy driven desalination technologies." *Renewable and Sustainable Energy Reviews* 41:1080–1118. doi: 10.1016/j.rser.2014.09.002

St-Onge, Joanie, Anne Carabin, Oumar Dia, Patrick Drogui, and Kamal El Haji. 2020. "Development of a solar electrocoagulation technology for decentralised water treatment." In *Proceedings of the Institution of Civil Engineers-Water Management.*

Vivar, M., M. Fuentes, J. Torres, and M. J. Rodrigo. 2021. "Solar disinfection as a direct tertiary treatment of a wastewater plant using a photochemical-photovoltaic hybrid system." *Journal of Water Process Engineering* 42:102196. doi: 10.1016/j.jwpe.2021.102196

12 Decentralized Sanitation Implementation in Different Countries

Mohamed S. Zaghloul[1], Abir Hamze[1], Farokh Laqa Kakar[2], Rania A. Hamza[1], and Elsayed Elbeshbishy[1]
[1] Toronto Metropolitan University (formerly Ryerson University), Toronto, Canada
[2] Brown and Caldwell, Burnaby, British Colombia, Canada

12.1 INTRODUCTION

One of the main challenges that developing communities across the globe are facing is sanitation. According to the United Nations World Water Development Report (UN Water 2017), 2.4 billion people lack access to basic sanitation infrastructure. The lack of sanitation infrastructure leads to the spread of waterborne diseases as well as limits the possibility of water reuse in water-scarce communities. Approximately 80% of domestic wastewater is released into the environment without any treatment leading to the degradation of open water, soil, and groundwater quality. In most developing countries, the crisis is further complicated by the rapid population growth and unplanned expansion of residential communities without proper sanitation infrastructure. An example of this crisis is in the African sub-Saharan urban communities where 62% of the population does not have sanitation infrastructure and dispose of their wastewater into open water bodies.

The developed countries aim to meet the United Nations 6th Sustainable Development Goal ensuring proper sanitation for all their population; however, the developing countries are still far behind in achieving this goal. Poor infrastructure regulation coupled with the struggling economic situation in most developing countries leads to improper wastewater collection and treatment. The conventional centralized wastewater treatment systems have proved unfeasible in many developing communities where most of the population lives in dispersed rural communities. Centralized treatment is quite expensive and requires continuous maintenance, which can be impossible to sustainably maintain in developing countries. Centralized systems also require robust process control and close monitoring by highly trained personnel which can be difficult to have in some developing countries. That is why decentralized treatment systems for separate households or household clusters present an attractive, sustainable solution due to their lower cost, operation, and maintenance requirements. While decentralized treatment is relatively simple to establish, it is yet to have an impact on sanitation quality as it is only beginning to be used in several developing countries. The lack of institutional regulation on their use and maintenance, as well as effluent quality standards, makes it difficult to effectively manage the treatment levels. Additionally, introducing the decentralized technologies to existing communities where no proper urban planning was done limits the versatility and effectiveness of the treatment technologies.

In this chapter, the decentralized wastewater treatment approaches in different developing countries are discussed. The commonly used treatment technologies are presented. Followed by the challenges in the implementation of decentralized treatment in developing countries, a few specific case studies of implementing decentralized treatment in residential communities, as well as on-site industrial wastewater treatment, are reviewed.

 DOI: 10.1201/9781003381419-14

12.2 DECENTRALIZED SANITATION TECHNOLOGIES IN DEVELOPING COUNTRIES

12.2.1 Septic Tanks

The septic tank is an anaerobic domestic wastewater treatment technology. It is one of the most common decentralized wastewater treatment technologies due to its simple design and construction, minimal maintenance and operation requirements, and relatively lower cost. Septic tanks are typically used for single households but are often also used for small clusters of households. They can be made using different materials, but the most common is on-site construction using concrete or pre-made plastic units for smaller units. Septic tanks are constructed underground and consist of two compartments that are separated by a vertical baffle. The baffle opening is located at the center or bottom of the wall to reduce wastewater shortcutting. Each compartment has a covered access port at the top of the septic tank. Wastewater enters and exits the tank through a T-shaped pipe where the inlet and outlet ports are facing downwards. The upward-facing ports are used for maintenance. Solids settle at the bottom of the first compartment, while the oils, fats, and grease float to the top to form an airtight scum layer. Anaerobic conditions are maintained as long as the scum layer is not disturbed, which is why the tank is optimally operated with the water level above the bottom port of the inlet T connection [1]. The tank sizing is designed so that the liquid wastewater has a retention time that is enough for the anaerobic decomposition of the soluble organics. The settled solids are also anaerobically digested for a period of up to three years. Non-biodegradable solids accumulate at the bottom of the tank and lead to the reduction of the space available for liquid wastewater, leading to the gradual shortening of the retention time. That is why regular desludging every one to three years is crucial for the successful operation of septic tanks. Desludging is done by removing the settled solids using sewage trucks and inspecting the tanks while it is empty for any needed structural repairs. Effluent wastewater from a successful septic tank contains minimal organic content and suspended solids. It flows into drainage fields that can be open channels lined with porous material or underground perforated pipes surrounded by coarse gravel. The drainage field disperses the effluent wastewater into the soil and allows it to percolate slowly into underground water or the nearest water body. Poorly managed septic tanks lead to effluent with high organic loading and suspended solids concentration which clog the drainage fields and require further expensive maintenance [2].

12.2.2 Pit Latrines

Pit latrines are an extremely low-cost but very effective solution to the open defecation problem. It is so effective that more than 1.7 billion people in developing countries use them. They protect the environment, resulting in clean communities, and reduce diseases spread by exposure to human waste. India was one of the most successful countries in using pit latrines to increase sanitation coverage. In 2014, the Clean India Mission took the initiative for social change to spread awareness and convince communities that practiced open defecation to use pit latrines instead. As a result, sanitation coverage rose from 39% in 2014 to 98% in 2019, with the installation of over 85 million pit latrines [3].

They are very simple in design, where a 3-m deep and 1-m diameter hole is dug into the ground with a floor at the top. The floor has a 25 cm hole, and the whole set-up is covered with an outhouse. When the hole is filled to about 0.5 m from the surface, it is dislodged and reused or covered, and a new hole is dug. The main consideration for this method is to ensure that it is at least 10 m away from wells or freshwater sources to prevent contamination due to soil percolation. Pit latrines that are used by large numbers of people and regularly emptied and reused are usually lined with brick or wood [4].

The main problem with pit latrines is that they can become a major health hazard if they leak wastewater into the groundwater. It is essential to know the water table level at locations where pit latrines will be used to ensure that the bottom of the pits is at least 2 m above the groundwater table. Otherwise, fully lined pits must be used where the lining must cover the sides and bottom of the pits. The lining material must be impervious such as concrete or plastic. Fully lined latrines have the disadvantage of not releasing urine and other liquids out of the pit leading to faster filling, more odors, and wet contents instead of solid sludge when being emptied. Ventilated pit latrines are modified by adding a ventilation pipe that leads from the pit to above the outhouse to drastically reduce the odors. The pipe also acts as a fly trap to attract flies using the spotlight from the pipe to the dark pit [5].

12.3 DECENTRALIZED TREATMENT IN VIETNAM

Vietnam, located in Southeastern Asia, is one of the most populous countries in the Asian Pacific Region. With an estimated population of 97 million, Vietnam has a low urbanization rate, and most Vietnamese live in rural areas and work in agriculture [6].

Over the past years, Vietnam has undergone a fast socio-economic development, increasing pressure on the urban infrastructure. Although national policies and strategies aimed at limiting urban environmental pollution, the infrastructure for waste treatment remains either absent or of poor quality. Consequently, most Vietnamese cities have rivers, lakes, and canals polluted with nutrients, mainly phosphorous and nitrogen. Also, population growth and rapid industrialization resulted in heavy water pollution; some places have high concentrations of chemical oxygen demand (COD) and biochemical oxygen demand (BOD) from 10 to 20 times higher than permissible values in Vietnamese standard limits for receiving water bodies of class A and from three to four times in comparison with water sources of type B [7].

Surface and natural water bodies are also contaminated, mainly because of the discharge of domestic and industrial wastewater into combined sewerage systems or open channels with no or minor pre-treatment [8].

The Sewerage and Drainage Company (SADCO) is the agency responsible for wastewater management in most urban areas; however, it does not have the resources or capacity to cover all parts. Accordingly, to overcome the limited resources in money, skills, time, and scientific knowledge, the Government adopted a decentralized approach to addressing wastewater management needs. It assigned key budget responsibilities for wastewater management to local authorities at the provincial level, leading to the adoption of a decentralized strategic plan to achieve sustainable wastewater management [9]. In Vietnam, decentralized wastewater facilities comprise some or all of the following stages:

a. wastewater pre-treatment
b. wastewater collection
c. wastewater treatment
d. effluent reuse or disposal
e. biosolids and septage management

Conventional septic tanks are the most common facilities in urban and peri-urban areas [10].

The quality of the obtained effluent depends on the composition and type of the treated wastewater, the temperature, the hydraulic retention time (HRT), the tank design (capacity and shape), and the state of maintenance of the tank itself. However, conventional septic tanks are mostly outdated and/or damaged, with no regular maintenance checkups. In addition to the septic tanks on-site installation, many Vietnamese towns rely on family and public pit or vault toilets and bucket latrines that are not connected to a sewer network [11].

The technology adopted in treating wastewater varies depending on the source itself. Accordingly, hospitals, hotels, and apartments in urban centers adopt activated sludge processes, trickling filter technologies, submerged aerated filters, and A_2O (anaerobic-anoxic-oxic treatment process). Several villages and administrative buildings rely on combined sewerage and drainage with combined sewerage with overflow chambers (CSO), baffled septic tanks with anaerobic filters (BASTAF), and horizontal flow-constructed wetlands (HF CW). The introduction of baffles and anaerobic filter material significantly improved the treatment performance of conventional septic tanks in terms of COD (58.0–76.1%) and total suspended solids (TSS) (61.0–78.0%) [12].

Pig farms adopt biogas digesters to manage their waste. Public amusement parks and a few hospitals use anaerobic baffled reactors and horizontal flow-constructed wetlands as sanitary facilities. Since conventional septic tanks are mostly outdated and/or damaged, with no regular maintenance checkups, Vietnam has significantly increased the number of urban sewerage and drainage projects implementing decentralized wastewater systems with secondary treatments [13].

These projects with promising outcomes are budgeted either from the central Government, external assistance (the World Bank or the Asian Development Bank), or loans from other countries such as Japan, France, and Denmark, with a cost of billions of US dollars.

12.4 DECENTRALIZED TREATMENT IN INDIA

The highest number of people practicing open defecation is in India, counting for 90% of the people in South Asia and 59% of the 1.1 billion people who rely on open defecation. This equals about 600 million people [14]. Around 140 kg of excreta each year is defecated by each human, and when multiplied by the 600 million people, it becomes challenging to equate access to sanitation with access to the toilet [15]. Open ground or water bodies get contaminated in such areas since 80% of this excreta finds its way to these water resources. Therefore, complete sanitation and treatment of human excreta are something to acknowledge.

Furthermore, India has been ranked the 46th highest risk country in the Water Stress Index 2019 by London-based Verisk Maplecroft [16]. There is a shortage in the supply for agricultural and drinking purposes as the groundwater level is dropping rapidly and quality is deteriorating. The insufficient water supply has raised conflict between farmers. In 2018, the Government of India's think tank, Niti Aayog, estimated that by 2020, 20 major cities, including Bengaluru, New Delhi, and Hyderabad, with over 100 million residents, have reached Day Zero when groundwater resources have fully depleted (https://indianexpress.com/article/lifestyle/life-style/resolving-india-sanitation-issues-sustainable-living-solutions-toilet-7697047/). Critical actions required in India are the treatment of wastewater and fecal sludge to provide clean water and sanitation, improve health and urban environment, as well as increase water supply through the reuse of treated water.

Various decentralized wastewater treatment technologies have been used in India to tackle the defecate management and water supply crisis. A list of the available technologies and systems is reported in Table 12.1. Given only one-third of India has toilets connected to the sewage system, on-site sanitation systems (OSS) such as septic tanks have become the widely used alternative. Septic tanks in India are emptied periodically by a mechanized desludging truck that collects the waste from an underground tank, transporting the sludge and septic to the fecal sludge treatment plant (FSTP). The by-product of the treatment process (water and biosolids) is then used mainly for agricultural purposes [17].

12.5 DECENTRALIZED TREATMENT IN SUB-SAHARAN AFRICA

The levels of sanitation in sub-Saharan Africa are some of the least developed in the world. Rural communities are rarely covered with any sanitation, with the average coverage across Africa less than 50% in 2010. Urban communities have better coverage, but poor urban planning in residential communities, the spread of illegal or unplanned communities, as well as the lack of authoritative

TABLE 12.1
Decentralized Wastewater Systems and Technologies [17]

Technology	Collection System	Type of Waste
Ecosan systems – Nepal	At toilet site	Black water
Compost toilets	At toilet site	Black water
Terra preta toilets	At toilet site	Black water
Blue diversion toilets	At toilet site	Black water
Septic tank and absorption field/trenches	Sewage pipe network	Gray water
Moving bed bioreactors (MBBR)	Sewage pipe network	Gray water
Membrane bioreactors (MBR)	Sewage pipe network	Gray water
Anaerobic baffled reactors	Sewage pipe network	Gray water
Upflow anaerobic sludge blanket (UASB)	Sewage pipe network	Gray water
Integrated fixed film activated sludge	Sewage pipe network	Gray water
Rotating biological contactors (RBC)	Sewage pipe network	Gray water
MBBR + activated sludge	Sewage pipe network	Gray water
MBBR + dissolved air floatation	Sewage pipe network	Gray water
Dynamic aerobic/anaerobic systems	Sewage pipe network	Gray water
Waste stabilization ponds	Sewage pipe network	Gray water

management of the wastewater sector lead to illegal or unmanaged waste disposal and inefficient implementation of wastewater treatment solutions. More than 60% of the urban communities in Africa are unplanned slums without reliable urban infrastructure. Around 830 million people lived in slums in 2012 compared to 760 million in 2000 [18]. Solid waste, including municipal, industrial, medical, and commercial, is regularly disposed of directly into the environment by dumping in unmanaged landfills or even on the side of roads. This leads to a constant stream of leachate and polluted runoff into open water sources such as rivers or lakes and underground water reservoirs. This contributes to the rising stress on freshwater availability in all African countries. On-site treatment facilities, such as pit latrines and septic tanks, have been installed in a small portion of the rural and semi-urban communities. But they remain poorly managed, leading to wastewater short-cutting and overflow without treatment. The overflowing municipal wastewater in poor, crowded residential communities is directly tied to the frequent outbreaks of waterborne diseases [19].

Compared to the global development target in sanitation led by the WHO and UNICEF (77% of the world population had access to proper sanitation in 2015), Africa was the second lowest region in sanitation growth which increased from 24% in 1990 to 30% in 2015, a 6% increase in 25 years. Out of the 69 countries that were not on track to meet the target, 36 countries (52%) were African. The rapid unplanned urban expansion also led to the decline in sanitation progress in countries that previously had better sanitation coverage, such as Nigeria, where the coverage declined from 37% in 1990 to 28% in 2012. Further, due to the lack of proper sanitation, 23% of the population in poor African communities practiced open defecation in 2012. Out of the top ten countries with open defecation, five were in Africa: Nigeria, Ethiopia, Sudan, Niger, and Mozambique [19].

On-site sanitation is much more popular than centralized treatment plants in African cities, mainly due to the capital and running costs of wastewater treatment plants. It was estimated by the WHO that the on-site sanitation coverage in African cities in 2000 was between 60 and 100%. The responsibility for the construction of on-site treatment facilities is taken by community developers. However, the long-term operation and maintenance often are left to the residents. Due to their cheaper capital investment and lower process complication, the most popular decentralized treatment technologies used in African communities are pit latrines, ventilated improved pit (VIP) latrines, and septic tanks. Septic tanks and VIP are considered capable technologies when properly operated, which includes the regular emptying, cleaning, and retrofitting of the underground chambers. This procedure is rarely performed in most African communities, mainly due to the associated costs and absence of skilled labor. Emptying and cleaning the tanks requires hiring sewage trucks that collect the sludge from septic tanks and transport it to the nearest treatment plant to safely stabilize and dispose of the biological and pathogenic sludge. Since this is the responsibility of the residents of poor communities, many are not able to afford this service. Instead, residents wait for the rainy season to release the contents of the septic tanks to drainage channels and rely on the rainwater to wash it away. This is one of the main reasons for the spread of waterborne diseases every year in the rainy season, e.g., cholera, dysentery, and other diseases caused by *Escherichia coli* such as pneumonia and neonatal meningitis, and urinary tract infections. Additionally, even in communities where residents can afford to have their septic tanks emptied by sewage trucks, there are no sludge treatment facilities nearby to manage their waste. This leads to the trucks dumping the sludge into open water sources, also contributing to the degrading water quality and spread of diseases [20].

One approach that is intended to encourage the wider spread and better management of decentralized wastewater treatment is resource recovery. Currently, wastewater treatment in Africa is considered a health and environmental protection measure. However, it can also be considered a source of revenue or at least alleviate some of the associated costs of implementation. The first challenge to resource recovery is that the majority of the existing on-site treatment infrastructure, such as pit latrines and septic tanks, is not suitable for resource recovery. The idea is becoming more popular now in African cities, which have the capability to implement them. The process starts with modifying existing facilities such as urine diversion and ecological sanitation (EcoSan) technology. Additionally, sludge treatment and resource recovery plants are being proposed in a cost-effective and sustainable way by researchers, as mentioned in Section 12.8.

In theory, adopting decentralized wastewater treatment across Africa can solve the sanitation crisis. However, practical experience has shown that the poor management of these facilities always resulted in effluents that did not meet the required quality standards. The lack of institutional monitoring and authority lets the situation continue without any structured improvement plans. Further, the poor-quality control in building the pit latrines and septic tanks, including the walls and lining, leads to serious pollution of the valuable groundwater reserves by the liquid wastewater that infiltrates the soil. During the rainy season, the wastewater trapped in the soil can also travel to the nearest water bodies.

12.6 DECENTRALIZED TREATMENT IN CHINA

China is the most populated country in the world, with a total population exceeding 1.4 billion [21]. The historical need to access water led to an uneven distribution of the Chinese population. Currently, 94% of the population lives East of the Heihe-Tengchong line, which occupies only 43% of the total land area, and 60.8% of the population is urban [22].

China has undergone fast urbanization accompanied by environmental challenges affecting the sustainable economic development in most cities. Most Chinese cities are encountering serious urban flooding and eutrophication in urban water bodies; 8 billion of sewage is generated yearly, and the primary source is agricultural non-point source pollution [23]. It has been reported that sewage is characterized by COD of about 387 mg/L, total nitrogen of 50.5 mg/L, and total phosphorus of 3.7 mg/L, while the rate of rural sewage treatment is only 13–34% [24].

Moreover, 96% of Chinese villages discharge directly or in open channels because they lack drainage and sewage treatment systems. Accordingly, the Chinese Government issued several policies. The "Five-Year Plan" for elevating Improvement of Rural Living Environment (2021–2025) states it is imperative to accelerate the treatment of rural domestic sewage. However, the advanced centralized technologies that adopt the biologically activated sludge processes (like anoxic oxic AO, oxidation ditch, and anaerobic-anoxic-oxic A_2O methods) are not feasible in rural areas due to their high costs and large space requirements. Accordingly, the adoption of decentralized could be more suitable, especially for rural areas, and presents more sustainable water management [25].

Decentralized wastewater treatment plant systems treat and dispose of domestic sewage on-site or near the source. The total cost includes the cost of wastewater treatment and transportation, which is road based. These systems provide several advantages, mainly by (1) protecting and promoting human health and (2) providing water quality and ecosystem protection [26].

With limited accessibility and remoteness, many Chinese areas found a solution by implementing a decentralized system. In China, decentralized wastewater treatment systems combine biological and ecological techniques. Currently, these systems exhibit three types of technical sewage treatment modes: biological treatment modes (8.4%), ecological treatment modes (16%), and combined treatment modes (75.6%) [23].

Septic tanks are the most common primary technology being inexpensive yet simple to operate. Septic systems consist of (1) the septic tank, which provides solids a long residence time to settle, and (2) the soil dispersion system [27]. Septic systems do not effectively remove total nitrogen (18–34%) or phosphorous compounds (26–36%) or the level of pathogenic organisms [28]. However, failure in designing septic tanks or their poor maintenance can cause an overflow of wastewater, causing detrimental health issues. The design of septic systems is based on the average daily flow, the HRT, the volume of wastewater, the sludge production rate, the emptying frequency, and the sludge retention time [27]. Since the soil infiltration system is challenging, the majority of septic systems adopt the three-chamber septic tank configuration, which provides quiescent and anaerobic conditions to separate sludge and grease [29].

Constructed wetlands, soil infiltration purification systems, and oxidation ponds are widely applied in decentralized Chinese systems [23, 30]. Constructed wetlands are engineered natural treatment systems characterized by a diversity in their designs and operational features that can be adapted to treat domestic, agricultural, and agro-industrial wastewater [26]. These systems are characterized by low capital costs, low operation costs, low infrastructure, and simplicity in their design. Constructed wetlands produce reclaimed water for irrigation, and non-potable urban applications like toilet flushing, street washing, and fire protection [31]. Lagoons are also popular in Chinese rural areas. These simple systems consist of a natural or constructed pond built on an impermeable layer. The high nutrient content of these systems enhances the application of the treated water for irrigation purposes.

Biological treatment techniques in decentralized systems differ from the ones in centralized systems because they should satisfy more stringent regulations but still enable contaminants removals (COD, suspended solids, NH_4-N, P, and pathogens).

National Rural Biogas Production Plan (2006–2010) encouraged the implementation of anaerobic digestion technology, which most of the cost was granted by the Government. By 2000, about 30% of the rural Chinese population was served by biogas bioreactors [32]. Anaerobic digestion produces biogas used in kitchens and bathrooms. However, these systems are poorly maintained and do not remove nutrients [29].

Electric aerobic treatment is implemented in areas of stringent regulations that require higher removal of COD, BOD, and ammonia. Although these systems are controlled automatically, they are considered complex and expensive for single families. Biofilters with ceramic carriers, characterized by high removal capacity for ammonia (84–98%) and COD (84.5–95%), are also used as a decentralized system [33].

To comply with stringent regulations concerning nutrient removal, anoxic-oxic systems have been used in decentralized technologies, mainly in Taihu Basin (http://en.cnki.com.cn/Article_en/CJFDTOTAL-SLZG200617007.htm). The geographical feature of this area allows the reduction of power consumption associated with the process.

Although membrane bioreactor technology is not appropriate in many Chinese areas mainly due to its operation and management requirements, it has been used, and the reclaimed water is used for irrigation purposes [34].

Recently, the most common decentralized systems use a combination of biological and ecological processes. This coupling produces high effluent quality with a balance of capital and operation costs. The most implemented modes consist of: (1) a combination of a septic tank and biofilm/sequencing batch reactor/activated sludge with the following configurations, (2) implemented modes consist of aerobic/ecological treatment combined with septic tanks, aerobic biofilms, and constructed wetlands with the following configurations [35]:

a. activated-sludge process + constructed wetlands
b. activated sludge + lagoon
c. bio-contact oxidation tank + constructed wetlands
d. bio-contact oxidation tank + soil treatment

12.7 CASE STUDIES OF DECENTRALIZED TREATMENT IN DEVELOPING COUNTRIES

12.7.1 Zimbabwe

Harare is the capital city of Zimbabwe. It is the largest city in the country, with busy urban activities and a population of around 2.1 million people. Wastewater treatment is currently mostly centralized; however, it is failing to meet the needed level of treatment. There are two activated sludge treatment plants in the city that were designed to serve 250,000 people, with a total capacity for the two plants of 208,000 m^3/day. The recent uncontrolled urban expansion without proper planning for the required infrastructure resulted in overloading the existing sewerage and wastewater treatment infrastructure where the influent wastewater flowrates exceed 300,000 m^3/day, which is more than 44% overload at the treatment plants. The remaining population does not have access to the sewer network and relies on open defecation and pit latrines. In Epworth, on the east side of Harare, only 1–3% of the population have access to toilets, 23–27% use pit latrines, and 2–13% have no access to toilets and practice open defecation. An expansion of the treatment plant was planned by the City of Harare, which would cost US$68 million. However, it was found that the expansion would only be able to cover 70% of the total wastewater generated in the city. The excess wastewater is discharged into local rivers, which lead to the two main lakes around Harare, Lakes Chivero and Manyame, leading to significant pollution in the two lakes. One of the reported reasons for excess wastewater is the higher water use in wealthy parts of the city that is covered with sewers. This leads to overloading the sewer lines as well, which results in sewage bursts and pools of raw sewage in the streets [36].

An analysis of the situation and plan for using decentralized treatment at Harare has been done by researchers. They found that decentralized treatment was more suitable for higher density communities with up to 1500 m^2 per residential plot. This is due to the generation of more wastewater than what can be utilized on-site for agricultural fertilizer purposes. Decentralized treatment also allowed the coverage of new residential expansions in areas that do not have sewer coverage. Septic tanks with urine separation provided a better solution in that case. However, the lack of authoritative management of decentralized treatment in Harare resulted in poor implementation of the decentralized methods that will lead to the same problems faced by centralized plants: overloading and poor treatment levels. That is why the City of Harare is focused on expanding the centralized treatment plants rather than the implementation of decentralized methods, serving only the main city. There are new trends in research at Harare where natural unmanaged treatment systems can be used to alleviate the sanitation problem in remote and rural communities with lower strength wastewater. Unaerated oxidation ponds, algae ponds, duckweed ponds, and unaerated wetlands are all proposed technologies where minimal management is needed beyond the construction of the ponds. The land restriction in the city is not a problem in remote areas, and new expansions allow using ponds. The situation in Harare calls for using different treatment approaches depending on the location and population density. The older and more established city core is already covered with sanitation infrastructure where the investment cannot be thrown away in favor of decentralized treatment. Therefore, centralized treatment is the solution there. But to alleviate some of the loading on the treatment plants, different methods of decentralized treatment can take care of the wastewater produced from the less densely populated outer city [36].

12.7.2 India

In 2013, a study was done on open defecation and the use of pit latrines in Uttarakhand, India, based on interviews with the local population to understand the geographical and social effects of the implementation of decentralized treatment. Uttarakhand is a mountainous area with 65% jungle coverage and a population of 10 million people. Around 70% of Uttarakhand is rural, where 0.8 million people live in poverty. Around 40% of the population had no access to roads, unreliable electricity supply, and poor access to goods that were hauled to the area by humans or mules. Less than 50% of the households have access to tap water, and the remaining population has to haul water from community tanks. Around 34% of the population (45% of the rural population) does not have access to toilets or latrines [4].

It was found that the population adoption of using pit latrines rather than practicing open defecation was related to multiple factors. The first factor was culture, where poor communities that practiced open defecation regarded it as the norm and resisted change due to associated costs and efforts. The sanitation intervention by multiple organizations was usually incomplete and left the communities without building any sanitation infrastructure. On the other hand, word of mouth about the costs of building pit latrines usually exaggerated the total costs due to the fact that the people who actually built the latrines were the wealthier families who built nicer and more advanced pit latrines. However, poorer families were not aware of the actual costs of a low-cost pit latrine, which can be within their means. The second factor was remoteness and access to paved roads, which directly correlated to the ability to acquire materials needed to build the pit latrines. Access to roads and the ability to travel outside of the community also resulted in better knowledge about the pit latrines and the impact they have on health, which leads to a higher desire to build them. The third reason was convenience, as the members of the community did not want to travel far to a pit latrine while working. The fourth reason was access to fresh water, which is necessary for toilets in Uttarakhand, areas with no access to water, or families that cannot afford to get tap water, usually did not have pit latrines. The fifth reason was that filled pits were difficult to empty and had to be closed, where they would attract insects and cause foul smells [4].

12.8 ENERGY RECOVERY AND REUSE OF TREATED WATER FROM DECENTRALIZED SYSTEMS

The ultimate objective of decentralized systems is to achieve a reduction in pollution rather than the generation of effluent with a fixed standard. However, centralized wastewater treatment plants require high capital costs and extra costs for operation and maintenance. Developing countries, suffering from restricted local budgets and a lack of local expertise found the decentralized system a preferable option, especially in low-density populated rural areas or improper zoning. Decentralized systems could be limited to simple primary treatment levels, such as septic tanks and anaerobic ponds, or more complex secondary treatment levels such as anaerobic filters, facultative ponds, or constructed wetlands. They might also implement a tertiary treatment level like maturation ponds or duckweed ponds [37].

Septic tanks are the most commonly used decentralized sanitary systems in developing countries. A conventional septic tank consists of one single cell which allows suspended solids to settle. This single cell also presents some degree of anaerobic digestion that stabilizes and reduces some of the organics. The quality of the treated effluent depends on several factors, including the composition of the influent, the weather conditions (temperature mainly), the HRT, and the design of the tank itself as its state of maintenance. Conventional septic tanks (2 m^3) perform about 70% and 50% removal efficiencies for TSS and COD respectively in tropical weather conditions. The majority of septic tanks confine the separation of black water and gray water to minimize the pathogenic content of the effluent. The generated sludge, hence, originating from the purely domestic origin is suitable for agricultural reuse [38].

Constructed wetlands are gaining popularity because of their cost and energy saving in addition to the advantage they have in terms of high-quality effluent, self-remediation, and self-adaptation to the surrounding environment. Waste stabilization ponds characterized by their simplicity and the long retention time enhance pathogen abatement. These ponds include anaerobic, facultative, and purely aerobic zones resulting in fair growth of algae concentrations enough to minimize the level of nitrates that infiltrate the groundwater [30].

Developing countries implementing anaerobic/anoxic/oxic or anoxic/oxic processes produce effluents to be reused for irrigation or directly meet the standards to be discharged into surface water bodies. The effluents from decentralized wastewater treatment plants can be reused when they match the appropriate environmental requirements of the ecological area. To protect the health of the local population effective effluent disinfection should be considered. The highest reuse potential for the generated effluent is for urban purposes and landscapes such as toilet flushing, irrigation, gardening, and street washing, followed by reusing water for energy recovery and/or agriculture [39].

12.9 DIFFICULTIES IN IMPLEMENTING DECENTRALIZED TREATMENT IN DEVELOPING COUNTRIES

To successfully implement decentralized wastewater treatment, the most important factor is proper management and regulation enforcement. In developing countries, the main difficulties facing the successful implementation of decentralized treatment are as follows:

The majority of urban expansion in the last 25 years occurred due to the migration of rural community residents to areas surrounding large cities. These cities, being already overpopulated, do not have the space or infrastructure to accommodate new occupants. This results in the development of slums and illegal communities that are characterized by a low standard of living due to low income, overpopulation, absence of necessary urban infrastructure such as water supply, sewerage, roads, and electricity. This situation makes sanitation a major social and environmental problem. These communities have existed for years, and installing new sewage collection infrastructure would be extremely difficult and expensive. Even if proper sewerage was achieved, the centralized wastewater treatment plants in the core city are usually not designed to accommodate the extra wastewater inflow. The installation of decentralized wastewater treatment facilities on a community level requires robust planning, coordination, and budgeting by the Government and the residents, which is difficult to do in developing countries [40].

In both rural communities and urban slums, the cost of installation of decentralized wastewater treatment often is required to be paid by the residents. This is almost always beyond the capability of the people needing these services, even though decentralized treatment facilities, are orders of magnitude cheaper than centralized plants. The upfront capital payment is carried by the Government in the case of centralized treatment, while it is carried by the individual in the case of decentralized treatment.

Common decentralized wastewater treatment facilities, such as pit latrines and septic tanks, are usually built by local contractors without much quality control. This leads to inefficient treatment as well as undetected pollution due to leaks and infiltration into the soil and groundwater (https://agwt.org/content/septic-systems).

It is very common to not have any operation or maintenance at all for existing decentralized wastewater treatment facilities, leading to the rapid deterioration of the treatment quality. The reason for this is both due to the high costs and the lack of knowledge from the residents who are responsible for their own decentralized sanitation. A very common example is the lack of maintenance of septic tanks. Accumulated sludge must be emptied every one to three years according to the tank design. However, septic tanks are often left for much longer without emptying. This results in the sludge accumulating beyond the maximum design level and occupying the space intended for the liquid wastewater, which leads to shortcutting and effluents that are not treated. It also exerts high pressure on the walls of the tanks leading to cracks and wastewater leakage. If the septic tanks are left further, the tank will be completely blocked, and wastewater will not be able to enter the subsequent planned effluent transportation system (which can be a pipe or channel). This leads to an overflow of wastewater on the surface of the ground and the exposure of the population to pathogenic hazards.

Finally, the key challenge facing decentralized sanitation is the absence of authoritative regulation. This can be due to the absence of actual standards and regulations for decentralized wastewater treatment, so the governments do not have the tools to manage the process. It can also be due to the absence of resources, whether financial, logistical, or technical, that are needed to manage decentralized wastewater treatment. If governments take the responsibility of installing the decentralized facilities in a well-planned approach and manage their operation and maintenance, as well as enforce the regulations regarding the effluent quality, then they may be able to successfully implement decentralized sanitation. This can be done by taking on the responsibility for the initial capital cost and facilitating a payment plan for the residents, providing regular technical maintenance service for a monthly fee, and establishing a system for reporting issues and rapid response. While every developing city has a different situation regarding sanitation, proper planning, regulation, and resource management can lead to a successful implementation of decentralized wastewater treatment [37].

12.10 CONCLUSIONS AND PERSPECTIVES

Developing countries find themselves significantly lagging in achieving sustainable sanitation primarily due to a confluence of factors, including inadequate infrastructure regulation, economic struggles, and deficiencies in wastewater collection and treatment systems. Despite the evident challenges, decentralized treatment systems offer a pragmatic, cost-effective, and straightforward solution. However, their potential to positively influence sanitation quality remains largely untapped. This can be attributed to a lack of comprehensive institutional regulations governing their utilization and upkeep, as well as the establishment of benchmarks for effluent quality.

The intricate challenge lies in introducing decentralized technologies into pre-existing communities that often lack proper urban planning. This limitation curtails the adaptability and efficacy of these treatment methodologies. Presently, decentralized wastewater treatment systems focus primarily on safeguarding human health and the environment. However, a paradigm shift could occur if revenue generation through resource recovery were integrated into these systems. This innovative approach could potentially mitigate some of the financial burdens associated with implementation.

A significant hurdle in realizing resource recovery is the unsuitability of much of the existing on-site treatment infrastructure, such as pit latrines and septic tanks, for such purposes. Nevertheless, a promising trend is emerging in African urban centers where modifications to existing facilities, like urine diversion and Ecological Sanitation (EcoSan) technology, are being adopted. These adaptations enable the integration of resource recovery practices.

Furthermore, researchers are actively proposing the establishment of sludge treatment and resource recovery plants that are cost-effective and align with sustainability principles. A growing imperative underpins this forward-thinking approach to adhere to stringent environmental regulations. For instance, the drive to meet exacting environmental standards in China has spurred the development of novel combinations of biological and ecological processes within decentralized wastewater treatment systems. This integration achieves a delicate equilibrium between capital and operational costs while yielding high-quality effluent.

The disparity in sanitation progress among developing nations stems from various complex factors. While decentralized treatment systems offer a promising solution, their impact on sanitation quality has yet to be fully realized. Overcoming regulation, infrastructure adaptation, and resource recovery challenges could transform these systems into viable, self-sustaining assets that safeguard public health and the environment and offer economic incentives for further expansion. This ongoing evolution reflects the dynamic nature of sustainable sanitation practices within the context of developing countries.

REFERENCES

1. J. A. Cotteral Jr, and D. P. Norris, "Septic tank systems," *J. Sanit. Eng. Div.*, vol. 95, no. 4, pp. 715–746, 1969.
2. T. R. Bounds, "Design and performance of septic tanks," *ASTM Spec. Tech. Publ.*, vol. 1324, pp. 217–234, 1997.
3. W. H. Organization, "State of the world's sanitation: An urgent call to transform sanitation for better health, environments, economies and societies." *UNICEF*, 2020.
4. K. O'Reilly, R. Dhanju, and A. Goel, "Exploring 'the remote' and 'the rural': Open defecation and latrine use in Uttarakhand, India," *World Dev.*, vol. 93, pp. 193–205, 2017.
5. J. P. Graham, and M. L. Polizzotto, "Pit latrines and their impacts on groundwater quality: A systematic review," *Environ. Health Perspect.*, vol. 121, no. 5, pp. 521–530, 2013.
6. P. T. M. Hanh, S. Sthiannopkao, K.-W. Kim, D. T. Ba, and N. Q. Hung, "Anthropogenic influence on surface water quality of the Nhue and day sub-river systems in Vietnam," *Environ. Geochem. Health*, vol. 32, no. 3, pp. 227–236, 2009. doi: 10.1007/s10653-009-9279-9
7. M. Lee, M. Kim, Y. Kim, and M. Han, "Consideration of rainwater quality parameters for drinking purposes: A case study in rural Vietnam," *J. Environ. Manage*, vol. 200, pp. 400–406, 2017. doi: 10.1016/j.jenvman.2017.05.072
8. East Asia & Pacific," *Global Economic Prospects*. The World Bank, pp. 107–117, 2015. doi: 10.1596/978-1-4648-0483-0_ch2_eap
9. V.-A. Nguyen, "Why DEWATS is still not popular in Vietnam?," *Water Pract. Technol.*, vol. 5, no. 4, 2010. doi: 10.2166/wpt.2010.117
10. N. V. Anh, P. T. Nga, T. H. Nhue, and A. Morel, "Potential decentralized wastewater management for sustainable development from Vietnamese experience," *Proc. Water Environ. Fed.*, vol. 2005, no. 5, pp. 917–946, 2005. doi: 10.2175/193864705783977989
11. F. Klingel, A. Montangero, and M. Strauss, "Nam Dinh–planning for improved faecal sludge management and treatment," in *Vietnam Water and Sanitation Association's yearly conference. December 6–7*, 2001.
12. A. V. Nguyen, N. T. Pham, T. H. Nguyen, A. Morel, and K. Tonderski, "Improved septic tank with constructed wetland, a promising decentralized wastewater treatment alternative in Vietnam," in *International Conference of the National On-site Wastewater Recycling Association (NOWRA) and International Water Association*, 2007.
13. N. V. Anh *et al.*, "Decentralized wastewater treatment-new concept and technologies for Vietnamese conditions," in *5th Specialised conference on small water and wastewater treatment systems*, Istanbul, Turkey, 2002, pp. 24–26.
14. V. G. Varma, S. Jha, L. H. K. Raju, R. L. Kishore, and V. Ranjith, "A review on decentralized wastewater treatment systems in India," *Chemosphere*, vol. 300, p. 134462, 2022. doi: 10.1016/j.chemosphere.2022.134462
15. M. Rath, T. Schellenberg, P. Rajan, and G. Singhal, "Decentralized wastewater and fecal sludge management: Case studies from India," 2020.
16. A. Singh *et al.*, "Performance evaluation of a decentralized wastewater treatment system in India," *Environ. Sci. Pollut. Res.*, vol. 26, no. 21, pp. 21172–21188, 2019, doi: 10.1007/s11356-019-05444-z
17. J. Arora, A. Yadav, and D. Saroj, "Potential of decentralised wastewater treatment systems applicable to India," *Curr. World Environ.*, vol. 11, no. 2, pp. 338–350, 2016. doi: 10.12944/cwe.11.2.01
18. H. Muzioreva, T. Gumbo, N. Kavishe, T. Moyo, and I. Musonda, "Decentralized wastewater system practices in developing countries: A systematic review," *Util. Policy*, vol. 79, p. 101442, 2022.
19. H. Wang *et al.*, "Water and wastewater treatment in Africa–current practices and challenges," *CLEAN–Soil, Air, Water*, vol. 42, no. 8, pp. 1029–1035, 2014.

20. I. Nansubuga, N. Banadda, W. Verstraete, and K. Rabaey, "A review of sustainable sanitation systems in Africa," *Rev. Environ. Sci. bio/technology*, vol. 15, pp. 465–478, 2016.
21. H. Li, J. Chen, Z. Chen, and J. Xu, "Urban population distribution in China: Evidence from internet population," *China Econ. Rev.*, vol. 74, p. 101808, 2022. doi: 10.1016/j.chieco.2022.101808
22. W. Tang, Y. Pei, H. Zheng, Y. Zhao, L. Shu, and H. Zhang, "Twenty years of China's water pollution control: Experiences and challenges," *Chemosphere*, vol. 295, p. 133875, 2022. doi: 10.1016/j.chemosphere.2022.133875
23. Y. Xu *et al.*, "Systematically assess the advancing and limiting factors of using the multi-soil-layering system for treating rural sewage in China: From the economic, social, and environmental perspectives," *J. Environ. Manage.*, vol. 312, p. 114912, 2022, doi: 10.1016/j.jenvman.2022.114912
24. Y. D. Xie *et al.*, "Towards the formulation of rural sewage discharge standards in China," *Sci. Total Environ.*, vol. 759, p. 143533, 2021, doi: 10.1016/j.scitotenv.2020.143533
25. N. Leigh, and H. Lee, "Sustainable and resilient urban water systems: The role of decentralization and planning," *Sustainability*, vol. 11, no. 3, p. 918, 2019. doi: 10.3390/su11030918
26. A. G. Capodaglio, A. Callegari, D. Cecconet, and D. Molognoni, "Sustainability of decentralized wastewater treatment technologies," *Water Pract. Technol.*, vol. 12, no. 2, pp. 463–477, 2017. doi: 10.2166/wpt.2017.055
27. N. Lourenço, and L. M. Nunes, "Review of dry and wet decentralized sanitation technologies for rural areas: Applicability, challenges and opportunities," *Environ. Manage*, vol. 65, no. 5, pp. 642–664, 2020. doi: 10.1007/s00267-020-01268-7
28. X. Hao, "Water infrastructure for sustainable communities: China And the world," *Water Intell. Online*, vol. 9, 2010, doi: 10.2166/9781780401485
29. H.-Y. Dong, Z.-M. Qiang, W.-D. Wang, and H. Jin, "Evaluation of rural wastewater treatment processes in a county of eastern China," *J. Environ. Monit.*, vol. 14, no. 7, p. 1906, 2012. doi: 10.1039/c2em10976j
30. V. Torrijos, O. G. Gonzalo, A. Trueba-Santiso, I. Ruiz, and M. Soto, "Effect of by-pass and effluent recirculation on nitrogen removal in hybrid constructed wetlands for domestic and industrial wastewater treatment," *Water Res.*, vol. 103, pp. 92–100, 2016. doi: 10.1016/j.watres.2016.07.028
31. Y. Bo, and W. Wen, "Treatment and technology of domestic sewage for improvement of rural environment in China," *J. King Saud Univ. – Sci.*, vol. 34, no. 7, p. 102181, 2022. doi: 10.1016/j.jksus.2022.102181
32. V. P. Aravani *et al.*, "Agricultural and livestock sector's residues in Greece & China: Comparative qualitative and quantitative characterization for assessing their potential for biogas production," *Renew. Sustain. Energy Rev.*, vol. 154, p. 111821, 2022. doi: 10.1016/j.rser.2021.111821
33. L. T. Pan, and Y. Han, "A novel anoxic–aerobic biofilter process using new composite packing material for the treatment of rural domestic wastewater," *Water Sci. Technol*, vol. 73, no. 10, pp. 2486–2492, 2016. doi: 10.2166/wst.2016.099
34. G. Schories, "IWAPIL — Innovative wastewater treatment applications for isolated locations," *Desalination*, vol. 224, no. 1–3, pp. 183–185, 2008. doi: 10.1016/j.desal.2007.02.091
35. X. Guo, Z. Liu, M. Chen, J. Liu, and M. Yang, "Decentralized wastewater treatment technologies and management in Chinese villages," *Front. Environ. Sci. Eng*, vol. 8, no. 6, pp. 929–936, 2014. doi: 10.1007/s11783-013-0623-z
36. I. Chirisa, E. Bandauko, A. Matamanda, and G. Mandisvika, "Decentralized domestic wastewater systems in developing countries: The case study of Harare (Zimbabwe)," *Appl. Water Sci*, vol. 7, no. 3, pp. 1069–1078, 2016. doi: 10.1007/s13201-016-0377-4
37. D. Bernal, I. Restrepo, and S. Grueso-Casquete, "Key criteria for considering decentralization in municipal wastewater management," *Heliyon*, vol. 7, no. 3, p. e06375–e06375, 2021. doi: 10.1016/j.heliyon.2021.e06375
38. J. C. L. Van Buuren, and G. Lettinga, "The role of septic tanks in future sewer systems in Vietnamese cities," in *Technical papers of International Seminar on Urban Sewerage in Vietnam, Hanoi, Vietnam, 2001.-[Sl]:[sn], 2001*, 2001, pp. 279–292.
39. A. G. Capodaglio, "Integrated, decentralized wastewater management for resource recovery in rural and peri-urban areas," *Resources*, vol. 6, no. 2, p. 22, 2017.
40. T. Makinde, "Problems of policy implementation in developing nations: The Nigerian experience," *J. Soc. Sci*, vol. 11, no. 1, pp. 63–69, 2005.

Index